Publications of the International Agricultural Research and Development Centers

1985 exhibition
at the Frankfurt Book Fair
sponsored by:

Deutsche Gesellschaft für
Technische Zusammenarbeit (GTZ) Gmb H
(German Agency for Technical Cooperation) (GTZ)

Consultative Group on
International Agricultural Research (CGIAR)

International Rice Research Institute (IRRI)

1985
INTERNATIONAL RICE RESEARCH INSTITUTE
Los Baños, Laguna, Philippines
P.O. Box 933, Manila, Philippines

ISBN 971-104-145-6

Foreword

Knowledge gained through international agricultural research is useless unless it is published and disseminated to its target audience of scientists, educators, extension specialists, and farmers.

Thus, there is an increasing interest by International Agricultural Research Centers and development organizations in exhibiting our educational materials at international book fairs.

The first International Agricultural Research Center Book Exhibition was in China in May 1982 and was cosponsored by the International Rice Research Institute (IRRI) and the China National Publications Import and Export Corporation. The German Agency for Technical Cooperation (GTZ) first exhibited at Frankfurt in 1979. In October 1982, IRRI also exhibited at the Frankfurt Book Fair.

The success of the 1982 Frankfurt exhibition led to GTZ-IRRI cooperation to sponor a joint exhibition of International Center and GTZ publications in Frankfurt in 1983 and 1984. Those exhibitions featured material on agricultural research and development for the Third World published by the International Centers supported by the Consultative Group on International Agricultural Research (CGIAR), and other International Centers, and the German Agency for Technical Cooperation (GTZ).

In 1985, the CGIAR, GTZ, and IRRI have again joined forces to sponsor the third Center exhibition at the Frankfurt Book Fair.

Worldwide information dissemination requires use of channels other than our own mailing lists. Therefore, we shall be seeking more contact and interaction with international publishers and distributors at the 1985 Fair.

We are interested in joint copublication agreements with other publishers. We particularly want to meet Third World publishers who might wish to publish translations of our materials in languages other than English. Since our aim is to increase knowledge about agricultural science and production, in most cases we would not expect royalties or payment for editions copublished in languages other than in English in the Third World.

We also would like to meet international publishers who want to copublish with us in English.

Last but not least we hope to establish during the Fair closer contacts with agricultural libraries everywhere.

We hope that participation in the 1985 Frankfurt Book Fair will bring us closer to the worldwide community of publishers and distributors in agricultural science. A wider exchange of knowledge on international agricultural research and production will help us achieve our ultimate objective — to make hunger a problem of past.

IRRI coordinated International Center contributions and activities. GTZ coordinated preparations in Frankfurt for this exhibition. IRRI compiled and published this catalog.

Dr. K. J. Lampe
Head, Department of Agriculture
Health and Rural Development
Deutsche Gesellschaft für Technische
Zusammenarbeit (GTZ) GmbH

Dr. M. S. Swaminathan
Director General
International Rice Research
Institute (IRRI)

Contents

Part II
Non CGIAR-Supported Centers

Part III

CGIAR

Consultative Group on International Agricultural Research

1818 H St., N.W. Washington, D.C. 20433 USA
Cable Address: INTBAFRAD - Telex Worldbank440098
Telephone: 202-477-5347

What is the CGIAR?
What do the international centers do?

The Consultative Group on International Agricultural Research (CGIAR) is a relatively new and exciting strategy to increase global food production. Established in 1971, the CGIAR is an association of 48 countries, including the Federal Republic of Germany, international and regional organizations, and private foundations supporting 13 international agricultural research institutes. The purpose of the Consultative Group is to bring the resources of modern biological and socioeconomic research to bear on the problems of improving agricultural productivity in the tropics and subtropics where most of the developing nations lie. The research and training programs undertaken by the centers and sponsored by the Group seek to arm the developing countries with superior varieties of essential crops and improved farming systems for the production of food plants and animals. The food crops and livestock on which the centers focus provide 75% of the food consumed in the developing countries. Located in developing countries, the international centers work together with scientists in national institutions to set research priorities, develop technologies suited to the specific needs of the region, and to evaluate and test the new technologies. There are currently 7,000 staff working at the international centers; 600 are senior scientists recruited from over 50 different countries. The CGIAR is formally cosponsored by the World Bank, the United Nations Development Program (UNDP) and the Food and Agricultural Organization (FAO). In 1984, the CGIAR will finance the 13 centers' collective budgets for a total of $180 million.

Contributions of the CGIAR System

Four of the institutes were initiated in the 1960's by the Ford and Rockefeller Foundations. The remainder were set up at different times since 1971 when the CGIAR was established. Despite their relatively short existence, the international centers have made remarkable contributions to world food production. As might be expected the impact of the two oldest institutes, the Centro Internacional de Mejoramiento de Maiz y Trigo (CIMMYT) in Mexico, and the International Rice Research Institute (IRRI) in the Philippines, has been the most dramatic. Their mandated crops, rice and wheat, are the two most widely consumed food crops in the world. High yielding, management responsive varieties of rice and wheat derived from those developed by CIMMYT and IRRI are grown on over 55 million hectares — one third of the area planted to these cereals in the developing world. It is estimated that the increased yields feed 300 million people and the economic value of this additional food supply is more than $5 billion annually. While less well known, the accomplishments of the newer centers are also making an impact on food production and hold even greater potential for the future. For example, in Latin America national programs have adopted and released more than 100 bean varieties developed by scientists at the Centro Internacional de Agricultura Tropical (CIAT) in Colombia. CIAT is also working on genetic improvements that can double and triple yields of cassava, an important food for some 400 million people in the tropics. In Lima, Peru, the Centro Internacional de la Papa (CIP) is concentrating on increasing production of the world's fourth most important crop. CIP houses the largest collection of potato germplasm in the world. A unique tissue culture technique makes possible the transmission of germ-free potato genetic resources. Improved lines of edible legumes and innovative soil and crop management systems for Asia and Africa are being developed by the International Crops Research Institute for the Semi-Arid Tropics (ICRISAT) based in India.

In addition to those international agricultural research centers supported by the CGIAR there are several funded independently of the Consultative Group. To further their objectives all the international centers are seeking more extensive exchange of knowledge on agricultural research and production among scientists and educators, particularly in developing countries. At the 1983 Frankfurt Bookfair they look forward to establishing new contacts and opening new channels of communication with publishers and book distributors from throughout the world.

Slide sets, Videotapes

International Service for National Agricultural Research for the Consultative Group on International Agricultural Research

CGIAR: Catalyst for Progress

June 1983. 158 slides. 23 minutes 35 seconds.

Narrated slide presentation describes 13 international agricultural research centers under the Consultative Group. Some of the stories of success are given in brief illustrated vignettes, along with accounts of the mandates, operational approaches, and collaboration between the centers and national agricultural research systems.

International Service for National Agricultural Research for the Consultative Group on International Agricultural Research

CGIAR: Catalyst for Progress

June 1983. Videotape presentation of narrated slide presentation. 23 minutes 35 seconds.

A videotape presentation of a narrated series of slides describing 13 international agricultural research centers under the Consultative Group. Some of the stories of success are given in brief illustrated vignettes, along with accounts of the mandates, operational approaches, and collaboration between the centers and national agricultural research systems.

Price for both to be determined.

Harnessing the monsoons: improved cropping systems in Asia

1982. Film. 27 minutes. (Also available in Spanish and French).

In a pilot project in Sri Lanka's dry zone, agricultural researchers have shown the way to double and triple food crops. This film documents how new rice varieties and a revised planting schedule take full advantage of the monsoon rains that fill the water reservoir or "tank" in the village of Walagambahuwa. The film records a rare partnership between modern science and ancient agricultural know-how, presenting a model of rural development that is being replicated in many other villages in Sri Lanka's dry zone. Also available in Sinhalese from the Department of Agriculture (Research), Peradeniya, Sri Lanka.

La Mousson Apprivoisée

1982. Film. 27 minutes.

Un projet pilote réalisé dans la zone sèche de Sri Lanka a montré le moyen de doubler et tripler les récoltes vivrières. Ce film explique comment l'utilisation de nouvelles variétés de riz et le choix de nouvelles dates de semis permettent de tirer pleinement profit des pluies de la mousson qui viennent remplir le lac-réservoir artificiel du village de Walagambahuwa. Il décrit un exemple remarquable de coopération entre chercheurs et agriculteurs — les uns apportant les connaissances scientifiques modernes et les autres le savoir agricole hérité de siècles d'expérience — et présente un modèle de dèveloppement rural qui est maintenant appliqué à de nombreux autres villages de la zone sèche de Sri Lanka. Disponible également en version cinghalaise. S'adresser au ministère de l'Agriculture (Recherches), Peradeniya, Sri Lanka.

Pods of Protein

This film is a coproduction of IDRC's Communications Division and IITA.

1979. Film. 23 minutes. (Also available in French and Spanish).

This film documents the work being done to improve cowpeas under the auspices of the International Institute of Tropical Agriculture (IITA) and IDRC in Nigeria and Upper Volta. An important food legume for the Third World, cowpeas originated somewhere in Sahelian Africa 3000-4000 years ago. Today they account for a large percent of the human protein intake in some regions. The film demonstrates that any attempt to improve this traditional crop must take into account local tastes and food processing methods.

L'Ostréiculture sous les tropiques

1979. 28 minutes.

Les huîtres sont généralement considérées en Occident comme un aliment de luxe, mais dans certains pays en développement elles representent avant tout une source précieuse de nourriture pour la collectivité. Des techniques permettant d'accroître la production ostréicole dans ces pays peuvent assurer un apport supplémentaire de protéines et satisfaire ainsi un besoin important. Ce film, conçu principalement comme outil éducatif, décrit les méthodes et les problèmes de l'élevage des huîtres tel qu'il est pratiqué dans plusieurs régions du monde. Il montre des projets d'ostreiculture entrepris à Sabah (Malaisie) et en Sierra Leone ainsi que des élevages situés au Japon et aux Philippines.

CIAT

Centro Internacional de Agricultura Tropical

Apartado Aéreo 6713 - Cali, Colombia
Cable: CINATROP - Telex 05769 CIAT Co
Telephone: 57-3-680111

General Information

Major research programs:
Cassava, field beans, rice, tropical pastures

Objectives:
To work with national organizations to develop improved agricultural technology to increase the quality and quantity of specific basic food commodities in the tropics. The main crops are cassava, field beans, and rice. CIAT also conducts a large program on developing suitable pasture technology for the infertile, acid soils of Latin America.

In this catalog CIAT publications are listed in the following criteria:

General; Beans (*Phaseolus vulgaris*); Cassava; Pastures (tropical pasture species); Rice; Seeds (seed technology).

Order information:
As a nonprofit organization, CIAT is primarily interested in making its research information widely available to the people who need it. Prices for developing countries are determined from direct production costs plus a small markup to cover warehouse and distribution costs. The "list price" for each book is its commercial sale price, which is based on actual production costs. This is also the developed country cost for the book. (Discounts are available for educational institutions and for distribution agents.)

Libraries can request copies of the CIAT Report, Miscellaneous Reports, CIAT International, commodity newsletters, and the abstracts journals through the Library exchange System at CIAT.

Many publications are available in English through the IADS Agri-bookstore, Rosslyn Plaza, 1611 Kent Street, Suite 600, Arlington, West VA 22209, USA. (Price list available on request.)

Payment:
Prices are listed in US dollars. Payment may be made with a personal or cashiers check, drawn *only* on banks that are affiliated with a US bank. PLEASE NOTE THAT PAYMENT MUST ACCOMPANY THE ORDER.

Payment may also be made in Colombian pesos, or in CIAT coupons, AGRINTER coupons, or UNESCO coupons.

Checks should be made payable to CIAT. No refunds can be made for checks exceeding the exact amount of the invoice, due to currency restrictions. An invoice can be requested, in order to verify availability of materials and total price. All refunds will be made in CIAT coupons.

Send orders to:
Distribution Office
CIAT
Apartado Aéreo 6713
Cali, Colombia

Como hacer pedidos:
Al ser una organizacion sin animo de lucro, el CIAT esta principalmente interesado en que la informacion sobre aus programas sea ampliamente conocida por aquellos que la necesitan. Los precios para paises en desarrollo reflejan los costos directos de publicacion, con un incremento pequeno para cubrir los costos de almacenamiento y distribucion.

Las bibliotecas pueden solicitar copias del Informe CIAT, de CIAT Internacional, de los boletines e, informes miscelaneos, y de las revistas de resumenes, mediante el sistema de canje del Centro.

Forma de pago:
Los precios anotados indican el precio en dolares estadounidenses por cada ejemplar, incluyendo el costo de su porte via correo aereo. SE REQUIERE PAGO ADELANTADO PARA CUALQUIER ENVIO DE PUBLICACIONES.

Se puede pagar con cheque personal o de gerencia en dolares estadounidenses y girado solamente contra bancos cuya oficina principal este localizada en los Estados Unidos. Si el pago se hace en pesos colombianos, por favor envie cheque giro o giro bancario.

El pago puede hacerse tambien con cupones del CIAT, cupones Agrinter, y cupones de la UNESCO.

Los cheques deben ser girados a nombre del CIAT. Favor tomar nota de que no es posible devolver dineros sobre cheques recibidos cuyo valor exceda el total de la factura.

Envie los pedidos a la siguiente direccion:
Distribucion de Publicaciones
CIAT
Apartado Aereo 6713
Cali, Colombia

Research Highlights

Centro Internacional de Agricultura Tropical

CIAT Report (Informe CIAT)

Annual (latest 1984). 96 pages. 18 × 23 cm. Perfect bound, paperback. ISBN/ISSN 0120-3150. Free.

This annual publication, written in popular language, presents highlights of the objectives, activities, and research results of CIAT's four basic crop commodity programs and international cooperation. Published every September for the preceding calendar year.

(Esta publicacion anual escrita en lenguaje sencillo, presenta aspectos sobresalientes de los objetivos, actividades y resultados de la investigacion de los programas y unidades de trabajo del CIAT. Se publica en septiembre para el ano calendario anterior.)

Available in Spanish and English. Limited number of back issues; some reports out of stock.

Annual Reports

Centro Internacional de Agricultura Tropical

CIAT Annual Report consisting of (a) Bean Program Report; (b) Cassava Program Report; and (c) Tropical Pastures Program Report
(Informes Anuales de los Programas: (a) Frijol, (b) Yuca; and (c) Pastos Tropicales)

Annual 1979-1984. Varies. 20.5 × 27 cm. Perfect bound, paperback. Free.

Specific program reports are available from the Coordinators of each commodity program. These detailed progress reports are working documents and are designed primarily for research collaborators.

(Estos informes pueder ser solicitados directamente a los Coordinadores de cada programa. Son documentos de trabajo principalmente diseñados para uso de colaboradores en investigación.)

Monographs

E. M. Rubinstein

La Economia de la Fiebre Aftosa: Analisis de sus Externalidades y Estrategias de Control en la Costa Norte de Colombia
(The Economics of Foot-and-Mouth Disease: Analysis of its Externalities and Control Strategies in the North Coast of Colombia)

1978. 54 pages. 17 × 24 cm. Saddle stitched, paperback. Prices: CIAT: $3.00, Colombia: $3.50, Other: $4.00 (Includes air mail postage).

Estudio economico de distintas alternativas de control de fiebre aftosa para el Departamento de Cordoba, Colombia; incorporando explicitamente los beneficios que tal control tiene sobre la produccion de carne.

(Economic evaluation of various alternatives for controlling foot-and-mouth disease in the Department of Cordoba, Colombia, including benefits on meat production.)

Available in Spanish only.

J. Doll

Manejo y Control de Malezas en el Tropico
(Management and Control of Weeds in the Tropics)

1979. 114 pages. 16 × 20.5 cm. Perfect bound, paperback. Prices: CIAT: $3.50, Colombia: $5.00, Americas: $6.50, Other: $7.00 (Includes air mail postage).

Esta monografia familiariza al agronomo con el manejo de malezas (malezas tropicales comunes en los cultivos, competencia y alelopatia, herbicidas y surfactantes) y le ayuda en el desarrolo de programas de produccion agricola.

(This monograph acquaints the agronomist with weed control and management (tropical weeds common in crops, plant interference, herbicides, and surfactants) and aids in the development of agricultural production programs.)

Available in Spanish only.

G. A. Morales and L. E. Beltran

Enfermedades Porcinas de Importancia en el Tropico Colombiano
(Important Swine Diseases of the Colombian Tropics)

1979. 71 pages. 17 X 23.5 cm. Perfect bound, paperback. Prices: CIAT: $2.50, Other: $3.50.

La información que se presenta proporciona a los veterinarios y zootecnistas una apreciación integrada de los problemas sanitarios que afronta la industria, y de las soluciones que mejor se ajustan a cada uno de los problemas encontrados. Discute enfermedades causadas por virus, bacterias, micoplasmas, parasitos y otros.

(An integrated look at sanitary problems that plague the industry and at solutions that may improve the situation. Discusses diseases caused by viruses, bacteria, mycoplasms, parasites, and others.)

Available in Spanish only.

A. Diaz-Duran and R. Narvaez D.

Revestimiento de Canales de Riego con una Mezcla de Suelo-Cemento
(Paving Irrigation Ditches with Soil-Cement)

Reprinted 1983. 24 pages. Illus. 17 X 21 cm. Saddle stitched, paperback. Prices: CIAT: $1.50. Other: $2.00 (Includes air mail postage).

Describe los materiales y los metodos para construcción de canales de riego de bajo costo con la mescla suelo-cemento.

(Contains materials and methods for paving irrigation ditches economically, using soil-cement.)

Available in Spanish only.

R. H. Howeler

Analisis del Tejido Vegetal en el Diagnostico de Problemas Nutricionales: Algunos Cultivos Tropicales
(Tissue Analysis for Diagnosis of Nutritional Problems: Some Tropical Crops)

1983. 28 pages. 15 X 23 cm. Saddle stitched, paperback. List price: $7.00. Developing country prices: CIAT: $1.50, Colombia: $2.50, Americas: $3.50, Other: $4.00 (Includes air mail postage).

Introduce la tecnica del analisis del tejido vegetal para determinar deficiencias, toxidades, o desbalances de nutrimentos de una planta, y por consiguiente de la disponibilidad de nutrimentos en el suelo. Informa sobre los metodos de muestreo, la distribución de nutrimentos dentro de la planta y los niveles criticos en los tejidos indicadores de ciertos cultivos como el arroz, frijol, soya, maiz, yuca, pastos, y forrajes.

(Introduces the techniques of tissue analysis for determination of deficiencies, toxicities, and poorly balanced levels of nutrients in plant tissue, which then point to similar problems of nutrient availability in the soil. Covers rice, beans, soybean, maize, cassava, and pasture and forage species.)

Available in Spanish only.

Beans (*Phaseolus vulgaris*)

N. de Londoño, P. Pinstrup-Andersen, J. H. Sanders, and M. A. Infante

Factores que Limitan la Productividad del Frijol en Colombia
(Limiting Factors to Bean Productivity in Colombia)

1978. 44 pages. 17 × 23 cm. Saddle stitched, paperback. Prices: CIAT: $1.50, Other: $2.00 (includes air mail postage).

Los autores encuentran que el factor mas importante en la reducción del rendimiento es el daño ocasionado por enfermedades e insectos y que la incorporación a la planta de resistencia a estos factores puede producir beneficios sustanciales en la producción de frijol.

(In an economic analysis, the authors find that the most important factors in bean yield reduction are disease and insect damage.)

Available in Spanish only.

J. H. Sanders, and C. Alvarez

Evolución de la Producción de Frijol en America Latina Durante la Ultima Decada
(Evolution of Bean Production in Latin America During the Last Decade)

1978. 48 pages. 16 × 21.5 cm. Saddle stitched. Prices: CIAT: $1.50, Other: $2.00.

Con base en el analisis economico de la producción de frijol, los autores llegan a la conclusion de que pueden obtener beneficios economicos sustanciales mediante investigaciones intensivas para producir o seleccionar nuevas variedades de alto rendimiento y con el establecimiento de programas de mercadeo.

(Based on economic analyses of on-going bean production, the authors conclude that substantial economic benefits can be obtained through intensified research to produce or select new varieties, together with the establishment of bean marketing programs.)

Available in Spanish only.

H. F. Schwartz and G. E. Galvez (Editors)

Bean Production Problems: Disease, Insect, Soil, and Climatic Constraints of *Phaseolus vulgaris* (Problemas de Producción de Frijol: Enfermedades, Insectos, Limitaciones Edaficas y Climaticas de *Phaseolus vulgaris*)

1980. 424 pages. 200 figures and color plates. 17 × 24 cm. Smythe-sewn, hard cover. ISBN:8489206-00-7 (84-89206-01-5 Spanish ed.). List price: $37.00. Developing country prices: CIAT: $16.00, Colombia: $20.00, Americas: $25.00, Other: $28.00 (Includes air mail postage).

A current review of bean production constraints in Latin America and other bean-growing regions of the world is contained in 20 chapters. Problems caused by diseases, nematodes, pests, nutritional disorders, and soil and climatic factors are described.

(Los 20 capitulos de este libro son una revision actualizada de los problemas que afectan la producción de frijol en America Latina y otras regiones del mundo. Se describen los problemas causados por enfermedades, nematodos, plagas, desordenes nutricionales y por factores edaficos y ambientales.)

Available in English and Spanish.

C. Cardona, C. A. Flor, F. J. Morales and M. A. Pastor-Corrales

Field Problems of Beans in Latin America (Prolemas de Campo en los Cultivos de Frijol en America Latina)

1982. 184 pages. 145 color plates. 24 × 17 cm. Perfect bound, paperback. List price: $20.00. Developing country prices: CIAT: $8.50, Americas: $13.00, Other: $15.50. (Includes air mail postage).

Describes diseases and pests attacking beans (*Phaseolus vulgaris*) and includes notes on clean seed production. Also discusses nutritional disorders due to toxicity or deficiency of chemical elements. (Shortened version of Bean Production Problems.)

(Se describen las enfermedades y plagas que atacan el cultivo de frijol (*Phaseolus vulgaris*) y se hacen algunas anotaciones sobre la producción de semilla limpia. También se identifican los desórdenes ocasionados por deficiencias o toxicidades nutricionales. [Versión resumida de Problemas de Producción de Frijol.])

Available in English and Spanish.

W. U. Jayasinghe

Chlorotic Mottle of Beans (*Phaseolus vulgaris* L.) (Moteado Clorotico del Frijol)

1982. 156 pages. 15.5 × 22.5 cm. Perfect bound, paperback. List price: $15.00. Developing country price: CIAT: $5.00, Colombia: $7.00, Americas: $9.50, Other: $12.00 (Includes air mail postage).

This report identifies and characterizes the causal agent of chlorotic mottle and recommends genetic breeding for resistance to the virus as the only method completely effective against the disease.

(Este estudio identifica y caracteriza el agente causal del moteado clorotico y recomienda el mejoramiento genetico por resistencia al virus como el unico metodo completamente efectivo contra esta enfermedad.)

Available in English only.

O. Voysest

Variedades de Frijol en America Latina y su Origen (Bean Varieties in Latin America and Their Origin)

1983. 87 pages. With poster of beans classified by colors. 15 × 22 cm. Perfect bound, paperback. Prices: CIAT: $3.00, Colombia: $3.50, Americas: $4.00, Other: $5.00 (Includes air mail postage).

Describe el sistema de clasificación de frijoles por medio de colores y tamaños. Resume los origines geneticos de las variedades de frijol latinoamericanas, sus nombres comunes y las preferencias en varias regiones.

(Describes the classification system for beans using colors and sizes. Summarizes the genetic origins of Latin American bean varieties, their common names, and the preferences of various regions.)

Available in Spanish only.

F. J. Morales

El Mosaico Común del Frijol: Metodología de Investigación y Técnicas de Control (Edición Revisada) (Bean Common Mosaic Virus: Methodology for Research and Techniques for Control [Revised Ed.])

1983. 26 pages. Illus. 18 color plates. 17 × 23 cm. Saddle stitched, paperback. Rices: CIAT: $2.00, Colombia $2.50, Americas: $3.00, Other: $3.50 (Includes air mail postage).

Proporciona a los investigadores una visión general del problema fitopatológico del mosaico común del frijol; define la metodología usada en el CIAT para evaluar germoplasma de frijol por su resistencia al virus causal (BCMV). Describe la sintomatología general, los diferentes medios de transmisión del virus, la identificación de cepas del virus según reacción de variedades diferenciales, los métodos de aislamiento del BCMV en laboratorio y su control integrado.

(Describes bean common mosaic virus, how to evaluate germplasm for resistance, and methods for integrated control of the virus.)

Available only in Spanish.

O. Voysest

Viveros Internacionales de Rendimiento de Frijol: Manual Descriptivo (International Bean Yield Nurseries: Descriptive Manual)

1983. 22 pages. 17 × 23 cm. Saddle stitched, paperback. Prices: CIAT: $1.50, Other: $2.00 (Includes air mail postage).

Explica la estructura general de los viveros de evaluacion de germoplasma del CIAT, su objeto, las pruebas a que se someten sus materiales, y los procedimientos que deben seguirse para desarrollar cada vivero. Comentarios historicos sobre su fundación y resultados acompañan las notas tecnicas.

(Explains the structure of the bean yield nurseries, and how they function. Comments regarding their history and results are included in the technical notes.)

Available only in Spanish.

F. Berti

**Synthese de Travaux Réalisés sur *Phaseolus* aux Zaire, Rwanda et Burundi de 1945 a nos jours
(Syntheses of Research Work carried out on *Phaseolus* beans in Zaire, Rwanda, and Burundi from 1945 to date)**

1985. 54 pages. 17 × 23 cm. Perfect bound, paperback. Price: $2.00.

Research carried out in *Phaseolus* beans by national agricultural research institutes in Zaire, Rwanda, and Burundi is highlighted.

Available only in French.

CIAT, Bean Program Staff

**The CIAT Bean Program. Research Strategies for Increasing Production
(El Programa de Frijol del CIAT. Estrategias de Investigación para Aumentar Produccion)**

88 pages. Illus. 15 × 23 cm. Perfect bound, paperback. ISBN 84-89206-11-2. Program distribution.

Outlines the CIAT Bean program's objectives, research strategies and methodologies as well as research achievements.

(Describe los objetivos, estrategias y metodologias de investigación del programa de Frijol del CIAT, asi como los logros de su investigacion.)

Available in English only.

Cassava (*Manihot esculenta* Crantz)

J. C. Lozano and R. H. Booth

Diseases of Cassava
(*Manihot esculenta* Crantz)

1976. 46 pages. 17 × 24 cm. Saddle stitched, paperback. Prices: CIAT $1.50. Other: $2.50.

Brief description of diseases attacking cassava and methods for their control.

Available in English only.

J. C. Lozano, J. C. Toro, A. Castro, and A. Bellotti

Production of Cassava Planting Material
(**Produçã**o de Material de Plantio da Mandioca)

1977. 28 pages. 16 × 21 cm. Saddle stitched, paperback. Prices: CIAT: $2.50. Other: $3.00.

Describes the quality, sanitation, and storage period of cassava cuttings along with the factors that affect the effectiveness of cassava "seed."

Available in English and Portuguese.

J. D. Doll and W. Piedrahita

Methods of Weed Control in Cassava
(Metodos de Control de Malezas en Yuca)

1978. 12 pages. 16 × 22.5 cm. Saddle stitched, paperback. Prices: CIAT: $1.50. Other: $2.00.

In experiments conducted at CIAT, it was demonstrated that weeds cause significant yield reductions in cassava. The critical period for weed control is defined.

(En experimentos realizados en el CIAT se demostro que las malezas causan reducciones significativas en el rendimiento del cultivo de la yuca y se definio el periodo critico para su control.)

Available in English and Spanish.

A. Bellotti and A. van Schoonhoven

Cassava Pests and their Control
(Plagas de Yuca y su control)

1978. 73 pages. 60 color plates. 17 × 24 cm. Perfect bound, paperback. List price: $17.00. Developing country prices: CIAT: $5.50, Colombia: $7.00, Americas: $8.50, Other: $9.00.

This monograph discusses host plants, pest distribution, mites and insects attacking the crop, and economic damage. An integrated cassava pest management program is described and recommended.

(Monografia sobre plantas hospederas, distribución de plagas, ácaros e insectos que atacan el cultivo y sus daños economicos. Programa integrado de manejo de plagas. 60 ilustraciones a color sobre plagas.)

Available in English only.

A. Diaz-Duran

A Cassava Harvesting Aid
(Un Implemento Para Cosechar la Yuca)

1979. 14 pages. Illus. 17 × 21 cm. Saddle stitched, paperback. Prices: CIAT: $1.50, Other: $2.00 (Includes air mail postage).

Contiene dibujos detallados para construir un implemento para cosechar la yuca, el cual esta diseñado para romper el suelo y aflojar las raices, facilitando asi la extracción manual del producto.

(Contains a detailed plan for building the harvesting implement, which is designed to break the soil and loosen the roots, making manual extraction easy.)

Available in Spanish only.

R. Best

Cassava Drying
(Secamiento de la Yuca)

1979. 26 pages. Illus. 15.5 × 21 cm. Saddle stitched, paperback. Prices. CIAT: $1.50, Other: $2.00.

Defines the physiological and commercial parameters for cassava drying methods designed principally for use by the small farmer.

(Define los parametros fisiologicos y comerciales del secamiento de la yuca y describe dos metodos de secamiento diseñados para el pequeño agricultor, con una utilizacion minima de insumos.)

Available in English and Spanish.

C. J. Asher, D. C. Edwards, and R. H. Howeler

Nutritional Disorders of Cassava (*Manihot esculenta* Crantz)
(Desordenes Nutricionales de la Yuca [*Manihot esculenta* Crantz])

1980. 48 pages. 26 color plates. 18 × 24 cm. Saddle stitched, paperback. ISBN 84-89206-03-1. Prices: $10.00. CIAT: $5.00, Colombia: $5.50, Americas: $6.00. Other: $6.00.

This monograph discusses the optimum conditions for cassava production and the nutrient requirements.

(Discute las condiciones optimas para la producción de yuca y los nutrimentos necesarios.)

Available in Spanish only.

R. Howeler

Mineral Nutrition and Fertilization of Cassava (*Manihot esculenta* Crantz)
(Nutrición Mineral y Fertilización de la Yuca) (*Manihot esculenta* Crantz)

1981. 55 pages. 17 × 21.5 cm. 10 color plates. Saddle stitched, paperback. ISBN 84-89206-05-8 (84-89206-06-X Spanish ed.). List price: $7.50, CIAT: $3.00, Colombia: $3.50, Americas: $4.00, Other: $4.00.

This report reviews the results of basic research on nutrient uptake and distribution in cassava, as well as the numerous fertilizer experiments conducted on various soils in Latin America, Africa and Asia.

(Revision de los resultados de las principales investigaciones sobre absorción y distribución de nutrimentos de la yuca, asi como los numerosos experimentos que se han realizado con fertilizantes en diferentes suelos de America Latina, Africa y Asia.)

Available in Spanish and English.

J. C. Lozano, A. Bellotti, J. A. Reyes, R. Howeler, D. Leihner, and J. Doll

Field Problems in Cassava (2nd edition)
(Problemas en el Cultivo de la Yuca - 2a. Ed.)

1981. 192 pages. 180 color plates. 11.5 × 17.5 cm. Smythe-sewn, soft cover. ISBN 84-85206-08-2 (84-89206-08-2 Spanish ed.). List price: $20.00. Developing country prices: CIAT: $7.00. Colombia: $9.00. Americas: $10.00. Other: $12.00.

Designed for the identification of field problems in cassava, this manual discusses important diseases, pests, and nutritional problems that diminish cassava yields.

(Diseñado para la identificación de problemas de campo en el cultivo de la yuca, este manual discute las enfermedades y plagas importantes y los problemas nutricionales que disminuyen el rendimiento de este cultivo.)

Available in English and Spanish.

J. C. Toro and C. B. Atlee

Practicas Agronomicas para la Producción de Yuca (Una Revision de la Literatura
(Agronomic Practices in Cassava Production
[Literature Review])

1981. 44 pages. 17 × 23.5 cm. Saddle stitched, paperback. List price: $8.50, CIAT: $2.00, Colombia: $2.50, Americas: $2.50, Other: $2.50.

Revision actualizada de la literatura que cubre las practicas agronomicas principales utilizadas actualmente para producir yuca en diversas regiones del mundo.

(Current literature review covering the main agronomic practices in cassava production used in various parts of the world.)

Available in Spanish only.

D. Leihner

Management and Evaluation of Intercropping Systems with Cassava
(Yuca en Cultivos Asociados: Manejo y Evaluacion)

1983. 79 pages. 15 × 22 cm. Perfect bound, paperback. ISBN (84-89206-28-7 Spanish ed.). List price: $11.00. Developing country prices: CIAT: $4.00. Colombia: $5.00. Americas: $6.00, Other: $6.50.

Guidelines for using cassava in intercropping systems. Improved technology includes selection of plant types; planting time, density, and pattern; nutrition and fertilization; and pest management. Methods of evaluation include biological efficiency, time factor, and crop competition.

(Guia para la utilización de la yuca en cultivos multiples. La tecnología mejorada descrita incluye selección de tipos de planta; tiempo, densidad, y patrones de siembra; nutrición y fertilización; y manejo de plagas. Los métodos de evaluación incluyen la eficiencia biológica, el factor tiempo, y la competencia del cultivo.)

Available in English and Spanish.

Rice

R. L. Cheaney and P. R. Jennings

Field Problems of Rice in Latin America
(Prolemas en Cultivos de Arroz en America Latina)

1975. 91 pages. 12 × 16 cm. 147 color plates. Smythe-sewn, soft cover. List price: $18.50, CIAT: $7.50, Colombia: $8.50, Americas: $9.50, Other: $10.00.

This manual is designed for identifying problems of rice. It includes the most important pests and diseases and problems caused by nutritional deficiencies or toxicities.

(Manual diseñado para ayudar en la identificación de los problemas del cultivo del arroz. Incluye las plagas y enfermedades mas importantes y los problemas ocasionados por deficiencias o toxicidades nutricionales.)

Available in English and Spanish.

A. Diaz-Duran & L. Johnson

A Continuous Rice Production System
(Sistema de Producción Continua de Arroz)

1979. 23 pages. 16 × 21.5 cm. Saddle stitched, paperback. Prices: CIAT: $1.00, Other: $1.50.

The authors suggest the implementation of the Asian flooding system of land preparation to achieve continuous rice production. The factors affecting selection of the farm and equipment, and the cultural practices associated with the system, are discussed.

(Los autores sugieren el empleo del sistema asiatico de preparación del suelo bajo agua para lograr una producción continua de arroz. Se tratan los factores que afectan la selección de la finca y del equipo y las practicas culturales asociadas con el sistema.)

Available in English and Spanish.

P. R. Jennings, W. R. Coffman, and H. E. Kauffman

**Mejoramiento del Arroz
(Rice Breeding)**

1981. 237 pages. 16 × 23 cm. 73 color figs. and plates. Smythe-sewn, soft cover. ISBN 84-89206-06-6. List price: $30.00. Developing country prices: CIAT: $8.50, Colombia: $12.00, Americas: $17.00, Other: $20.00.

Manual practico para los cientificos que deseen consultar cada fase del desarrollo de variedades mejoradas de arroz. Escrito desde un punto de vista interdisciplinario, este libro ilustra la solución de problemas, metodos especificos de fitomejoramiento, y procedimientos operacionales de campo. Incluye descripción detalladas sobre como seleccionar, cruzar y evaluar variedades por sus caracteristicas agronomicas, por su resistencia a enfermedades e insectos; y por su tolerancia a condiciones ambientales desfavorables.

(Practical manual for scientists interested in each phase of the development of improved rice varieties. Written from an interdisciplinary point of view, the book is oriented toward production problem-solving, specific breeding methods, and operational field procedures. It includes detailed descriptions on how to select, cross, and evaluate varieties for their agronomic characteristics, for their resistance to diseases and pests, and for their tolerance to unfavorable climatic conditions.)

Available in Spanish only.

G. M. Scobie and R. Posada T.

The Impact of High-yielding Rice Varieties in Latin America with Special Emphasis on Colombia (El Impacto de las Variedades de Arroz con Altos Rendimientos en America Latina con Enfasis Especial en Colombia)

Reprinted 1983. Orig. pub 1977. 165 pages. 14 × 22 cm. Perfect bound, paperback. List price: $10.00. Developing country prices: CIAT: $2.50, Colombia: $3.00, Americas: $4.00, Other: $5.00 (Includes air mail postage).

This economic analysis provides an overview of the economic impact of high-yielding varieties on rice production and trade in Latin America, with emphasis on Colombia, from 1950 to 1974.

(Este analisis economico brinda una vision general del impacto de las variedades de alto rendimiento sobre la producción de arroz y su comercio desde 1950 hasta 1974.)

Available in English and Spanish.

M. Rosero (Translator and Adaptor)

Sistema de Evaluación Estandar para Arroz (Standard Evaluation System for Rice)

1983. 61 pages. 11 × 17 cm. Perfect bound, paperback. Prices: CIAT: $2.00, Other: $2.50 (Includes air mail postage).

Traducción y adaptación de Standard Evaluación System for Rice, publicado en 1981 por IRRI (International Rice Research Institute). Presenta la metodologia para calificar carateristicas agronomicas, reacción a plagas, enfermedades y problemas fisico-quimicos de los materiales en evaluacion mediante una escala uniforme, y para determinar sus caracteristicas morfologicas.

(Translation and adaptation of Standard Evaluation System for Rice, published in 1981 by the International Rice Research Institute (IRRI). Presents the methodology for describing agronomic characteristics; quantifying reaction to pests, diseases, and physiochemical problems in planting materials being evaluated, using a uniform scale; and determining morphological characteristics.)

Available from CIAT in Spanish only. English edition is available from IRRI.

Tropical Pastures

J. Doll and P. Argel

Guía Práctica para el Control de Malezas en Potreros (Practical Guide for Weed Control in Pastures)

Orig. pub. 1978. Reprinted 1984. 30 pages. 16×22 cm. Saddle stitched, paperback. Prices: CIAT: $1.50. Other: $2.00 (Includes postage.)

Los autores consideran que el control de malezas es un paso muy importante al disenar un plan de manejo de potreros. Trata los factores que favorecen la invasion de malezas, metodos para controlarla, selección de herbicidas, y precauciones necesarias.

(The authors consider that weed control is a very important step in a pasture development plan. This guide considers factors favoring weed invasion, weed control methods, herbicides selection, and necessary precautions.)

Available in Spanish only.

G. O. Mott (Editor)

Handbook for the Collection, Preservation and Characterization of Tropical Forage Germplasm Resources (Manual para la Colección, Preservación y Characterización de Recursos Forrajeros Tropicales)

1979. 106 pages. 17 × 24 cm. Perfect bound, vinyl cover. List price: $16.00. CIAT: $0.00, Columbia. $7.00, Americas: $8.00, Other: $8.50.

Guide for establishing a coordinated and standardize system for the collection, classification, preservation, distribution, and evaluation of various ecotypes of native tropical grass and legume forage species.

(Guia para el establecimiento de un sistema coordinado y estandarizado para la colección de ecotipos de especies de gramineas y leguminosas forrajeras nativas de los tropicos.)

Available in English and Spanish.

J. M. Toledo (Editor)

Manual para la Evaluación Agronómica: Red Internacional de Evaluación de Pastos Tropicales (Handbook for Agronomic Evaluation: International Network for the Evaluation of Tropical Pastures)

1982. 168 pages. 17 × 24 cm. 75 color plates. Smythe-sewn, hard cover. ISBN: 84-89206-12-0. CIAT: $12.00, Colombia: $14.00, Americas: $16.50, Other: $19.00.

Summarizes the methodologies agreed upon at the first workshop of the Tropical Pastures Network held in 1979, for agronomic evaluations in regional trials A and B. This publication is a guide for participants in the network to standardize techniques in evaluating materials in these trials and to improve evaluation of germplasm adaptable to the various ecosystems in the region.

(Resumen de las metodologias acordadas en la primera reunión de trabajo de la Red Internacional de Pastos Tropicales efectuada en 1979, para las evaluaciones agronómicas en los ensayos regionales A y B. Es una guia que permite a los participantes en la Red uniformar sus técnicas de evaluación de los materiales en dichos ensayos, logrando conocer mejor el germoplasma más susceptible de adaptación a los distintos ecosistemas de la región.)

Available in Spanish only.

L. G. Reca and J. M. Frogone

Rasgos Characterísticos de la Ganaderia Vacuna Argentina (Characteristic Traits of the Cattle Industry of Argentina)

1982. 84 pages. 20 × 27 cm. Perfect bound, paperback. ISBN: 84-89206-20-1. Prices: CIAT: $5.00, Colombia: $6.00, Americas: $8.00, Other: $10.00 (Includes air postage).

Estudio económico de la industria: caracterización del sector ganadero, demanda de carne vacuna, el mercado de leche, y la politica ganadera.

(Economic study of the industry: characterization of the beef cattle sector, demand for beef, market for milk and dairy products, and policies that influence the industry.)

Available in Spanish only.

L. Rivas and J. L. Cordeu

Potencial de Producción de Carne Vacuna en América Latina: Estudio de Caso
(Potential for Beef Production in Latin America: a Case Study)

1983. 93 pages. 15 × 23 cm. Perfect bound, paperback. List price: $12.00. Prices for developing countries: CIAT: $3.00, Colombia: $3.50, Americas: $4.50, Other: $6.00 (Includes air mail postage).

Estudio hecho en conjunto con la FAO, analiza el desarollo histórico del sector ganadero en América Latina. Describe la producción, consumo, demanda, y productividad de la carne, y estima en términos ecónomicos el impacto potencial de la nueva tecnología de pasturas en tierras marginales de los trópicos de América Latina.

(This study, done in conjunction with FAO, analyzes the historical development of the beef cattle industry in Latin America. It describes the production, consumption, and demand for beef, and estimates the potential, in economic terms, of the new pasture technology for marginal lands in the Latin American tropics.)

Available in Spanish only.

O. Paladines and C. Lascano (Editors)

Germoplasma Forrajero bajo Pastoreo en Pequeñas Parcelas
(Forage Species under Grazing on Small Plots)

1983. 185 pages. 17 × 23 cm. Smythe-sewn, hard cover. ISBN: 84-89206-29-5. List Price: $19.00. Developing country prices: CIAT: $7.00, Colombia: $9.00, Americas: $11.50, Other: $13.00 (Includes air mail postage).

Contiene la contribución de 53 especialistas de 20 países sobre producción animal y manejo de praderas. En sus varios capítulos, se analiza el efecto de los factores edáficos y climáticos sobre las pasturas; la estructura y la descarga de éstas; la evaluación de pequeños potreros por corte o por pastoreo; y diversas relaciones entre los sistemas de manejo del pastoreo y la selección, temprana o avanzada, de especies forrajeras.

(The contributions of 53 specialists representing 20 countries analyze the effects of soil and climatic factors on pastures, the structure and capacity of pastures, the evaluation of small plots for feed and forage, and the relationships between management systems and selection of species for pastures.)

Available in Spanish only.

D. García and J. Ferguson

Cosecha y Beneficio de Semillas de *Andropogon gayanus* (Harvesting and Processing of *Andropogon gayanus* Seed)

1984. 35 pages. 15 × 23 cm. Perfect bound, paperback. ISSN: 0120-5943. Prices: CIAT: $1.50, Other: $2.00.

Primero en la serie de Boletines Técnicos del Programa de Pastos Tropicales del CIAT. Describe métodos de cosecha y beneficio de la semilla.

(First of a series of Technical Bulletins from CIAT's Tropical Pastures Program. Describes seed harvesting and processing methods.)

Available in Spanish only.

R. Vera and C. Seré (Editors)

Sistemas de Producción Pecuaria Extensiva. Brasil, Colombia, Venezuela (Extensive Cattle Production System. Brazil, Colombia, Venezuela)

1984. 530 pages. 15 × 23 cm. Perfect bound, paperback. ISBN: 84-89206-41-4. List price: $26.00. Developing country prices: CIAT: $8.00, Colombia: $10.00, Americas: $13.00, Other: $16.00.

Estudio biológico y económico de los sistemas de producción ganadera en ecosistemas de sabana y cerrado en América Latina. Resultados del Proyecto de Estudio Técnico-Económico de Sistemas de Producción Pecuaria (ETES) con patrocinio de la Agencia Alemana para la Cooperación Técnica (GTZ), Universidad Técnica de Berlín (TUB) y Centro Internacional de Agricultura Tropical (CIAT).

(Biological and economic study of cattle production systems in savanna and "cerrado" ecosystems of Latin America. Results from the Technical-Economic Study of Cattle Production Systems Project sponsored by the German Agency for Technical Cooperation (GTZ), Technical University of Berlin (TUB) and Centro Internacional de Agricultura Tropical [CIAT]).

Available in Spanish only.

Seeds

J. Douglas (Compiler, Editor)

Programa de Semillas: Guía para su Planeación y Manejo (Successful Seed Programs: a Planning and Management Guide)

1982. 357 pages. 15 × 23 cm. Perfect bound, paperback. ISBN (84-89206-16-3 Spanish ed.). List price: $28.50, CIAT: $6.50, Colombia: $8.50, Americas: $11.00, Other: $13.50.

Guia internacional que trata sobre cómo planear, ejecutar y manejar programas de semillas, asi cómo sobre los requisitos y responsabilidades al hacer las multiplicaciones iniciales de semillas de nuevas variedades; los sistemas de control de calidad; y los métodos de extensión y mercadeo para ayudar a la transferencia de tecnología y al suministro de semillas a los agricultores.

(International guide dealing with the requirements and responsibilities involved in initial seed multiplication from new varieties; quality control systems; and extension and marketing to assist in the transfer of technology and supply of seed to the farmer. [English edition of a book originally published by Westview Press, Boulder, Colo., ISBN 0-89158-872-8.])

Available in Spanish only. (English edition available from Westview Press.)

Conference, Symposia, and Workshop Proceedings

General

CIAT, Communication and Information Support Unit

Proceedings of the 10th Anniversary (Memorias 10° Aniversario)

1984. 90 pages. 21.5 × 28 cm. Illus. Perfect bound, paperback. Free.

Includes complete texts of the messages presented during the Symposium and Formal Acts of Commemoration on the occasion of CIAT's 10th anniversary of the inauguration and dedication of its facilities, as well as the program, summary of activities, a photo collage of memories, and several examples of the articles that appeared in the Colombian press.

(Incluye la versión completa de los mensajes presentados durante el Simposio y Actos Formales de Conmemoración durante la celebración del 10° aniversario de la inauguración y dedicación de las instalaciones del CIAT, así como el programa, resumen de actividades, composición fotográfica de los eventos y varios recortes de artículos aparecidos en la prensa nacional.)

Available in English and Spanish.

C. J. Lozano (Editor)

Proceedings of the Fifth International Conference on Plant Pathogenic Bacteria

1982. 640 pages. 15 × 23 cm. Perfect bound, paperback. ISBN: 84-89206-21-X. List price: $40.00. Developing country prices: CIAT: $20.00, Colombia: $24.00, Americas: $30.00, Other: $35.00.

Contains the texts and illustrations of 80 papers presented by scientists from 29 countries who participated in the 1981 international conference. The papers report on research dealing with new bacterial diseases, recent developments in microbial identification and taxonomy, ecology of bacterial pathogens, host-pathogen interactions, and disease management of bacterial pathogens.

Available in English only.

S. B. Hecht (Editor)

Amazonia: Agriculture and Land Use Research (Amazonia: Investigación Sobre Agriculturas y Uso de Tierras)

1982. 430 pages. 16 × 23 cm. Perfect bound, paperback. ISBN: 84-89206-14-7 (84-89206-15-5 Spanish ed.). List price: $28.00. Developing country prices: CIAT: $11.00, Colombia: $16.00, Americas: $22.50, Other: $26.50.

(Reune los trabajos científicos y técnicos presentados por 21 autores en la conferencia celebrada en 1980 en el CIAT. Los asistentes se ocuparon del tema de las políticas nacionales en la cuenca amazónica, revisaron el estado actual de la investigación científica en curso como la ya realizada, e identificaron nuevos temas de investigación. El propósito general de la conferencia fue ampliar el acervo de conocimientos sobre nuevas technologías agrícolas y de uso de la tierra para los trópicos húmedos de América del Sur.)

Complete scientific and technical papers presented by 21 authors at the conference held at CIAT in 1980. The participants addressed the issue of national policies in the Amazon basin; reviewed the current state of research knowledge, both completed and in progress; and identified new topics for research. The overall purpose of the conference was to enlarge the knowledge base about new agriculture and land use technologies for the humid tropics of South America.

Available in English and Spanish.

P. H. Graham and S. C. Harris (Editors)

Biological Nitrogen Fixation Technology for Tropical Agriculture

1982. 768 pages. 16 × 23 cm. Smythe-sewn, hard cover. ISBN: 84-89206-22-8. List price: $42.00. Developing country prices: CIAT: $20.00, Colombia: $25.00, Americas: $32.00, Other: $37.00.

Contains 84 papers presented at the workshop held at CIAT in 1981, covering all aspects of BNF technologies as they are applied to agricultural productivity. Includes reports on inoculant production; critiques of methodologies used in BNF research; discussions of the plant, soil, microbial, and environmental factors that affect BNF; consideration of the nitrogen economies of tropical cropping systems; and discussion of the present research programs underway in this area of research.

Available in English only.

Beans

S. Amaya and F. Motta (Editors)

Potential for Field Beans in Eastern Africa: Proceedings of a Regional Workshop Held in Lilongwe, Malawi

1981. 226 pages. 15 × 23 cm. Perfect bound, paperback. ISBN: 84-89206-07-4. List price: $15.00. Developing country prices: CIAT: $5.00, Colombia: $7.00, Americas: $9.50, Other: $12.00.

The proceedings of this seminar include reports on bean production in each of the seven participating countries, followed by a summary of panel discussions and the concluding recommendations for increasing bean production in Eastern Africa.

Available in English only.

S. Amaya and C. Dagnon (Editors)

Potential for Field Beans (*Phaseolus vulgaris* L.) in West Asia and North Africa: Proceedings of a Regional Workshop in Aleppo, Syria

1985. 143 pages. 15 × 22 cm. Perfect bound, paperback. ISBN: 84-89206-45-7. List price: $15.00. Developing country prices: CIAT: $5.00, Colombia: $7.00, Americas: $9.50, Other: $12.00.

This publication compiles papers by bean researchers from 12 countries in West Asia, North Africa, and Spain describing the situation of bean cultivation in each country, with emphasis on research activities and prospects.

Available in Spanish only.

Cassava

T. Brekelbaum, A. Bellotti, and J. C. Lozano

Proceedings of the Cassava Protection Workshop

1978. 244 pages. 17 × 24 cm. Perfect bound, paperback. List price: $25.00, CIAT: $10.50, Colombia: $13.50, Americas: $18.00, Other: $21.40.

Participants at this workshop advocated pest management through host plant resistance, biological control, and phytosanitary practices.

(Los participantes en el taller apoyaron del control de plagas mediante la resistencia de la planta hospedante, control biólogico y medidas fitosanitarias.)

Available in English only.

W. H. Roca, C. D. Hershey, and O. S. Malamud

**Primer Taller Latinoamericano sobre Intercambio de Germoplasma de Papa y Yuca — Memorias, Cali, Colombia
(First Latin American Workshop on Exchange of Potato and Cassava Germplasm — Proceedings, Cali, Colombia)**

1982. 295 pages. 15 × 22 cm. Perfect bound, paperback. ISBN: 84-89206-23-6. List price: $18.00, CIAT: $6.00, Colombia: $8.00, Americas: $10.50, Other: $13.00.

Recoge los aspectos netamente técnicos sobre las plagas y enfermedades que afectan los cultivos de yuca y papa, y los métodos más refinados para disminuir el riesgo de contaminarlos cuando se introduce material vegetal foráneo a un país o región.

(Reviews technical aspects of pests and diseases limiting potato and cassava crops and describes the most refined methods to diminish the risk of contamination when introducing foreign vegetative material into a country or region.)

Available in Spanish only.

J. C. Toro

Evaluación de Variedades Promisorias de Yuca en América Latina y el Caribe
(Evaluation of Promising Cassava Varieties in Latin America and the Caribbean)

1983. 184 pages. 25 × 22 cm. Perfect bound, paperback. ISBN: 84-89206-24-4. List price: $15.00, CIAT: $5.00, Colombia: $6.50, Americas: $7.50, Other: $8.00.

Memorias del primer taller sobre el tema, llevado a cabo en el CIAT en 1982, patrocinado por el Centro Internacional de Investigaciones para el Desarollo (CIID). Participaron nueve paises cuyos trabajos describen el estado del cultivo de la yuca en ellos, asi como los progresos obtenidos en investigación y extensión rural.

(Proceedings of the first workshop on this subject, held at CIAT in 1982 and coordinated by the International Development Research Centre (IDRC). The papers presented by the nine countries participating describe the current state of cassava culture in their respective countries as well as the progress achieved in research and rural extension.)

Available in Spanish only.

J. A. Reyes (Compiler)

Yuca: Control Integrado de Plagas
(Cassava: Integrated Pest Control)

1983. 362 pages. 17 × 23 cm. Perfect bound, paperback. List price: $18.00. Developing country prices: CIAT: $7.00, Colombia: $8.50, Americas: $10.00, Other: $12.00 (Includes air mail postage).

Una referencia para los cursos de capacitación sobre control integrado de plagas de la yuca dictados por el CIAT. Contiene artículos sobre fundamentos del control integrado, pérdidas en rendimiento causadas por insectos y ácaros, biología, morfología, y ecología de varios insectos, resistencia varietal, control biólogico, y control integrado.

(A reference book for the training courses at CIAT on integrated pest control.)

Available in Spanish only.

J. Cock (Editor)

Global Workshop on Root and Tuber Crops Propagation (Seminario Mundial sobre Propagación de Raíces y Tubérculos)

1986. Expected in the first trimester.

Proceedings of a global workshop held in CIAT in September 1983 covering description of crops (potato, cassava, sweet potato, yam and cocoyam) and emphasis on the various aspects of seed production. Sponsorship from CIAT, CIP and IITA.

(Memorias del seminario mundial celebrado en CIAT en septiembre 1983, que comprendió descripción de los cultivos (papa, yuca, batata, ñame y taro) a hizo énfasis especial en los varios aspectos de producción de semilla. Patrocinado por CIAT, CIP y IITA.)

Available in English only.

Carlos E. Domínguez (Compiler)

Yuca: Investigación, Producción y Utilización (Cassava: Research, Production and Utilization)

Reprinted 1985. Orig. Pub. 1983. 660 pages. 16 × 23 cm. Perfect bound, paperback. List price: $30.00. Developing country prices: CIAT: $15.00, Colombia: $20.00, Americas: $22.00, Other: $25.00.

Referencia de los cursos de capacitación sobre yuca dictados por el CIAT. Contiene artículos sobre el conocimiento morfológico y fisiológico de la planta, mejoramiento, prácticas agronómicas y manejo del cultivo, plagas y enfermedades, utilización y evaluación agronómica de las variedades mejoradas.

(This compilation of conferences given during CIAT training courses on cassava includes articles on morphological and physiological aspects of the plant, breeding, agronomic practices and crop management, pests and diseases, utilization, and agronomic evaluation of improved varieties.)

Available only in Spanish.

Pastures

J. C. Thompson (Editor)

Proceedings of a Workshop on the Ecology and Control of Ectoparasites on Bovines in Latin America

1978. 186 pages. 17 × 24 cm. Perfect bound, paperback. List price: $15.00, CIAT: $6.50. Colombia: $8.50, Americas: $11.00, Other: $13.50.

This workshop deals principally with the problem of ticks and disease vectors on bovines. Topics include tick distribution in Latin America, design of tick control programs, and current research on the control of these ectoparasites.

Available in English only.

E. A. Wells (Editor)

Workshop on Hemoparasites (Anaplasmosis and Babesiosis)
(Reunión de Discusión sobre Hemoparásitos — Anaplasmosis y Babesiosis)

1978. 137 pages. 17 × 24 cm. Perfect bound, paperback. List price: $15.00, CIAT: $5.50, Colombia: $7.50, Americas: $10.00, Other: $12.50.

Considered in this workshop, held at CIAT in 1975, are the disease epidemiology, diagnosis, and preliminary control methods for two tick-vectors, hemoparasites in cattle.

(En esta reunión celebrada en el CIAT en 1975 sobre dos hemoparásitos transmitidos por garrapatas a los bovinos, se consideró la epidemiología de las dos enfermedades asi como su diagnóstico y métodos preliminares de control.)

Available in English and Spanish.

L. E. Tergas and P. A. Sanchez (Editors)

Pasture Production in Acid Soils of the Tropics (Producción de Pastos en Suelos Acidos de los Trópicos)

1979. 521 pages. 17 × 23.5 cm. Perfect bound, paperback. List price: $35.00. CIAT: $12.00, Colombia: $16.00, Americas: $20.00, Other: $25.00.

Proceedings of a seminar conducted at CIAT on production, management, and forage utilization of pastures under acid soil conditions for beef production in Latin America.

(Memorias de un seminario llevado a cabo en el CIAT sobre producción, manejo y utilización de forrajes bajo condiciones de suelos acidos para la producción de ganado de carne en América Latina.)

Available in English and Spanish.

Rice

M. Rosero

Report of the Fifth Conference of the IRTP for Latin America
(Informe de la Quinta Conferencia del IRTP para América Latina) August 9-13, 1983

1984. 182 page. 21 × 27 cm. Illus. Perfect bound, paperback. Free.

Proceedings of the fifth IRTP conference containing rice research highlights, needs and future IRTP plans for Latin America, technical reports on hoja blanca in Latin America and rice leafhoppers in Asia, germplasm collection in Latin America, upland rice production systems in Latin America and Asia, as well as descriptions of current rice production in Latin America.

(Memorias de la quinta conferencia del IRTP con avances de la investigación, necesidades y planes futuros del IRTP, informes sobre saltahojas en Asia y hoja blanca en América Latina, colección de germoplasma de arroz en América Latina, sistemas de producción de arroz de secano en América Latina y Asia, así como descripción de la producción actual de arroz en América Latina.)

Available in English and Spanish.

Seeds

CIAT Seed Unit

Memorias de las Reuniones de Trabajo sobre Semillas. 5 volúmenes
(Proceedings of the Seed Workshop. 5 volumes)

1983. Pages vary. 15 X 21 cm. Perfect bound, paperback. Prices: CIAT: $3.00, Colombia: $4.00, Americas: $5.00, Other: $6.00 (Includes air mail postage).

Una serie de memorias de reuniones de trabajo patrocinadas por la Unidad de Semillas del CIAT. Contiene: Semilla Mejorada para el Pequeño Agricultor; Estrategias, Planeación, y Ejecución de un Programa de Semillas; Estrategias para la Capacitación en Tecnología de Semillas (dos vols.); y Administración y Mercadeo en Empresas de Semillas.

(A series of proceedings, in abstract form, of meetings convened by the CIAT Seed Unit. Contains: Improved Seed for Small Farmers; Planning and Implementing a Seed Program; Training Strategies in Seed Technology (2 vols.); and Administration and Marketing in Seed Enterprises.)

Available in Spanish only.

Miscellaneous Reports and Brochures

Unless otherwise indicated, the following reports and working documents are available from the Distribution Office at a charge of US$2.00/copy, or directly from the Coordinator of each Program.

Si no se indica la contrario, los siguientes informes y documentos de trabajo pueden solicitarse en la Oficina de Distribución de Publicaciones, o directamente a los coordinadores de los programas del CIAT.

CIAT

CIAT in the 1980s: a Long-Range Plan for the Centro Internacional de Agricultura Tropical
(CIAT en la Década de los 80: Un Plan a Largo Plazo para el Centro Internacional de Agricultura Tropical)

1981. 182 pages. 17 × 23 cm. Perfect bound, paperback.

As a basis for future planning, this document examines CIAT's role in relation to specific needs of the developing countries, the total global agricultural research and development framework, and the family of CGIAR-supported international research centers.

(Planeamiento del CIAT para la decado de los 80.)

Available in English only.

O. Voysest and J. García

Vivero Internacional de Rendimiento y Adaptación de Frijol: IBYAN 1976-1980
(International Yield and Adaptation Nursery)

1976-1981. Pages vary. 20 × 27 cm. Perfect bound, paperback.

Describe el desarrollo y los resultados del IBYAN en el cual se evaluaron variedades en diversos países del mundo para medir su: hábito de crecimiento, altura de la planta; volcamiento, dehisencia, número de vainas, peso calidad de la semilla, días a la madurez fisiológica y resistencia a plagas y infermedades.

(Describes the development and results of the IBYAN field trials. Evaluates growth habit, architecture, seed weight and quality, days to maturity, resistance to diseases and pests, and other key characteristics.)

Available in Spanish only.

CIAT Bean Program Staff

Ensayos Preliminares, EP 1982-1983
(Preliminary Trials, 1982-1983)

1983. 37 pages. 20 × 27 cm.

Resultado de la evaluación de 304 materiales seleccionados entre 1101 líneas evaluadas en el Vivero del Equipo de Frijol (VEF). Se consideraron los siguientes parámetros: resistencia a las plagas y enfermedades, respuesta al fotoperíodo y a la sequía, calidad del grano, rendimiento y adaptación.

(Results of the evaluation of 180 lines with different growth habits and different grain types.)

Available in Spanish only.

CIAT Bean Program Staff

Vivero Internacional de Roya del Frijol: Resultados
(International Rust Nursery: Results)

1983. 22 pages. Published every two years. 79-80 results available.

Describe el desarrollo y los resultados de este vivero, incluyendo cuadros en los que se clasifica la reacción al ataque de la roya de diferentes accesiones de frijol de acuerdo al sitio donde se llevaron a cabo las evaluaciones.

(Describes the development and results of this nursery, including tables classifying the reaction to attack by rust of different bean accessions.)

Available in Spanish only.

G. Gálvez, S. H. Orozco et al.

VICAR 1981 y 1982
(VICAR 1981 and 1982)

1984. 92 pages. 21 × 28 cm. Perfect bound, paperback. Free.

Resultados de pruebas agronómicas, de rendimiento y tecnológicas de frijol en cooperación con agencias nacionales de países centro-americanos.

(Results from agronomic, yield and technological bean trials in cooperation with national agencies from Central American countries.)

Available in Spanish only.

E. Pizarro (Editor)

Red Internacional de Evaluación de Pastos Tropicales: Resultados 1979–1982
(Results of the International Network for Pasture Evaluation, 1979–1982)

1983. 460 pages. 20 × 27 cm. Perfect bound, paperback. ISSN: 0120-4882. Prices: CIAT: $6.00, Colombia: $7.00, Americas: $8.00, Other: $10.00 (Includes air mail postage).

Este informe presenta los resultados de investigaciones colaborativas de la RIEPT, organizados por ecosistemas y tipo de ensayo.

(Presents the results of collaborative research done by the International Network for Pasture Evaluation [RIEPT], organized by ecosystem and type of trial.)

Available in Spanish only.

M. Calderón and G. Arango

Insectos Asociados con Especies Forrajeras en América Tropical
(Insects Associated with Forage Species in Tropical America)

1985. 44 pages. 17 × 23 cm. Perfect bound, paperback. ISBN: 84-89206-42-2. List price: $12.00. Developing country prices: CIAT: $4.00, Colombia: $4.50, Americas: $5.00, Other: $7.00.

Available in Spanish only.

CIAT Rice Program Staff

Informe Final de los Viveros Internacionales de Arroz para América Latina
(Final Report on the International Rice Nurseries for Latin America)

1983 results available.

Estos informes presentan los resultados de los ensayos de los siguientes viveros: VIRAL-P (variedades precoces), VIRAL-S (variedades de secano) VIAVAL (añublo de la vaina), y VIRAL-T (variedades tempranas).

(These reports give the results of field trials from the various rice nurseries.)

Available in Spanish only.

Bibliographies, Directories, Abstracts

General

CEDEAL Staff (Documentalists)

Resúmenes Analiticos en Economia Agricola Latinoamericana
(Abstracts on Latin American Agricultural Economics)
(Vol. I, 1976; Vol. II, 1977; Vol. III, 1978; Vol. IV, 1979; Vol. V, 1980; Vol. VI, 1981)

Yearly 1976-1981. Varies (approx. 275 pages/vol.). 16.5 × 23 cm. Perfect bound, paperback. ISSN 0120-2936. $3.00.

Cada volumen compila resúmines analiticos en economia agrícola latinoamericana e incluye un listado de palabras claves para solicitar la búsqueda de teams especificos al Centro de Documentación Económica para la Agricultural Latinoamericana (CEDEAL).

(Each volume constitutes a compilation of abstracts on agricultural economics in Latin America and includes a list of key words used by CIAT's Documentation Center for Latin American Agricultural Economics [CEDEAL] to facilitate topic searches.)

Available in Spanish only.

Beans

Bean Information Center Staff

Abstracts on Field Beans
(Resúmenes Analiticos sobre Frijol)
Vol. I, 1976; Vol. II, 1977; Vol. III, 1978; Vol. IV, 1979; Vol. V, 1980; Vol. VI, 1981; Vol. VII, 1982; Vol. VIII, 1983; Vol. IX, 1984; Vol. X, 1985

Yearly 1976-1981; 3/year 1982+. Varies (approx. 325/vol.). 15×22 cm. Perfect bound, paperback. ISSN: 0120-2928. List price: $25.00/yr.

Developing country price: $15.00. Back issues: $3.00 (1970's); $7.00 (1980's).

Subject specialists analyze hundreds of articles in scientific and technical journals, manuals, bulletins, mimeographed works, and other sources and abstract the contents, augmenting the citation with keywords for quick retrieval. These abstracts are arranged by subject heading in periodic volumes that correspond to three key areas of CIAT research. All abstracts are subject and author indexed. Every title listed in the abstracts is available through the Documentation Service Unit, at CIAT.

(Documentalistas especializados analizan cientos de articulos publicados en revistas cientificas y técnicas, manuales, boletines, trabajos mimeografiados y de otras fuentes, y resumen sus contenidos ampliando la cita bibliográfica con palabras claves que permiten una recuperación rápida. Estos resúmenes se organizan por encabezamientos de materia, en volúmenes periódicos que corresponden a las 3 áreas claves de la investigación del CIAT. Todos los resúmenes se indizan por autor y por materia, y cada titulo indicado en los resúmenes esta disponible por intermedio de la Unidad de Documentación del CIAT.)

Available in English and Spanish.

J. López S. (Compiler)

Bibliography on Bean Research in Africa

1983. 177 pages. 17 × 23 cm. Perfect bound, paperback. ISBN: 84-89206-38-8. List price: $10.00. Developing country price: $6.00 (Includes air mail postage).

Contains 813 references to African bean literature. These documents, most difficult to obtain, give the history of research on *Phaseolus vulgaris* on the African continent. More than 50% of the documents are abstracted, and many are available in their entirety from the Bean Information Center at CIAT.

Titles in language of original publication. All abstracts in English only.

Bean Information Center

Bibliography on Bean Research in Africa. Supplement 1984

1984. 164 pages. 17 × 23 cm. Perfect bound, paperback. ISBN: 84-89206-43-0. List price: $10.00. Developing country price: $6.00.

This supplement includes 358 new references. Documents cited in the first issue and acquired by the Bean Information Center during 1984 are listed and summaries are given of another 138 documents.

Available in English only.

Pastures

Pastures Information Center Staff

Resúmenes Analiticos sobre Pastos Tropicales (Abstracts on Tropical Pastures) Vol. I, 1979; Vol. III, 1981; Vol. VI, 1984, Vol. VIII, 1985

Yearly 1979-1981; 3/yr 1982+. Varies (approx. 400 pages/vol.). 15×22 cm. Perfect bound, paperback. ISBN/ISSN: 0120-2944. List price: $25.00. Developing country price: $15.00. Back issues: $3.00 (1970's), $7.00 (1980's).

Especialistas en información analizan cientos de articulos publicados en revistas cientificas y técnicas, manuales, boletines, trabajos mimeografiados y de otras fuentes, y resumen sus contenidos ampliando la cita bibliográfica con palabras claves que permiten una recuperación rápida. Estos resúmenes se organizan por encabezamiento de materia, en volúmenes periódicos que corresponden a las 3 áreas claves de la investigación del CIAT. Todos los resúmenes se indizan por autor y por materia y cada titulo indicado en los resúmenes está disponible por intermedio de la Unidad de Servicios de Documentación del CIAT.

(Subject specialists analyze hundreds of articles in scientific and technical journals, manuals, bulletins, mimeographed works, and other sources and abstract the contents, augmenting the citation with keywords for quick retrieval. These abstracts are arranged by subject heading in periodic volumes that correspond to three key areas of CIAT research. All abstracts are subject and author indexed. Every title listed in the abstracts is available through the Documentation Services Unit, at CIAT.

Available in Spanish only.

M. Mejía M. (Compiler)

Scientific and Common Names of Tropical Forage Species
(Nombres Científicos y Vulgares de Especies Forrajeras Tropicales)

1984. 75 pages. 17 × 23 cm. Perfect bound, paperback. ISBN: 84-89206-40-6. List price: $5.00. Developing country price: $3.00.

Compiles the scientific name and all the common names, in English, Spanish and Portuguese, of the major tropical forage species of Latin America.

(Esta publicación presenta los nombres científicos de las especies forrajeras tropicales más importantes en América Latina con sus nombres vulgares en inglés, español, y portugués.)

English/Spanish publication.

M. Mejía (Compiler)

Andropogon gayanus Kunth: Bibliografía Analítica

1984. 176 pages. 17 × 23 cm. Perfect bound, paperback. List price: $10.00. Developing country price: $6.00.

Contiene 393 referencias organizadas en 13 categorías temáticas. Estos documentos presentan los resultados de estudios realizados en esta especie en América Latina y otras regiones del mundo.

Títulos en el lenguaje original de la publicación. Todos los resúmenes en español.

(A total of 393 references to the results of studies carried out on this species in Latin America and other areas are organized under 13 thematic categories. Titles are in the language of the original publication and all abstracts are in Spanish.)

Cassava Information Center Staff

Abstracts on Cassava
(Resúmenes Analíticos sobre Yuca)
Vol. II, 1976; Vol. III, 1977; Vol. IV, 1978; Vol. V, 1979; Vol. VI, 1980; Vol. VII, 1981; Vol. VIII, 1982; Vol. IX, 1983; Vol. X, 1984

Yearly 1976-1981; 3/yr 1982 +. Varies (approx. 300/vol.). 15 × 22 cm. Perfect bound, paperback. ISSN: 0120-288X. List price: $25.00. Developing country price: $15.00. Back issues: $3.00 (1970's), $7.00 (1980's).

Subject specialists analyze hundreds of articles in scientific and technical journals, manuals, bulletins, mimeographed works, and other sources and abstract the contents, augmenting the citation with keywords for quick retrieval. These abstracts are arranged by subject heading in periodic volumes that correspond to three key areas of CIAT research. All abstracts are subject and author indexed. Every title listed in the abstracts is available through the Documentation Service Unit, at CIAT.

(Especialistas en información analizan cientos de articulos publicados en revistas cientificas y técnicas, manuales, boletines, trabajos mimeografiados y de otras fuentes, y resumen sus contenidos ampliando la cita bibliográfica con palabras claves que permiten una recuperación rápida. Estos resúmenes se organizan por encabezamiento de materia, en volúmenes periódicos que corresponden a las 3 áreas claves de la investigación del CIAT. Todos los resúmenes se indizan por autor y por materia y cada titulo indicado en los resúmenes está disponible por intermedio de la Unidad de Servicios de Documentación del CIAT.)

Available in English and Spanish.

J. López S. and J. H. Cock (Compilers)

Cassava Workers Directory (Second edition)
(Directorio de Investigadores de la Yuca [Segunda edición])

Expected in the third semester.
1984. 200 pages. 21 × 27 cm.

Periodicals

H. Ferrerosa (Writer)
J. Reeves (Writer)
S. Amaya (Editor)

CIAT International
(CIAT Internacional)

3/yr. 8 pages. 21 × 27.5 cm. Saddle stitched. ISSN: 0110-4084.

Newsletter highlighting CIAT's current research activities and results and its impact in international development.

(Este boletin destaca las actividades y los resultados de la investigación actual del CIAT, y su impacto sobre el desarrollo internacional.)

Available in English and Spanish.

A. L. de Román (Ed.) F. Gonzales (Doc.)

**Hojas de Frijol
(Bean Newsletter)**

3/yr. 4 pages. 21 × 27.5 cm. Saddle stitched. ISSN: 0120-2480.

Este boletin ofrece notas técnicas, noticias sobre eventos a realizarse, e informes de programas de investigaciones en frijol de todas partes del mundo.

(This newsletter provides interchange of technical notes, news of upcoming events, and progress reports from bean researchers throughout the world.)

Available in Spanish only.

A. L. de Román and M. de West (Ed. & Trans.)

**Cassava Newsletter
(Yuca-Boletin Informativo)**

2/yr. 8 pages. 21 × 27.5 cm. Saddle stitched. ISSN: 0120-1824.

This newsletter provides interchange of technical notes, news of upcoming events, and progress reports from cassava researchers throughout the world.

(Esta serie de boletines ofrece notas técnicas, noticias sobre eventos a realizarse a informes de progreso de los investigadores en yuca de todas partes del mundo.)

Available in English and Spanish.

A. Ramírez (Editor)

**Pasturas Tropicales, Boletín
(Tropical Pastures Bulletin)**

3/yr. pages vary. 21 × 27.5 cm. Saddle stitched.

Publicado tres veces al año, reune artículos científicos, notas de investigación y comentarios sobre investigación en pasturas en el trópico.

(Published three times a year, this bulletin provides interchange of scietific articles, research advances and comments of tropical pastures research.)

Available in Spanish only with abstracts in English.

A. L. de Román (Editor)

Arroz en las Américas

2/yr. pages vary. 21 × 27.5 cm. Saddle stitched. ISSN: 0120-2634.

Boletin divulgativo que ofrece notas técnicas, noticias sobre eventos a realizarse, e informes de los resultados de la investigación en arroz en América Latina.

(This newsletter provides interchange of technical notes, news of upcoming events, and progress reports from rice researchers throughout Latin America.)

Available in Spanish only.

J. Fernández de Soto y Constanza Anzola (Editors)

Semillas para América Latina

4/yr. 4 pages. 21 × 27.5 cm. Saddle stitched. ISSN: 0120-2979.

Boletin que ofrece información sobre cursos, reuniones y otras actividades de la Unidad de Semillas del CIAT y de organizaciones semillistas de América Latina.

(Newsletter offering information on courses, meetings, and other activities of CIAT's Seed Unit and of seed procedures in various regions of Latin America.)

Available in Spanish only.

Unidades Audiotutoriales

Catálogo disponible.

Envíe los pedidos a la siguiente dirreción:

Distribución de Publicaciones
CIAT
Apartado Aéreo 6713
Cali, Colombia

Slide Sets (Audiotutorials)

Catalogue available in Spanish only.

Send orders to:

Distribution Office
CIAT
Apartado Aéreo 6713
Cali, Colombia

Training Materials (Audiotutorials)

Price information for all Audiotutorial Units:

Standard format (slides, tape, script, 5 study guides).

CIAT: US$90.00 plus airmail postage.

Special package (slides in carousel, tape, script, 25 study guides)

CIAT: US$130.00 plus airmal postage.

10% discount for universities and nonprofit institutions.

Study guides for all units: US$2.00 each.

NOTE: Airmail postage prices may change without previous notice.

All units are available in the language in which it is described (Spanish, English or French)

Beans

Descripción y Daños de las Plagas que Atacan el Frijol

Code: 04SB-05.01. 1980 (2nd ed.). 97 slides. 32 min.

Es muy importante conocer no solo las características morfológicas de las plagas en los cultivos de frijol, lo cual permite su identificación, como también sus hábitos y los daños que causan a la planta durante los diferentes estados de desarrollo.

Principales Insectos que Atacan el Grano de Frijol Almacenado y su Control

Code: 04SB-05.03. 1981 (2nd ed.). 80 slides. 30 min.

El frijol almacenado está expuesto al ataque de los insectos denominados comunmente bróquidos, los cuales, en algunos casos, causan pérdidas irremediables. Esta unidad presenta métodos de control a nivel doméstico y del pequeño productor, mediante baja temperatura, mezcla de frijol con ceniza, uso de aceites vegetales, y otras medidas de control. A escala comercial se habla de desinfestación y protección.

El Lorito Verde (*Empoasca kraemeri* Ross y Moore) y su Control

Code: 04SB-05.04. 1980. 84 slides. 30 min.

El lorito verde es una plaga directa de amplia distribución en América Latina, que influye en el crecimiento y desarrollo de la planta de frijol; como consecuencia de su ataque resultan afectados los principales componentes del rendimiento, por lo tanto la disminución de la producción es drástica.

Principales Crisomélidos que Atacan el Frijol y su Control

Code: 04SB-05.05. 1982. 80 slides. 20 min.

Esta unidad incluye la distribución geográfica e importancia económica de los crisomélidos: su biología y hábitos, la importancia de los daños causados por larvas y adultos, y su control.

Enfermedades del Frijol Causadas por Hongos y su Control

Code: 04SB-06.01. 1980. 140 slides. 52 min.

Esta unidad discute las enfermedades que afectan el follaje, las vainas, el tallo, y la raíz y además presenta las principales estrategias de control.

Bean Diseases Caused by Fungi and their Control

Code: 04EB-06.01. 1981. 140 slides. 52 min.

Bean crops are attacked by many fungal diseases, which cause significant differences between actual and potential yields. The unit discusses diseases that attack different plant parts and their control.

Enfermedades del Frijol Causadas por Virus y su Control

Code: 04SB-06.02. 1980 (2nd ed.). 125 slides. 62 min.

Del complejo de enfermedades que afectan al frijol y ocasionan pérdidas de importancia económica las causadas por virus son especiales, debido a las características biológicas muy particulares de los virus, lo cual hace que el mejoramiento genético sea la alternativa para su control. Esta unidad discute siete enfermedades importantes causadas por virus.

Control Genético del Mosaico Común del Frijol (BCMV)

Code: 04SB-06.03. 1983. 55 slides. 22 min.

El mosaico común del frijol es una enfermedad causada por un virus denominado universalmente BCMV (Bean Common Mosaic Virus). Esta virosis del frijol es la de mayor distribución en el mundo y causa disminuciones en el rendimiento del orden del 50%, en promedio, según la susceptibilidad de la variedad. En esta unidad se describe la metodología diseñada en el CIAT para la obtención de germoplasma resistente al mosaico común.

Técnicas para el Aislamiento, Identificación y Conservación de Hongos Patógenos del Frijol

Code: 04SB-06.04. 1981 (2nd ed.). 97 slides. 27 min.

El proceso de identificación de una enfermedad se puede resumir en los siguientes pasos: observación de los síntomas, el aislamiento del patógeno, las pruebas de patogenicidad; la consulta bibliográfica.

Enfermedades Bacterianas del Frijol: Identificación y Control

Code: 04SB-06.05. 1981. 104 slides. 39 min.

Las enfermedades del frijol causadas por bacterias son menos numerosas que las producidas por hongos y por virus. Sin embargo, por su importancia económica y por su amplia distribución geográfica, se hace necesario un estudio que incluya el proceso de identificación y las formas de control, especialmente la obtención de resistencia varietal.

La Roya del Frijol y su Control

Code: 04SB-06.06. 1980. 91 slides. 33 min.

La roya del frijol es una enfermedad importante, no solo por su amplia distribución geográfica sino por la considerable reducción que causa en las cosechas.

Pudriciones Radicales del Frijol y su Control

Code: 04SB-06.07. 1981. 111 slides. 40 min.

Existe un complejo de hongos patógenos al frijol presentes en el suelo, causantes de las enfermedades comunmente denominadas pudriciones. Este complejo está constituído por las especies *Sclerotium rolfsii, Fusarium oxysporum, Fusarium solani, Rhizoctonia solani, Thielaviopsis basicola, Pvmatotrichum omnivorum, Macrophomina phaseolina* y varias especies de *Pythium y Phytophthora.*

La Antracnosis del Frijol y su Control

Code: 04SB-06.08. 1981. 77 slides. 26 min.

La antracnosis causada por el hongo *Colletotrichum lindemuthianum* (Sacc. y Magn.) es una enfermedad que puede presentarse durante todo el ciclo del cultivo de frijol. Además de la reducción del rendimiento, las lesiones en las vainas y en los granos perjudican su calidad y por consiguiente el mercadeo.

La Mancha Angular del Frijol y su Control

Code: 04SB-06.09. 1982. 79 slides. 24 min.

La mancha angular del frijol es una enfermedad que se caracteriza por causar lesiones típicamente angulares en el follaje de la planta de frijol. El agente causante es el hongo *Isariopsis griseola* Sacc. y su presencia ha sido comprobada en muchos países de las zonas templadas y tropicales en las cuales el frijol es un cultivo importante.

Principales Nematodos que Atacan el Frijol y su Control

Code: 04SB-06.10. 1982. 94 slides. 35 min.

Los nematodos son microorganismos multicelulares que pueden ser parásitos de las plantas. Su importancia no solo se debe a su amplia distribución y al daño que causan directamente a la planta, sino también a que el punto de la planta en donde se inicia el ataque se convierte en puerta de entrada para hongos, bacterias y virus.

La Mustia Hilachosa del Frijol y su Control

Code: 04SB-06.12. 1983. 75 slides. 17 min.

Esta enfermedad causada por el hongo *Thanatephorus cucumeris,* es uno de los principales factores limitativos de la producción de frijol en las zonas hómedas y cálidas. Causa defoliación rápida y drástica, provocando en la mayoría de los casos la pérdida de las cosechas. Esta unidad muestra las etapas de control de la enfermedad.

Diseases of Beans (*Phaseolus vulgaris*) in Africa

Code: 04EB-06.13.

The estimated losses of rust in East Africa is 100%, anthracnose in Malawi is 92%, scab in Kenia causes 70% of estimated loss. This unit is a description of the principal diseases in field beans in Africa.

(It will be available in 1986)

Cruzamiento del Frijol (*Phaseolus vulgaris* L.)

Code: 04SB-08.02. 1982 (2nd ed.). 140 slides. 40 min.

La hibridación o cruzamiento constituye el principal medio para el mejoramiento genético de las plantas. A través de la hibridación se logran formas cultivadas superiores a las existentes. Describe las técnicas de hibridación en frijol.

Bean Crossing

Code: 04EB-08.02.

Hybridization or crossbreeding is the principal method of plant breeding. Through hybridization cultivated forms superior to those already existing are obtained. This unit discusses the techniques that are applied to beans.

(It will be available in 1986)

Mejoramiento del Frijol por Introducción y Selección

Code: 04SB-08.03.

Para el desarrollo de plantas eficientes los fitomejoradores utilizan diversos sistemas de mejoramiento genético, uno de los cuales es la introducción y selección de variedades y líneas.

(In press)

Bean Breeding by Introduction and Selection

Code: 04EB-08.03. 1985.

To develop high yielding varieties, bean programs use various breeding methods, among which are the introduction and selection of varieties and lines.

(In press)

Amelioration du Haricot par Introduction et Selection

Code: 04FB-08.03.

Pour le développement des plantes plus productives, les améliorateurs utilisent diverses sistèmes d'amélioration génétique, parmi lesquelles on peut mentionner l'Introduction et sélection de variétés et de lignés.

(It will be available in 1986)

Morfología de la Planta de Frijol Común (*Phaseolus vulgaris* L.)

Code: 04SB-09.01. 1981 (2nd ed.). 123 slides. 36 min.

El género *Phaseolus* incluye aproximadamente 35 especies de las cuales 4 se cultivan. Desde el punto de vista de la taxonomía *Phaseolus vulgaris* L. es el prototipo del género *Phaseolus.*

Morphology of the Bean Plant (*Phaseolus vulgaris* L.)

Code: 04EB-09.01. 1985. 123 slides. 36 min.

The genus *Phaseolus* includes approximately 35 species of which four are cultivated. From a taxonomic point of view *Phaseolus vulgaris* L. is the prototype of the genus *Phaseolus.*

Morphologie de la Plante du Haricot Comun (*Phaseolus vulgaris* L.)

Code: 04FB-09.01. 1985. 123 slides. 36 min.

Le genre *Phaseolus* comprend approximativement 35 espèces parmi lesquelles quatre sont cultivées. Du point de rue taxonomique cette espèce est le prototype du genre *Phaseolus*.

(In press)

Diversidad Genética de las Especies Cultivadas del Género *Phaseolus*

Code: 04SB-09.02. 1980. 115 slides. 54 min.

Dentro del género *Phaseolus* hay cuatro especies principales que se consideran como formas cultivadas. De estas cuatro especies la más importante es el frijol común (*Phaseolus vulgaris* L.). Las otras tres especies cultivadas desde hace 3000 años son: *Phaseolus lunatus* o frijol lima; *Phaseolus coccineus* o frijol ayacote y *Phaseolus acutifolius* o frijol tepari.

Genetic Diversity of Cultivated Species of the Genus *Phaseolus*

Code: 04EB-09.02. 1985. 115 slides. 54 min.

Within the genus *Phaseolus* there are four major cultivated species. Of these four species, the most important is dry beans (*Phaseolus vulgaris* L.). The other three species that have been cultivated for 3000 years are: *Phaseolus lunatus* or lima bean; *Phaseolus coccineus* or runner bean, and *Phaseolus acutifolius* or moth bean.

(In press)

Diversité Génétique des Espèces Cultivées du Genre *Phaseolus*

Code: 04FB-09.02. 1985. 115 slides. 54 min.

Dans le genre *Phaseolus* il y a quatre espèces principales. Parmi ces quatre espèces, le haricot commun est le plus important (*Phaseolus vulgaris* L.). Les autres trois espèces chacune cultivée depuis 3000 années, sont: *Phaseolus lunatus* ou haricot beurre (lima); *Phaseolus coccineus*, ou haricot d'Espagne; et *Phaseolus acutifolius* ou haricot tepari.

(It will be available in 1986)

Etapas de Desarrollo de la Planta de Frijol Común

Code: 04SB-09.03. 1983. 76 slides. 29 min.

Es importante elaborar una escala basada en el conocimiento mor-fológico de la planta de frijol y los cambios fisiológicos que se suceden durante su desarrollo, lo cual permite referir las observaciones y prácticas de manejo a etapas de desarrollo fisiológico. Con este propósito se dan a conocer en esta unidad las etapas de desarrollo de la planta de frijol común.

Stages of Development of the Common Bean Plant

Code: 04EB-09.03. 1985. 76 slides. 29 min.

The developmental stages of beans are shown. Observations and management practices can thus be related to stages of physiological development across research institutions.

(In press)

Etapes de Développement du Haricot (*Phaseolus vulgaris* L.)

Code: 04FB-09.03. 76 slides. 29 min.

Les étapes de développement du haricot sont présentés ici. Les observations et les practiques agronomiques peuvent se rapporter aux étapes du développement physiologique à travers des institutions de recherche.

(It will be available in 1986)

Semilla de Frijol de Buena Calidad

Code: 04SB-12.03. 1980. 122 slides. 52 min.

Un programa efectivo de semillas comprende diversos elementos y actividades que deben estar coordinadas para alcanzar su objetivo específico: la producción y distribución de semilla de buena calidad, en el tiempo requerido, a un costo razonable y en el lugar donde aquella se necesite.

Good Quality Bean Seed

Code: 04EB-12.03. 1981. 122 slides. 52 min.

An effective seed program is comprised of diverse elements and activities that should be coordinated to achieve these objectives: the production and distribution of good quality seed, with prompt delivery at a reasonable cost, and in the location where it is needed.

Bean Production Systems in Africa

Code: 04EB-12.04.

The development of improved bean production technology in Africa requires a knowledge of current bean production systems in the region including the principal production constraints and the potential for the improvement of these production systems. This unit give a brief description of current bean production systems.

Conceptual Basis of Intercropping Bean

Code: 04EB-12.05

Intercropping system is important for small farmers because it provides diversity of diet; a greater stability of production and income source; often reduces insect and disease incidence; and more efficient use of family labour and intensive production with limited land resources. The purposes of this unit are to: emphasize the importance of intercropping systems in bean production, develop a common scientific language and discuss the genetic, agronomic and socio-economic factors involved in intercropping.

Cassava

Desordenes Nutricionales de la Planta de Yuca

Code: 04SC-01.01. 1981. 100 slides. 28 min.

En la yuca los síntomas de desordenes nutricionales claramente reconocibles generalmente están asociados con condiciones severas. En casos menos críticos, particularmente cuando la deficiencia es de elementos mayores, no se presentan síntomas específicos sino una reducción del vigor y del rendimiento del cultivo; por otra parte, los diferentes desordenes nutricionales pueden producir algunas veces síntomas muy similares. En casos como éstos, los análisis del suelo y de la planta son una gran ayuda para hacer un diagnóstico correcto.

Nutritional Disorders of Cassava

Code: 04EC-01.01. 1985.

In cassava, clearly recognizable symptoms of nutritional disorders are associated with severe conditions. Under less critical conditions particularly when the deficiency is of macro elements, there are no specific symptoms. Instead there is a reduction in plant vigor and yield. Additionally, the different nutritional disorders can produce very similar symptoms. In these cases soil and plant analyses help in making a correct diagnosis.

(In press)

Un Tipo Ideal de Planta de Yuca para Rendimiento Máximo

Code: 04SC-02.01. 1981 (2nd ed.). 77 slides. 29 min.

El Programa de Fisiología de Yuca del CIAT ha elaborado un modelo matemático del desarrollo de la planta de yuca mediante el cual es posible predecir los efectos que produce en el rendimiento la variación de uno o mas de sus componentes. El tipo ideal de planta de yuca se describe en esta unidad.

El Cultivo de Meristemas de Yuca y sus Aplicaciones

Code: 04SC-02.02. 1980. 54 slides. 24 min.

El cultivo de tejidos, que consiste en cultivar *in vitro* cualquier parte de la planta, o sea una célula, un grupo de células o un órgano, es una de las técnicas que están siendo investigadas en la actualidad para tratar de solucionar los problemas encontrados en la propagación vegetativa, y facilitar el intercambio de germoplasma para poder llevar a cabo los programas de fitomejoramiento en los diferentes países productores de yuca.

Morfología de la Planta de Yuca (*Manihot esculenta* Crantz)

Code: 04SC-02.03. 1981. 98 slides. 30 min.

Uno de los generos más importantes dentro de la familia de las euforbiaceas es el género Manihot, al cual pertenecen tanto las variedades de yuca amarga como las de yuca dulce. El conocimiento detallado de la morfología de cada una de las partes de la planta es de gran importancia no sólo para la investigación sino también directamente en el proceso de la producción.

Morphology of the Cassava Plant (*Manihot esculenta* Crantz)

Code: 04EC-02.03. 1983. 98 slides. 30 min.

An important genus within the family of the Euphorbiaceae is the genus Manihot to which both bitter and sweet cassava varieties belong. Detailed information of the morphology of each part of the cassava plant is of great importance in cassava and in the cassava production process.

Acaros Presentes en el Cultivo de la Yuca y su Control

Code: 04SC-02.04. 1982. 99 slides. 33 min.

La yuca puede ser atacada por un gran número de artropodos (aproximadamente 200 especies identificadas) siendo los acaros, probablemente, los de mayor importancia económica. La descripción de las especies de acaros plaga del cultivo, como también los métodos de control, se describen en esta unidad.

El Cultivo de Meristemas para el Saneamiento de Clones de Yuca

Code: 04SC-02.04. 1982. 61 slides. 26 min.

Adelantos logrados en el cultivo de meristemas *in vitro* permiten ahora propagar la yuca y simultaneamente eliminar las enfermedades que se diseminan con el material propagativo comercial.

El Coquito (*Cyperus rotundus* L.): Biología y Posibilidades de Control

Code: 04SC-02.06. 1982. 127 slides. 39 min.

El coquito (*Cyperus rotundus* L.) ha sido reconocido como la maleza mas dañina del mundo. Esta ciperácea, de orígen asíatico (India), se encuentra ampliamente distribuída en el trópico y subtrópico. Su agresividad, gran capacidad de competencia, adaptación a diversos medios y condiciones y su difícil control, hacen de esta maleza uno de los peores problemas en la agricultura.

The Biology and Control of Purple Nutsedge (*Cyperus rotundus* L.)

Code: 04EC-02.06. 1985. 127 slides. 39 min.

Purple nutsedge (*Cyperus rotundus* L.) is known as the most damaging weed worldwide. Of Asiatic origin (India), it is widely distributed in the tropics and subtropics. Its aggressiveness, high competitive capacity, adaptability to different environments and conditions, and its difficult control make this weed one of the most severe problems in agriculture.

Descripción de las Enfermedades de la Yuca

Code: 04SC-03.01. 1980 (2nd ed.). 94 slides. 29 min.

La yuca puede ser atacada por mas de 30 agentes bacterianos, fungosos, virales y micoplasmas. Algunos patógenos atacan el tallo, otros el tejido foliar y las partes tiernas del tallo causando manchas, defoliación, marchitez, añublos y otros.

El Control de *Erinnyis ello* Gusano Cachón de la Yuca

Code: 04SC-04.01. 1982 (2nd ed.). 71 slides. 26 min.

El crecimiento del área cultivada y el uso indiscriminado de insecticidas han alterado el equilibrio ecológico de las poblaciones insectiles en muchas regiones, dando lugar a que algunos insectos que aparecían espordicamente se conviertan en plagas causantes de daños de importancia económica. Uno de ellos es *Erinnyis ello* llamado comunmente gusano cachón, el cual causa severos daños especialmente en plantas jóvenes, encontrándose en la mayoría de las zonas yuqueras del continente americano.

Descripción de las Plagas que Atacan el Cultivo de la Yuca y Características de sus Daños

Code: 04SC-04.02. 1980. 93 slides. 39 min.

Aunque ha sido considerado como un cultivo libre de plagas artropodas, hasta el presente se han identificado aproximadamente 200 especies de insectos y acaros que atacan la planta de yuca, algunos de los cuales están distribuídos en todo el mundo. La descripción de las principales plagas del cultivo permite identificarlas, evitar las pérdidas económicas que ocasionan, e impedir su diseminación.

Mosca Blanca de la Yuca: Biología y Control

Code: 04SC-04.05. 1985.

La yuca es atacada por varias especies de moscas blancas (*Aleyro-didae*) con diferentes grados de importancia económica y distribuídas en América, Africa y ciertas regiones de Asia. Estos insectos son especialmente importantes como vectores de virus.

(In press)

Intercambio Internacional de Clones de Yuca *in vitro*

Code: 04SC-05.02. 1982. 72 slides. 30 min.

Actualmente, para propósitos de intercambio internacional de germo-plasma de yuca se está empleando el método de cultivos meriste-máticos, ya que éstos, por su condición aséptica, permiten cumplir las regulaciones cuarentenarias que restringen dicho intercambio.

El Cultivo de Meristemas para la Conservación de Germoplasma *in vitro*

Code: 04SC-05.03. 1984. 99 slides. 38 min.

Existe una gran diversidad potencial de tipos de tejidos de una planta que se podría utilizar para la conservación *in vitro*, siempre y cuando el sistema cumpla con los siguientes requisitos; 1) establecimiento in vitro y regeneración de plantas de rangos amplios de germoplasma y 2) máxima estabilidad genética de los materiales. La habilidad para regenerar plantas en forma consistente a partir de un cultivo de células somáticas *in vitro* a un no está disponible para muchas especies de cultivos importantes. Sin embargo, existen técnicas para hacer crecer plantas de numerosas especies hasta la madurez a partir de meris-temas apicales, garantizando el mínimo de pérdidas y el máximo de estabilidad genética.

Sistema de Propagación Rápida de la Yuca

Code: 04SC-06.01. 1980 (2nd ed.). 62 slides. 19 min.

Una de las principales limitaciones para incrementar el área de cultivo de yuca ha sido la insuficiente disponibilidad de material de siembra, debido a que el sistema tradicional para la obtención de las estacas de siembra presenta una tasa de multiplicación muy baja. Con el fin de solucionar el problema planteado, el CIAT desarrolló un sistema de propagación rápido, sencillo y barato, el cual aumenta el factor de multiplicación por planta por año.

Selección y Preparación de Estacas de Yuca para Siembra

Code: 04SC-06.02. 1980 (2nd ed.). 65 slides. 27 min.

En la yuca, la selección y preparación del material de siembra exige una serie de recomendaciones que es necesario seguir cuidadosamente para obtener una buena germinación y plantas vigorosas.

Selection and Preparation of Cassava Cuttings for Planting

Code: 04EC-06.02. 1984. 73 slides. 21 min.

The selection and preparation of planting material for cassava should carefully follow a series of recommendations to achieve a high percentage of germination and vigorous plants.

Multiplicación Acelerada de Material Genético Promisorio de Yuca

Code: 04SC-06.06. 1983. 75 slides. 26 min.

El CIAT ha desarrollado un método de propagación de la yuca rápido, sencillo y barato, con base en estacas caulinares de dos yemas, y mejoro otro que fue desarrollado en Filipinas, que utiliza como propágulos esquejes de una hoja con una yema. Esta unidad muestra la técnica.

Secamiento Natural de la Yuca para Alimentación Animal

Code: 04SC-07.01. 1983. 54 slides. 25 min.

Uno de los mercados potenciales de la yuca es su utilización como sustituto de los cereales en las dietas balanceadas destinadas a la alimentación de aves y cerdos. Esta unidad discute el secado de la yuca, el cual podría constituirse en una importante agroindustria para países tropicales productores de este cultivo.

La Utilización de las Raíces de Yuca en la Alimentación de Cerdos

Code: 04SC-07.04. 1983. 37 slides. 19 min.

A pesar de que el mercado potencial pecuario más importante que existe hoy día en los paises tropicales de América Latina lo constituye el sector de la producción avícola, la mayor información sobre la utilización de las raíces de yuca en la alimentación animal se refiere a la producción porcina. El potencial presente y futuro de este último sector también es considerable, teniendo en cuenta que la mayoría de los productores de yuca a nivel de fincas pequeñas y medianas también cría cerdos.

Almacenamiento de Raíces Frescas de Yuca

Code: 04SC-07.05. 1983. 85 slides. 36 min.

Una de las mayores limitaciones para aumentar el consumo de yuca en la alimentación humana es la dificultad en conservar las raíces después de que se cosechan; las raíces se deterioran rápidamente, volviéndose inaceptables para el consumo humano y otros usos.

Durante los ultimos años, se han realizado investigaciones para determinar las causas del deterioro de las raíces de yuca en poscosecha y se han efectuado trabajos para disminuir su perecibilidad mediante tratamiento en pre y poscosecha y desarrollar sistemas o prácticas de almacenamiento de las raíces. En esta unidad se exponen los aspectos relacionados con el deterioro de las raíces de yuca en poscosecha, los factores que la afectan y los principales sistemas de almacenamiento que se han utilizado.

Almacenamiento de Raíces de Yuca en Bolsas de Polietileno

Code: 04SC-07.06.

El almacenamiento de raíces de yuca en bolsas de polietileno es un sistema sencillo y práctico, que ha dado excelentes resultados tanto en tiempo de almacenamiento como en la conservación de la calidad de las raíces. En esta unidad se descubre el procedimiento y se discuten los resultados del almacenamiento en bolsas de polietileno.

(In press)

Secado Natural de Raíces de Yuca en Pisos de Concreto

Code: 04SC-07.02.

El secado de la yuca se puede realizar en pisos de concreto, que es un método de secado natural que aprovecha la energía solar como el factor energético de secado. Esta unidad discute los requerimientos para establecer una planta de secado de este tipo, y describe las instalaciones y las herramientas utilizadas, así como el procedimiento a seguir.

(In press)

Rice

Selección y Adecuación de Lotes para la Producción Continua de Arroz

Code: 04SR-01.01. 1980. 119 slides. 30 min.

Al utilizar la técnica asitica de preparación de suelos bajo agua, es posible sembrar y cosechar semana tras semana, y usar mas eficientemente la mano de obra, la tierra, el agua y los equipos. La unidad presenta los factores que deben tenerse en cuenta en la selección de lotes para la producción continua de arroz de riego utilizando el sistema de preparación de suelos mediante inundación, y describe la adecuación de los lotes seleccionados.

Preparación de Suelos Mediante el Sistema de Inundación (Fangueo) para el Cultivo del Arroz

Code: 04SR-01.02. 1981 (2nd ed.). 78 slides. 18 min.

Para el cultivo de arroz en el trópico se utilizan diferentes sistemas de preparación del suelo; uno de estos sistemas es el de inundación, que consiste en arar y rastrillar la capa superior del suelo inundado. Este sistema ayuda a controlar las malezas, disminuye el movimiento del agua a través del suelo y aumenta significativamente la eficiencia del nitrógeno aplicado.

Siembra de Arroz Mediante Transplante

Code: 04SR-01.04. 1980. 98 slides. 22 min.

El arroz se siembra mediante siembra directa y transplante. En cualquier sistema que se adopte, los factores que afectan el crecimiento inicial de la planta también influyen en su rendimiento. Esta unidad presenta estos factores para la siembra mediante transplante.

Producción y Beneficio de Semilla Certificada de Arroz

Code: 04SR-01.08. 1979. 125 slides. 35 min.

La certificación de semillas garantiza la conservación y el mantenimiento de la pureza e identidad genética de las variedades mejoradas obtenidas por los fitomejoradores. En esta unidad se expondrán los mecanismos legales y técnicos que sirven de base a los programas de producción y beneficio de las semillas certificadas de arroz.

Control y Normas de Calidad de las Semillas Certificadas de Arroz

Code: 04SR-01.09. 1982. 110 slides. 25 min.

Para que una entidad certificadora de semillas pueda garantizar la pureza genética y física, la sanidad y la buena germinación de las semillas, debe establecer normas y determinar si las semillas las cumplen o no. En esta unidad se expone el procedimiento que se debe seguir para el análisis de la semilla en el laboratorio, y se informa sobre las normas de calidad.

Principales Malezas en el Cultivo del Arroz en América Latina

Code: 04SR-03.01. 1983. 105 slides. 27 min.

En términos generales se puede afirmar que la presencia de las malezas en el cultivo del arroz causa una pérdida superior a la producida por las enfermedades y las plagas en conjunto.

Por ser las malezas un problema tan grave, su manejo y control es de gran importancia para la producción. El primer paso en la estrategia para combatir las malezas es conocerlas. El poder identificarlas, saber de su distribución, sus características botánicas y su agresividad permitirá diseñar los mejores sistemas de manejo.

Descripción y Daño de los Insectos que Atacan el Arroz en América Latina

Code: 04SR-04.01. 1983. 111 slides. 31 min.

Esta unidad incluye información sobre la identificación de los insectos, lo mismo que el conocimiento de sus hábitos, el daño que ocasionan a la planta, la edad del cultivo en que atacan y la época del año en que aparecen.

Barrenadores del Tallo del Arroz en América Latina y su Control

Code: 04SR-04.02. 1980. 120 slides. 27 min.

Los insectos barrenadores del tallo atacan al arroz y ocasionan pérdidas en la producción en algunas regiones de América Latina. Esta unidad presenta la biología de los barrenadores *Diatraea saccharalis* (Fabricius) y *Rupela albinella* (Cramer), describe los daños característicos de cada uno de ellos y los métodos para su control, citando sus enemigos naturales. Las prácticas culturales y algunas recomendaciones sobre el uso de insecticidas son descritos.

La Sogata, *Sogatodes oryzicola*, y el Virus de la Hoja Blanca en Arroz

Code: 04SR-04.03. 1983. 66 slides. 17 min.

La hoja blanca es un de las enfermedades que puede ocasionar mayor perjuicio al arroz en el trópico del hemisferio occidental. El agente causante de la hoja blanca (VHB) es un virus transmitido por los insectos *Sogatodes oryzicola* Muir (Sogata) y *S. cubanus* Crawf. Estos insectos pueden además causar daños al alimentarse del cultivo y al ovipositar en él.

Latencia y Pregerminación de la Semilla de Arroz

Code: 04SR-05.01. 1979. 78 slides. 30 min.

Entre el período de madurez y el de germinación, en algunos casos, se presenta la latencia, la cual es de especial importancia porque evita la germinación de la semilla en la panícula o en la bodega, pero constituye un problema cuando se desean utilizar las semillas para su siembra inmediata. Se presentan causas, ventajas, y desventajas duración e intensidad, y métodos para romper la latencia.

Morfología de la Planta de Arroz

Code: 04SR-05.02. 1981 (2nd ed.). 87 slides. 18 min.

El conocimiento de la planta de arroz y de su morfología es importante en la investigación por ser la base para la diferenciación de las variedades y de los estudios de fisiología y de mejoramiento.

Crecimiento y Etapas de Desarrollo de la Planta de Arroz

Code: 04SR-05.04. 1980. 128 slides. 31 min.

La unidad describe el crecimiento y las etapas de desarrollo de la planta de arroz a través de sus características morfológicas y expresiones cuantitativas con referencia a una escala numérica de 0 a 9, y además, determina la relación entre las diferentes etapas y la acumulación de materia seca.

Los Macronutrimentos en la Nutrición de la Planta de Arroz

Code: 04SR-09.06. 1983. 75 slides. 21 min.

La siembra de las variedades de alto rendimiento y el manejo adecuado de los cultivos permite obtener mayor cantidad de granos por cada kilogramo de fertilizante aplicado. El conocimiento de cómo los nutrimentos contribuyen a aumentar los rendimientos a través de las etapas de desarrollo es importante porque permite un uso eficiente de los fertilizantes.

Enfermedades del Arroz en América Latina y su Control

Code: 04SR-06.01. 1983. 102 slides. 26 min.

La búsqueda de sistemas de producción que permitan reducir los costos y elevar los rendimientos de arroz constituye la actual estrategia de investigación en el Programa de Arroz del CIAT. Esta unidad contiene información general sobre las enfermedades del arroz en América Latina, especifica sus agentes causantes, describe los síntomas, el desarrollo de la enfermedad y menciona las medidas de control.

Evaluación de la Calidad del Arroz

Code: 04SR-07.01. 1980 (2nd ed.). 106 slides. 29 min.

Dos de los criterios para el fitomejorador en la evaluación de la calidad de arroz son la calidad de molinería y la calidad culinaria. La explicación de cómo estos criterios determinan y predicen la calidad y la descripción de algunos de los procedimientos e instrumentos que se usan para ésto son el tema de esta unidad.

Química de los Suelos Inundados

Code: 04SR-09.01. 1981. 85 slides. 30 min.

Las propiedades físicas y químicas de un suelo cambian drásticamente cuando se inunda y por lo tanto la planta de arroz responderá a la fertilización y al encalamiento en forma diferente que cuando se cultiva en suelos no inundados. Los cambios químicos y la influencia que ejercen en la planta de arroz son presentados.

Fertilización Nitrogenada del Arroz

Code: 04SR-09.02. 1982. 101 slides. 30 min.

El insumo agrícola que más influye en los rendimientos del arroz es el nitrógeno; en la mayoría de los casos su deficiencia es un factor que limita la producción. La respuesta de la planta a la aplicación de nitrógeno y el manejo del fertilizante se discute.

Fertilización Fosfórica del Arroz

Code: 04SR-09.04. 1981. 77 slides. 25 min.

En América Latina la deficiencia de fósforo es común. Es necesario hacer fertilización fosfórica en las regiones sin suficiente fósforo porque esta insuficiencia limita significativamente la producción. Esta unidad presenta información sobre la disponibilidad de fósforo, el manejo de fertilizantes, y una estrategia de baja energía.

Tropical Pastures

Establecimiento de Praderas en Sabanas Bien Drenadas de los Llanos Orientales de Colombia

Code: 04SP-01.01. 1981. 96 slides. 34 min.

La investigación en el establecimiento de pastos ha dado soluciones parciales a la acidez, la baja fertilidad y la erosión de los suelos, el alto costo de los insumos y la acción de plagas y enfermedades.

Síntomas de Deficiencia de Macronutrimentos y Nutrimentos Secundarios en Pastos Tropicales

Code: 04SP-02.01. 1981. 86 slides. 26 min.

Los principales factores que limitan el establecimiento de pasturas en las regiones de Oxisoles y Ultisoles de América tropical son la marcada acidez e infertilidad de estos suelos. Una de las manifestaciones de las deficiencias minerales en las plantas es la aparición de síntomas visibles especialmente en las hojas, típicos para cada nutrimento. Muestras de estos síntomas se encuentran en esta unidad.

Síntomas de Deficiencia de Micronutrimentos y de Toxicidades Minerales en Pastos Tropicales

Code: 04SP-02.02. 1981. 78 slides. 26 min.

Aunque los principales problemas nutricionales de las pasturas de América tropical se relacionan con las deficiencias de macronutrimentos y nutrimentos secundarios, las deficiencias de micronutrimentos a menudo son un factor limitativo para el establecimiento de especies forrajeras. De mayor importancia a un son los problemas causados por las toxicidades de Al y Mn. El conocimiento de los síntomas visibles y de la forma en que se manifiestan estos desórdenes nutricionales se encuentra en esta unidad.

Oxisoles y Ultisoles en América Latina. I. Distribución, Importancia y Propiedades Físicas

Code: 04SP-02.03. 1983. 79 slides. 30 min.

La vasta región de tierras bajas de América del Sur que incluye la Amazonía, los Llanos de Colombia y Venezuela, El Cerrado brasileño, las pampas bolivianas y otras áreas, representa una de las mayores reservas de tierras inexplotadas del mundo, con más de 800 millones de hectáreas.

El objetivo de esta unidad audiotutorial es mostrar la distribución de los Oxisoles y Ultisoles en la región central de tierras bajas de América tropical y describir sus propiedades físicas mas importantes, con sus ventajas y limitaciones.

Oxisoles y Ultisoles en América Latina. II. Mineralogía y Características Químicas

Code: 04SP-02.04. 1984. 79 slides. 30 min.

Los Oxisoles y Ultisoles que predominan en la extensa región de tierras bajas de América del Sur caen dentro de la categoría de "suelos ácidos e infértiles," debido a que han alcanzado un grado intenso de meteorización y lavado y tienen una mineralogía dominada por arcilla de baja actividad (arcillas caoliníticas y sesquioxidos de Fe y Al).

Selección y Evaluación de Pastos Tropicales en Condiciones de Alta Concentración de Aluminio y Bajo Contenido de Fósforo Disponible

Code: 04SP-02.06. 1984. 88 slides. 32 min.

La selección de especies de pastos tolerantes a las condiciones edáficas adversas es una de las soluciones posible para la producción de ganado de carne en la mayoría de los suelos ácidos de América tropical.

Manejo de la Fertilización Fosfatada de Pastos Tropicales en Suelos Acidos de América Latina

Code: 04SP-02.07. 1985. 87 slides. 28 min.

Esta unidad tiene el propósito de formular conceptos y factores a tener en cuenta en el manejo de la fertilización fosfatada de pastos en ecosistemas de América Latina con suelos ácidos. Como una base conceptual para el manejo de la fertilización fosfatada la unidad precisa características de los ecosistemas de América tropical para la producción de ganado y analiza alternativas para el manejo del fósforo.

(In press)

Descripción de las Plagas que Atacan los Pastos Tropicales y Características de sus Daños

Code: 04SP-03.01. 1982. 106 slides. 41 min.

El conocimiento de las diferentes plagas que atacan las especies forrajeras tropicales, su biología y sus hábitos, el daño que causan y los métodos adecuados de manejo y control se presentan en esta unidad.

Cercópidos Plagas de los Pastos de América Tropical: Biología y Control

Code: 04SP-03.02. 1983. 79 slides. 37 min.

Los insectos conocidos como "salivitas," "salivazos" o "saliveros," "miones" o "meones" de los pastos, "candelillas," "moscas pintas," "cigarrhinas," etc. (Homoptera:Cercopidae) constituyen una plaga de importancia creciente de las pasturas de gramíneas en América tropical. Esta unidad presenta una descripción detallada de la biología y hábitos de los cercópidos, de su importancia económica, y una serie de recomendaciones para su control, con énfasis en el establecimiento de gramíneas resistentes o tolerantes a esta plaga.

Descripción de las Enfermedades de las Principales Leguminosas Forrajeras Tropicales

Code: 04SP-03.03. 1983. 113 slides. 42 min.

Las enfermedades, junto con las plagas, son uno de los principales factores que dificultan el desarrollo de cultivares estables de forrajes tropicales. En América tropical, estas enfermedades afectan con mayor severidad a las leguminosas, debido a que en su mayoría son especies nativas que han evolucionado por miles de años en el mismo medio que sus patógenos especializados.

Manejo de la Sabana Nativa en los Llanos Orientales de Colombia y Venezuela

Code: 04SP-04.01. 1983. 72 slides. 21 min.

La extrema acidez e infertilidad de los suelos de estas sabanas clasificadas como Oxisoles y Ultisoles constituye la principal barrera para su desarrollo agrícola; alli la actividad económica predominante es la ganadería extensiva basada en los pastos naturales.

Prácticas de Manejo del Hato de Cría en Sabanas Bien Drenadas de Colombia y Venezuela

Code: 04SP-04.02. 1984. 88 slides. 24 min.

Las explotaciones ganaderas extensivas de los Llanos Orientales de Colombia y Venezuela se caracterizan por una productividad animal muy baja, causada principalmente por la pobre calidad nutricional de las especies nativas y por la falta de disponibilidad de forraje durante las épocas secas que duran entre cuatro y seis meses del año.

El objetivo de esta unidad audiotutorial es presentar un serie de recomendaciones sobre el establecimiento de la infraestructura mínima necesaria y un programa de manejo y sanidad animal encaminados a mejorar la productividad animal.

Cosecha de Semilla de *Andropogon gayanus*

Code: 04SP-05.01. 1984. 80 slides. 26 min.

La cosecha de semilla de *A. gayanus* es difícil y poco eficiente por la altura del cultivo, la desuniformidad en la floración y maduración de las espiguillas y el amplio rango de distribución de las espiguillas en los tallos florales.

En esta unidad audiotutorial se describen los cuatro métodos básicos para la cosecha de semilla, y se discuten las ventajas y desventajas de cada uno de ellos.

Seeds

Desarrollo y Morfología de la Semilla

Code: 04SSe-01.01. 1983. 80 slides. 26 min.

El hombre ha utilizado las semillas principalmente para dos propósitos; para la reproducción de las plantas y como alimento. En esta unidad se hace énfasis en la capacidad reproductora de la semilla. El mayor énfasis se hace en la terminología y en la relación de la estructura de la flor con la estructura de la semilla. El conocimiento de esta relación permitirá al entender e interpretar las técnicas de producción y las pruebas para evaluar la calidad de la semilla.

Seed Development and Morphology

Code: 04ESe-01.01. 1984. 80 slides. 26 min.

Man uses seed of many different plants for two primary purposes: reproduce the plant which produced the seed and as a source of food. The development of the seed and its reproductive capacity are the emphasis of this unit. Emphasis is placed on the terminology and structural relationships between flower and seed. An understanding of this relationship will enable to understand and interpret production techniques and the tests for evaluating seed quality.

Evaluación de la Calidad de la Semilla de Maiz

Code: 04SSe-02.01. 1981. 80 slides. 19 min.

La calidad de la semilla no se puede evaluar mediante un exámen visual, es necesario aplicar métodos objetivos. Se hace el muestreo de los lotes de semillas y se obtiene una muestra representativa, la cual se analiza en el laboratorio para determinar la humedad, efectuar el análisis de pureza y el ensayo de germinación.

Principios del Acondicionamiento de Semillas

Code: 04SSe-03.01. 1981. 79 slides. 19 min.

Para la obtención de semilla de buena calidad es necesario someter la semilla cosechada a una serie de procesos y controles, entre los cuales el acondicionamiento es uno de los más importantes. Los objetivos, principios, equipos y métodos se encuentran en esta unidad.

Técnicas de Muestreo

Code: 04SSe-03.02. 1983. 95 slides. 21 min.

Una forma de garantizar la calidad de la semilla que se siembra es mediante la evaluación de muestras de esta semilla antes de ser usadas; pero a veces el muestreo se hace incorrectamente, o no se usa equipo apropiado. Esta unidad muestra la técnica correcta.

Elementos Esenciales para el Exito de un Programa de Semillas

Code: 04SSe-04.01. 1981. 93 slides. 32 min.

Esta unidad presenta los elementos que contribuyen al éxito de los programas de semillas a los administradores y líderes de los mismos.

Essential Elements for Successful Seed Programs

Code: 04ESe–04.01. 1983. 93 slides. 32 min.

For many managers and leaders, who are responsible for making decisions that could affect program success, this unit identifies the fundamental elements that contribute to an efficient seed program.

Other Subjects

Prácticas de Manejo de las Cerdas Lactantes y sus Lechones

Code: 04SSE–01.02. 1981 (2nd ed.). 101 slides. 40 min.

Entre las causas más importantes, no solo de la muerte de lechones síno también del precario estado general y poco peso con que llegan al destete, están las deficiencias de las prácticas sanitarias, nutricionales y de manejo de la cerda y su camada.

Elementos Estructurales de un Experimento Agrícola

Code: 04SB–03.01. 1982. 80 slides. 31 min.

Aunque se pueden establecer múltiples clasificaciones de los experimentos agrícolas con relación al lugar de ejecución, la disciplina o el cultivo involucrado, en todos ellos existen elementos comunes los cuales se denominan en esta unidad elementos estructurales. El objetivo de la unidad es la caracterización de dichos elementos

La Heterogeneidad del Suelo y los Ensayos de Uniformidad

Code: 04SB–03.02. 1983. 76 slides. 28 min.

Es practicamente imposible obtener un campo para experimentación que sea homogéneo o uniforme, puesto que todos los suelos presentan, en mayor o menor grado, heterogeneidad derivada de sus diferentes características físicas, químicas y biológicas. A fin de caracterizar las tendencias en la fertilidad y cuantificar el grado de heterogeneidad del suelo se muestra cómo hacer los ensayos de uniformidad.

Efectividad Agronómica de las Rocas Fosfóricas

Code: 04SR-09.05. 1983. 80 slides. 31 min.

El objetivo general del Proyecto Fósforo del International Fertilizer Development Center (IFDC) en el Centro Internacional de Agricultura Tropical (CIAT) es ayudar a desarrollar una estrategia para el manejo de los fertilizantes fosfatados en los suelos ácidos e infértiles de América Latina utilizando, cuando sea posible, fuentes nativas de fosfatos.

Principios Básicos para el Manejo y Control de las Malezas en los Cultivos

Code: 04SW-01.01. 1980 (2nd ed.). 85 slides. 26 min.

Factores como fertilización, manejo de agua, y control de plagas, organismos patógenos y malezas, están relacionados entre sí y son dependientes el uno del otro. Experiencias en el campo han demostrado que los estragos causandos por las malezas son de igual o mayor magnitud que los ocasionados por las plagas y las enfermedades. Un adecuado manejo y control de las malezas en los cultivos es el tópico de esta unidad.

Información Básica sobre la Competencia entre Malezas y los Cultivos

Code: 04SW-01.02. 1980 (2nd ed.). 104 slides. 44 min.

Para lograr un crecimiento normal, tanto el cultivo como las malezas compiten por el agua, los nutrimentos, la luz, el espacio y el CO_2. Es así como las malezas afectan directa o indirectamente el rendimiento de los cultivos, lo cual se conoce con el nombre de "Competencia."

Principios Básicos sobre la Selectividad de los Herbicidas

Code: 04SW-01.03. 1981 (2nd ed.). 110 slides. 39 min.

Un herbicida selectivo es aquel que causa toxicidad a unas plantas (las malezas), las elimina o retarda su crecimiento y no el a otras (el cultivo). Para que un herbicida realice su acción tóxica es necesario que haya contacto y penetración en la planta, movilización al sitio donde ejerce su acción, y acción tóxica que afecte procesos vitales.

Los Herbicidas: Modo de Actuar y Síntomas de Toxicidad

Code: 04SW-01.04. 1982 (2nd ed.). 156 slides. 60 min.

El éxito del control químico de las malezas en forma selectiva radica en el poder que tienen los herbicidas de cambiar el metabolismo de las malezas de manera tal que impidan su crecimiento y desarrollo, sin afectar los cultivos.

Factores que Condicionan la Eficacia de los Herbicidas

Code: 04SW-01.05. 1980 (2nd ed.). 105 slides. 38 min.

El éxito del control de las malezas mediante el uso de los herbicidas no depende únicamente del producto en sí. Un alto porcentaje de los problemas se presentan al utilizar los herbicidas ya sea por desconocimiento del manejo del equipo o del producto, por formas inadecuadas de aplicación, o por deficiencia en el mantenimiento de los equipos.

Equipos para la Aplicación Terrestre de Herbicidas

Code: 04SW-01.06. 1980. 128 slides. 36 min.

El control de las malezas puede realizarse mediante diferentes métodos; uno de ellos es el químico, o sea la utilización de herbicidas, para cuya aplicación se pueden usar diferentes equipos.

Formulaciones de Herbicidas

Code: 04SW-01.07. 1981 (2nd ed.). 109 slides. 32 min.

Un herbicida es seleccionado teniendo en cuenta diversos factores, entre ellos las especies de malezas, el cultivo, y las condiciones del suelo. El siguiente paso es la selección de la formulación apropiada, de la cual muchas veces puede depender el éxito o fracaso de la aplicación del herbicida.

Recomendaciones Básicas Sobre el Manejo de Agroquímicos

Code: 04SW-01.08. 1981. (2nd ed.). 137 slides. 34 min.

Esta unidad discute el manejo apropiado de agroquímicos; preparación de la mezcla, tipos de incompatibilidad que pueden presentarse, procedimientos, y precauciones que deben tenerse al usar plaguicidas.

Los Surfactantes: Clases, Propiedades y Uso con Herbicidas

Code: 04SW-01.09. 1980. 83 slides. 33 min.

Los surfactantes son sustancias químicas que se le suelen adicionar a los pesticidas para mejorar o facilitar su acción; su principal característica es que actúan sobre las superficies. Esta unidad informa sobre lo que es un surfactante, los tipos que existen y sus propiedades, y capacita al interesado para que pueda usarlos correctamente con los herbicidas.

Manejo y Control de las Malezas en el Cultivo del Frijol

Code: 04SW-02.02. 1980. 112 slides. 41 min.

El diseño de una estrategia de manejo y control de malezas debe tener en cuenta factores económicos, sociales y ecológicos de la zona donde va a aplicarse. El control manual es usual en zonas de ladera donde los sistemas de cultivos múltiples son comunes y se dispone de suficiente mano de obra.

Principios Básicos para el Manejo y Control de las Malezas en los Potreros

Code: 04SW-03.01. 1981 (2nd ed.). 129 slides. 43 min.

Aunque no se tienen datos precisos sobre la reducción en la cantidad de forraje ocasionado por el complejo de malezas presentes en los potreros, es bien conocido que las praderas mas productivas son aquellas en donde, además de otros factores, existe un bajo porcentaje de malezas. Esta unidad discute métodos de control y sus ventajas.

Guía Práctica para el Control de las Malezas en los Potreros

Code: 04SW-03.02. 1977. 106 slides. 37 min.

El control químico en algunos casos es más eficaz que otros métodos pues con el uso de herbicidas selectivos se puede lograr un control rápido y eficaz de las malezas, sin perjudicar los pastos y las leguminosas presentes en los potreros.

Manejo y Control de Malezas en el Cultivo de la Yuca

Code: 04SW-02.01. 1979. 86 slides. 35 min.

Dada su rusticidad se ha creído que la yuca puede tolerar, sin mayor perjuicio, la competencia de las malezas. En los trabajos realizados por el CIAT se ha demostrado que la presencia de malezas durante los primeros 60 días del ciclo del cultivo causa una reducción en los rendimientos de aproximadamente un 50%, en comparación con cultivos de yuca libres de malezas durante todo el ciclo.

Preparación de Suelos en Zonas Mecanizables

Code: 04ST-01.01. 1980. (2nd ed.). 69 slides. 18 min.

Existen diferentes tipos de suelos con distintas propiedades físicas; cada uno de ellos requiere su propia preparación, uso y conservación, los cuales pueden variarse de acuerdo con el cultivo seleccionado para la siembra y las condiciones ambientales de la zona.

CIP

International Potato Center
Centro Internacional de la Papa
Centre International de la Pomme de Terre

Apartado Postal 5969, Lima, Peru
Cables: CIPAPA-LIMA
Telex: 25672 PE
Telephone: 354354-366920

Foreword

The International Potato Center (CIP) is a scientific institution established for the purpose of developing and disseminating knowledge for greater use of the potato as a basic food. The following publications are designed to support research and communication among agricultural programs around the world. Several CIP publications are out of print and no longer listed. Copies may be available for consultation at an agricultural library.

We recommend contacting your National Potato Program. It is authorized to reproduce all technical information of CIP, and may also respond to other needs you may have concerning potato research or training.

Prefacio

El Centro Internacional de la Papa (CIP) es una entidad cientifica establecida para desarrollar y diseminar conocimientos sobre la papa, con el propósito de lograr su mayor utilización como alimento básico. Las siguientes publicaciones han sido hechas para apoyar la investigación y comunicación entre los programas agrícolas en el mundo. Algunas publicaciones del CIP están agotadas y ya no son incluides en esta lista. Podrían encontrarse disponibles en una biblioteca agrícola.

Le recomendamos que se comunique con su Programa Nacional de Papa. Este está autorizado para reproducir toda la información técnica del CIP y quizás también pueda responder a otras necesidades que usted tenga en relación con la investigación y la capacitación en papa.

Avant-propos

Le Centre International de la Pomme de Terre (CIP) est une institution scientifique qui a été établie dans le but de développer et disséminer les connaissances sur la Pomme de terre pour stimular son emploi en qualité d'aliment de base. Les publications présentées ci-après ont été faites afin d'encourager la recherche et la communication entre les programmes agricoles dans le monde entier. Quelques-unes des publications du CIP sont épuisées et ne sont plus distribuées par le CIP. Cependant, il se peut qu'elles soient disponibles dans d'autres bibliothequès specialisées.

Aussi, nous vous recommandons de contacter votre Programme National de Pomme de terre. Celui-ci est autorisé à reproduire toute information technique venant du CIP et pourrait peut-être répondre à d'autres besoins que vous pourriez avoir concernant la recherche et la formation en matiére de Pomme de terre.

Instruction for ordering

1 Mail your order form to the address most convenient to you (see note below and next page).
2 CIP cannot assure that all publications are available at any time. Before ordering, we suggest you contact CIP for information on availability, prices, and form of payment.
3 If you need publications urgently, include your payment (US$ check on a US bank, or bank draft in US$) with your order form. *Do not request* publications that are not listed or "in revision".

Procedimiento de compra

1 Envíe su orden de compra a la dirección que más le convenga (vea la nota abajo y la pagina siguiente).
2 El CIP no puede asegurarle que siempre tenga todas las publicaciones disponibles. Antes de hacer su pedido le sugerimos que solicite al CIP información sobre disponibilidad, precios y forma de pago.
3 Si necesita las publicaciones urgentemente, adjunte su pago (cheque en US$ de un banco de los EE.UU., o giro bancario en US$) a la orden de compra. *No solicite* material que no esté en la lista o que esté "en revisión".

Instructions pour faire votre demande

1 Envoyez votre formulaire de demande à l'adresse qui vous est la plus appropriée (voir la note ci-dessous et la page suivante).
2 Le CIP ne peut pas vous assurer que toutes les publications soient toujours disponibles. Avant de faire votre demande, il vaudrait mieux contacter le CIP afin d'être informé(e) sur la disponibilité, les prix et le mode de paiement.
3 Si vous avez besoin de ces publications de facon urgente, joignez votre paiement (chèque en US$ d'un banque des Etats Unis, ou virement bancaire en US$) au formulaire de demande. *Ne demandez pas* des publications non listées ou marquées "en révision".

Note

You can order certain CIP publications also through:
Usted puede pedir ciertas publicaciones del CIP tambien a:
Vous pouvez commander certaines publications du CIP aussi a:

Agribookstore, IADS Inc.
1611 North Kent Street
Arlington, Virginia 22209
USA
Phone 703-525-9455

Josef Margraf
Eichendorfstr. 9
8074 Gaimersheim
West Germany
Tel. 08458/2207

Mail your order form to the address most convenient to you.
Envíe su formulario de pedido a la dirección que más le convenga.
Envoyez votre formulaire de demande a l'adresse la plus appropriée.

REGION I Latinoamérica Andina	CIP Apartado Aéreo 92654 Bogotá 8, D.E. *Colombia*
REGION II Latinoamérica no Andina	CIP c/o CNPH - EMBRAPA Caixa Postal (11)1316 70 000 Brasilia D.F., *Brasil*
REGION III Tropical Africa	CIP P.O. Box 25171 Nairobi *Kenya*
REGION IV Middle East & North Africa	CIP P.O. Box 2416 Cairo *Egypt*
REGION V Afrique du Nord et Afrique de l'Ouest	CIP 11 Rue des Orangers 2080 Ariana - Tunis *Tunisia*
REGION VI South Asia	CIP c/o NBPGR IARI Campus New Delhi, 110012, *India*
REGION VII Southeast Asia	CIP c/o IRRI P.O. Box 933 Manila *Philippines*
CIP Main Headquarters Oficina Principal del CIP Direction Génerale du CIP	CIP Apartado 5969 Lima *Peru*

Research Administration	Administración de la Investigación	Administration de la Recherche

General Publications — Publicaciones Generales

CIP Decennial Anniversary Post-Congress
Workshops: a Summary of Recommendations
International Potato Center. 1982. 65 pp.   US $ 2.-

Research for the potato in the year 2000   US $ 10.-
International Potato Center. 1983. 199 pp.
Proceedings International Congress.
   More than 120 papers on potato production,
   utilization and research presented during CIP's
   tenth anniversary, including reports from
   countries in which the potato is becoming an
   important food crop.

Abstracts of the Sixth Symposium of the   US $ 5.-
International Society for Tropical Root Crops, 20-
25 February, 1983
International Potato Center. 1983. 113 pp.

Symposium of the International Society for   US $ 20.-
Tropical Root Crops, 6th, Lima, 1983
Proceedings
International Potato Center. 1984. 672 pp.
   About 120 papers on the recent progress on
   improvement of root and tuber crops which are
   staple foods in many developing countries.

Potatoes for the developing world   US $ 12.-
International Potato Center. 1984. 150 pp
   Results of research and training conducted by
   the International Potato Center after 12 years of
   permanent collaboration with national potato
   programs ("Impact Study").

Annual Reports — Informes Anuales (80-170 pp.)   free - gratuito

CIP's overall research and training activities during the corresponding year, with emphasis on contributions made by national potato programs. List of CIP's board of directors, staff, donors, publications, funding.

Resume las actividades globales de investigación y capacitación del CIP durante el año respectivo, con énfasis en las contribuciones de los

Research Administration	Administración de la Investigación	Administration de la Recherche

programas nacionales de papa. Incluye listas de los miembros de la junta directiva, el personal, los donantes, las publicaciones científicas, y fuentes de financiación.

Latest issues — Ultimos números

English	Español
1981	1981
1982	1982
1983	1983

CIP Circular — Circular del CIP

Distribution on mailing list   free
Research reports, training activities, new
publications, etc.
Quarterly
Size: 22 × 28 cm, folded, 4 to 12 pages

Distribución por subscripción   gratuita
Informes de investigación, actividades de
capacitación, nuevas publicaciones, etc.
Trimestral
Tamaño: 22 × 28 cm, doblado, 4 a 12 páginas

Main topics of the past issues:
Temas principales de los últimos números:

Vol. 11, No. 1, 1983
— The use of mulch for potatoes in the hot tropics
— Coberturas protectoras del suelo para papa en
   clima cálido

Vol. 11, No. 2, 1983
— Pollination of potatoes under natural conditions
— Polinización de papa en condiciones naturales

Vol. 11, No. 3, 1983
— Breeding of potato populations at the
   International Potato Center
— Mejoramiento de poblaciones en el Centro
   Internacional de la Papa

Vol. 11, No. 4, 1983
— Control of Rhizoctonia damping-off in
   seedlings grown from true potato seed

Research Administration	Administración de la Investigación	Administration de la Recherche

— Control de la rizoctoniosis (Rhizoctonia solani) en plántulas de papa provenientes de semilla botánica

Vol. 12, No. 1, 1984
— Tissue culture: past, present, and future
— El cultivo de tejidos: su pasado, presente y futuro

Out of print

Vol. 12, No. 2, 1984
— China: Research on potato breeding and use of true seed
— China: Investigación en mejoramiento de papa y utilización de semilla botánica

Vol. 12, No. 3, 1984
— Breeding virus-resistant potato cultivars for developing countries
— Mejoramiento de cultivares de papa por resistencia a virus

Vol. 12, No. 4, 1984
— Follow-up: An essential element for effective training
— El seguimiento: Elemento fundamental para una capacitación efectiva

Planning Conference Reports   US $ 5.-

Comprehensive reviews of aspects relevant for CIP's research, based on CIP internal planning conferences to evaluate past research results and to determine future strategies.

Note: Many reports are out of print. Copies may be available for consultation at an agricultural library.

16   Developments in the control of nematode pests of potatoes II
1978. 193 pp.

22   Strategy for virus management in potatoes II
1980. 163 pp.

23   Utilization of the genetic resources of the potato III
1982. 235 pp.

Research Administration	Administración de la Investigación	Administration de la Recherche

24 Social science research at the International
    Potato Center
    1982. 196 pp.

26 Present and future strategies for potato
    breeding and improvement
    1985. 210 pp.

27 Integrated pest management
    1984. 257 pp.

Technology Evaluation Series   US $ 1.-
Serie de Evaluación de Tecnología

Description of technologies that have been generated at the International Potato Center (CIP) and that are ready for evaluation under local conditions by scientists of national potato programs.

Descripción de tecnologías generadas por el Centro Internacional de la Papa (CIP), para que éstas sean evaluadas bajo condiciones locales por los científicos de los programas nacionales de papa.

1 Evaluation of technology for production of
   seed tubers from true potato seed
   Wiersema, S.G. 1982. 14 pp.

2 Evaluación de campo para clones del CIP
   mejorados por resistencia a marchitez
   bacteriana
   French, E.R. 1982. 9 pp.

3 Evaluation of agronomic technology for
   potato production from true potato seed
   Malagamba, J.P. 1982. 19 pp.

4 Evaluation of agronomic technology for
   potatoes in the hot tropics
   Midmore, D.J. 1982. 11 pp.

5 Field screening procedures to evaluate   In revision
   resistance to late blight
   Henfling, J.W. 1982. 16 pp.

6 Field-screening CIP clones developed for   In revision
   resistance to bacterial wilt
   French, E.R. 1982. 9 pp.

Research Administration	Administración de la Investigación	Administration de la Recherche
7 Evaluación de tecnología agronómica para cultivar papa en climas cálidos y húmedos Midmore, D.J. 1983. 12 pp.		En revisión
8 Evaluación de tecnología para la producción de tubérculos-semillas de semilla botánica de papa Wiersema, S.G. 1983. 14 pp.		En revisión
9 Evaluación de tecnología agronómica para producción de papa a partir de semilla botánica Malagamba, J.P. 1983. 20 pp.		En revisión
10 Evaluation of technology for integrated control of potato tuber moth in field and storage Raman, K.V., Booth, R.H. 1983. 18 pp.		In revision
11 Evaluación de clones del CIP por resistencia al nematodo del quiste de la papa (Globodera pallida) Franco, J., Scurrah, M. 1985.		

Germplasm Management	Manejo de Germoplasma	Gestion du Germoplasma	
Present and future strategies for potato breeding and improvement International Potato Center. 1985. 210 pp. Planning Conference Report 26 1)		US $	5.-
Systematic botany and morphology of the potato Huamán, Z. 1980. 19 pp. Technical Information Bulletin 6 3)		US $	1.-
Botánica sistemática y morfología de la papa Huamán, Z. 1980. 20 pp. Boletín de Información Técnica 6 3)		US $	1.-

see - ver - voir: 1) - p. 86   3) - p. 108-110

Germplasm Management	Manejo de Germoplasma	Gestion du Germoplasma
Tissue culture micropropagation, conservation, and export of potato germplasm Espinoza, N., Estrada, R., Tovar, P., Bryan, J., Dodds, J.H. 1984. 20 pp. Specialized Technology Document 1 　Advantages, methodology and materials of tissue culture techniques applied at CIP, such as meristem isolation, micropropagation, long-term storage and *In vitro* export of germplasm.	US $	1.-

Disease and Pest Management	Manejo de Enfermedades y Plagas	Maladies et Animaux Prédateurs
Posters - Afiches - Affiches (56 × 83 cm)	US $	1.-

- The potato plant (description in English)
- La planta de la papa (descripción en español)
- La plante de promme e terre (description en francais)

General Publications - Publicaciones Generales - Publications Génerales

The potato: major diseases and nematodes International Potato Center. 1977. 68 pp. 　Description and color photographs of symptoms caused by 30 diseases and nematode pests.	US $	2.50
La pomme de terre; maladies et nématodes Centre International de la Pomme de Terre. 1979. 68 pp. 　Description et photographies en coleur des symptomes causés par 30 maladies et nématodes.	US $	2.50
Major potato diseases, insects, and nematodes International Potato Center. 1982. 95 pp. 　Description and color photographs of symptoms caused by 50 diseases, insects, and nematode pests.	US $	5.-

Disease and Pest Management	Manejo de Enfermedades y Plagas	Maladies et Animaux Prédateurs
Principales enfermdades, nematodes e insectos de la papa Centro Internacional de la Papa. 1983. 95 pp. Descripción y fotos de color de síntomas causados por 50 enfermedades y plagas de insectos y nematodes.		US $ 5.-
Compendio de enfermedades de la papa Hooker, W.J. (ed.). 1980. 166 pp. Recopilación de la información existente sobre las enfermedades y desórdenes de la papa para científicos, estudiantes y agricultores.		US $ 10.-
Compendium of potato diseases Hooker, W.J. (ed.). 1981. 144 pp. An English version of the publication mentioned before is available at: American Phytopathological Society (APS), 3340 Pilot Knob Road, St. Paul, Minnesota 55121, USA.		Contact APS
Vocabulario de enfermedades y plagas de la papa Crossby, M. 1981. 22 pp. Guía de la terminilogía de enfermedades de la papa con nombres en latín (nombre científico), inglés y español.		US $ 1.-
General Slide Set - Juego General de Diapositivas 4)		US $ 20.-

IV-1 — Major potato diseases, nematodes, and insects
    International Potato Center. 1983. 100 slides. 39 pp.
 	
    — Principales enfermedades, nematodos, insectos y ácaros de la papa
    Centro Internacional de la Papa. 1983. 100 Diapositivas. 44 pp.		

see - ver - voir: 4) - p. 113

Disease and Pest Management	Manejo de Enfermedades y Plagas	Maladies et Animaux Prédateurs

Virology - Virología

		US $	1.-

Transmission of potato viruses by aphids
Raman, K. V. 1985. 23 pp.
Technical Information Bulletin 2
3)

Transmisión de virus de papa por áfidos
Raman, K. V. 1985. 23 pp.
Boletín de Información Técnica 2
3)
   US $ 1.-

Virus detection in potato seed production
Salazar, L. F. 1982. 14 pp.
Technical Information Bulletin 18
3)
   US $ 1.-

Detección de virus en la producción de semilla de papa
Salazar, L. F. 1982. 14 pp.
Boletín de Información Técnica 18
3)
   US $ 1.-

Virus diseases of potato
Hooker, W. J. 1982. 17 pp.
Technical Information Bulletin 19
3)
   US $ 1.-

Enfermedades virosas de la papa
Hooker, W. J. 1982. 17 pp.
Boletín de Información Técnica 19
3)
   US $ 1.-

Virus testing of seed potatoes
Hunnius, W. 1982. 16 pp.
   Review on basis of more than 160 scientific publications.
   US $ 2.-

Manual de enfermedades virosas de la papa
Salazar, L. F. 1982. 111 pp.
   Revisión de conocimientos sobre enfermedades virosas de la papa.
   US $ 5.-

see - ver - voir: 3) - p. 108-110

Disease and Pest Management	Manejo de Enfermedades y Plagas	Maladies et Animaux Prédateurs

Slide Sets and Guidebooks on Virology
Juegos de Diapositivas y Guías de Virología
4)

Slide Set plus Guidebook		see below
Guidebook alone		US $ 1.-
Juego de Diapositivas y Guía		ver abajo
Guía solamente		US $ 1.-
II-1 Fribourg, C., Nakashima, J. 1981 — Latex test for detecting potato viruses 31 slides, 12 pp. — Prueba de látex para detectar virus de papa 31 diapositivas, 12 pp.		US $ 15.-
II-2 Salazar, L. F. 1981 — Detection of PSTV by gel electrophoresis 38 slides, 10 pp.		US $ 25.-
II-3 Salazar, L. F. 1983 — Detection with ELISA of potato viruses 32 slides, 12 pp. — Detección con ELISA de virus de papa 32 diapositivas, 12 pp.		US $ 15.-
IV-2 Raman, K. V. 1984 — Monitoring aphid populations 35 slides, 12 pp. — Estudio de poblaciones de áfidos 35 diapositivas, 13 pp.		US $ 15.-

Bacteriology - Bacteriología

Bacterial wilt of potato; Pseudomonas solanacearum Martin, C. 1981. 15 pp. Technical Information Bulletin 13 3)		US $ 1.-

see - ver - voir: 4) - p. 112-113   3) - p. 109

Disease and Pest Management	Manejo de Enfermedades y Plagas	Maladies et Animaux Prédateurs
La marchitez bacteriana de la papa; Pseudomonas solanacearum Martin, C. 1981. 15 pp. Boletín de Información Técnica 13 3)		US $ 1.-
Evaluación de campo para clones del CIP mejorados por resistencia a marchitez bacteriana French, E. R. 1982. 9 pp. Serie de Evaluación de Tecnología No. 2 2)		US $ 1.-
Field-screening CIP clones developed for resistance to bacterial wilt French, E. R. 1982. 9 pp. Technology Evaluation Series No. 6 2)		In revision
Marchitez bacteriana (Pseudomonas solanacearum) de la papa en América Latina Centro Internacional de la Papa. 1984. 120 pp. Dieciseis trabajos presentados durante el seminario "Avances en el control de la marchitez bacteriana de la papa en América Latina: Reunión de planeamiento", Brazilia, Agosto 31-Setiembre 3, 1982.		US $ 5.-

Micology - Micología

Late blight of potato; Phytophthora infestans Henfling, J. W. 1979. 13 pp. Technical Information Bulletin 4 3)		US $ 1.-
El tizón tardío de la papa; Phytophthora infestans Henfling, J. W. 1980. 16 pp. Boletín de Información Técnica 4 3)		US $ 1.-
Early blight of potato; Alternaria solani Zachmann, R. 1982. 14 pp. Technical Information Bulletin 17 3)		US $ 1.-

see - ver - voir: 2) - p. 86   3) - p. 108-110

Disease and Pest Management	Manejo de Enfermedades y Plagas	Maladies et Animaux Prédateurs
El tizón temprano de la papa; Alternaria solani Zachmann, R. 1982. 14 pp. Boletín de Información Técnica 17 3)		US $ 1.-
Field screening procedures to evaluate resistance to late blight Henfling, J. W. 1982. 16 pp. Technology Evaluation Series No. 5 2)		In revision

Nematology/Entomology — Nematologia/Entomología

Developments in the control of nematode pests of potatoes II International Potato Center, 1978. 193 pp. Planning Conference Report 16 1)		US$ 5.-
Integrated pest management International Potato Center. 1984. 257 pp. Planning Conference Report 27 1)		US$ 5.-
Annotated bibliography of nematode pests of potato, Jensen, H. J., Armstrong, J., Jatala, P. 1979. 315 pp. International Potato Center, Lima, Peru, and Oregon State University Agricultural Experiment Station, Corvallis, Oregon Nomenclature and synopsis of principal nematode pests of potato. Cross index among subject areas. Nearly 1800 references.		US$ 6.-
Parasitic nematodes of potatoes. Jatala, P. 1981. 17 pp. Technical Information Bulletin 8 3)		US$ 1.-
Nematodos parásitos de la papa. Jatala, P. 1981. 17 pp. Boletín de Información Técnica 8 3)		US$ 1.-
Potato cyst nematodes; Globodera spp. Franco, J. 1981. 21 pp. Technical Information Bulletin 9 3)		US$ 1.-

see - ver - voir: 1) p. 85-86   2) - p. 86   3) - p. 108-110

Disease and Pest Management	Manejo de Enfermedades y Plagas	Maladies et Animaux Prédateurs	
Nematodos del quiste de la papa; Globodera spp. Franco, J. 1981. 21 pp. Boletín de Información Técnica 9 3)		US$	1.-
Evaluating resistance to potato cyst nematodes Scurrah, M. M. 1981. 16 pp. Technical Information Bulletin 10 3)		US$	1.-
Evaluación de la resistencia en papa a los nematodos del quiste Scurrah, M. M. 1981. 16 pp. Boletín de Información Técnica 10 3)		US$	1.-
Transmission of potato viruses by aphids Raman, K. V. 1985. 23 pp. Technical Information Bulletin 2 3)		US$	1.-
Transmisión de virus de papa por áfidos Raman, K. V. 1985. 23 pp. Boletín de Información Técnica 2 3)		US$	1.-
Potato tuber moth Raman, K. V. 1980. 14 pp. Technical Information Bulletin 3 3)		US$	1.-
La polilla de la papa Raman, K. V. 1980. 14 pp. Boletín de Información Técnica 3 3)		US$	1.-
Evaluation of technology for integrated control of potato tuber moth in field and storage Raman, K. V., Booth, R. H. 1983. 18 pp. Technology Evaluation Series No. 10 2)		In revision	

see - ver - voir: 2) - p. 87   3) - p. 108-110

Disease and Pest Management	Manejo de Enfermedades y Plagas	Maladies et Animaux Prédateurs

Nematology/Entomology — Nematología/Entomología

Evaluación de clones del CIP por resistencia al nematodo del quiste de la papa (Globodera pallida) Franco, J., Scurrah, M. 1985. Serie de Evaluación de Technología No. 11 2)	US$	1.-

Slide Set — Juego de Diapositivas
4)

IV-2 Raman, K. V. 1984
    • Monitoring aphid populations
        35 slides, 12 pp.
    • Estudio de poblaciones de áfidos
        35 diapositivas, 13 pp.
	US$	15.-

Curso de estadística experimental avanzado Centro Internacional de la Papa. 1979. 2 Tomos, 444 pp. Conceptos estadísticos, metodologías y diseños experimentales al alcance de un curso básico para la investigación alimentaria.	US$	6.-
Effect of stem density on potato production Wiersema, S. G. 1981. 15 pp. Technical Information Bulletin 1 3)	US$	1.-
Efecto de la densidad de tallos en la producción de papa Wiersema, S. G. 1981. 15 pp. Boletín de Información Técnica 1 3)	US$	1.-
Planting potatoes Cortbaoui, R. 1981. 17 pp. Technical Information Bulletin 11 3)	US$	1.-
Siembra de papa Cortbaoui, R. 1981. 17 pp. Boletín de Información Técnica 11 3)	US$	1.-

see - ver - voir: 2) - p. 87   3) - p. 108-110   4) - p. 113

Disease and Pest Management	Manejo de Enfermedades y Plagas	Maladies et Animaux Prédateurs
Soil fertility requirements for potato production Vander Zaag, P. 1981. 20 pp. Technical Information Bulletin 14 3)		US$ 1.-
Necesidades de fertilidad de suelos para la producción de papa Vander Zaag, P. 1981. 20 pp. Boletín de Información Técnica 14 3)		US$ 1.-

Agronomy	Agronomía	Agronomie
Water management in potato production Haverkort, A. J. 1982. 22 pp. Technical Information Bulletin 15 3)		US$ 1.-
Manejo del agua en la producción de papa Haverkort, A. J. 1982. 22 pp. Boletín de Información Técnica 15 3)		US$ 1.-
Physiological development of potato seed tubers Wiersema, S. G. 1985. 16 pp. Technical Information Bulletin 20 3)		US$ 1.-
Desarrollo fisiológico de tubérculos-semillas de papa Wiersema, S. G. 1985. 16 pp. Boletín de Información Técnica 20 3)		US$ 1.-
Evaluation of agronomic technology for potatoes in the hot tropics Midmore, D. J. 1982. 11 pp. Technology Evaluation Series No. 4 2)		US$ 1.-
Evaluación de tecnología agronómica para cultivar papa en climas cálidos y húmedos Midmore, D. J. 1983. 12 pp. Serie de Evaluación de Technología No. 7 2)		En revisión

see - ver - voir: 2) - p. 86-87   3) - p. 108-110

Post-harvest Technologies	Tecnologías de Poscosecha	Tecnologie d'après-récoltes	
Principles of potato storage Booth, R. H. and Shaw, R. L. 1981. 105 pp. Information for scientists and technologists to understand and solve potato storage problems. Socioeconomic factors, biological and engineering aspects.		US$	5.-
Simple processing of dehydrated potatoes and potato starch Shaw, R. L., Booth, R. H. 1982. 32 pp. How to construct and operate a simple plant to process dehydrated potatoes and potato starch, including social and economic factors to be considered.		US$	3.-

Seed Production	Producción de Semilla	Production de Semences	

General Publications - Publicaciones Generales

Roguing potatoes Cortbaoui, R. 1984. 13 pp. Technical Information Bulletin 5 3)		US$	1.-
Descarte de plantas de papa Cortbaoui, R. 1984. 13 pp. Boletín de Información Técnica 5 3)		US$	1.-
On-farm seed improvement by potato seed plot toohniquo Bryan, J. E. 1981. 11 pp. Technical Information Bulletin 7 3)		US$	1.-
Técnica de parcelas de semillla de papa a nivel de agricultor Bryan, J. E. 1980. 11 pp. Boletín de Información Técnica 7 3)		US$	1.-
Clonal selection in potato seed production Bryan, J. E. 1981. 15 pp. Technical Information Bulletin 12 3)		US$	1.-

see - ver - voir: 3) - p. 108-110

Seed Production	Producción de Semilla	Production de Semences

Selección clonal en producción de semilla de papa
Bryan, J. E. 1981. 15 pp.
Boletín de Información Técnica 12
3)

True Potato Seed - Semilla Botánica

Potato production from true seed
Accatino, P., Malagamba, J. P. 1982. 20 pp.
    Technology and results of the first three
    years of research at CIP to produce a
    potato crop from true seed.
	US$	1.-

Evaluation of technology for production of seed tubers from true potato seed Wiersema, S. G. 1982. 14 pp. Technology Evaluation Series No. 1 2)	US$	1.-
Evaluatión de tecnología para la producción de tubérculos-semillas de semilla botánica de papa Wiersema, S. G. 1983. 14 pp. Serie de Evaluación de Tecnología No. 8 2)	En revisión	
Evaluation of agronomic technology for potato production from true potato seed Malagamba, J. P. 1982. 19 pp. Technology Evaluation Series No. 3 2)	US$	1.-
Evaluación de tecnología agronómica para producción de papa a partir de semilla botánica Malagamba, J. P. 1983. 20 pp. Serie de Evaluación de Tecnología No. 9 2)	En revisión	

Rapid Multiplication - Multiplicación Rápida

Rapid multiplication techniques for potatoes
Bryan , J. E., Jackson, M. T., Meléndez, N. 1981.
20 pp.
    Introduction to be used in conjunction with
    available slide sets which provide more
    specific details.
	US$	1.-

see - ver - voir: 2) - p. 86-87   3) - p. 109

Seed Production	Producción de Semilla	Production de Semences

Técnicas de multiplicación rápida de papa
Bryan, J. E., Jackson, M. T., Meléndez, N. 1981.
20 pp.
   Introducción al tema para ser utilizada en
   conjunto con series de diapositivas que
   muestran detalles más específicos.
	US$	1.-

Tissue culture micropropagation, conservation,
and export of potato germplasm
Espinoza, N., Estrada, R., Tovar, P., Bryan, J.,
Dodds, J. H. 1984. 20 pp.
Specialized Technology Document 1
   Advantages, methodology and materials of
   tissue culture techniques applied at CIP,
   such as meristem isolation, micropropagation,
   long-term storage and *in vitro* export of
   germplasm.
	US$	1.-

Slide Training Series and Guidebooks
Series de Diapositivas Didácticas y Guías Didácticas
Jeux de Diapositives Didactiques
4)

Slide Set plus Guidebook (except in French)	US$	15.-
Guidebook alone	US$	1.-
Juegos de Diapositivas con Guía (excepto en francés)	US$	15.-
Guía solamente	US$	1.-
Jeux de Diapositives (il n'y a pas de guide en francais)	US$	15.-

I-1   Bryan, J. E., Meléndez, N., Jackson, M. T. 1981

    • Sprout cuttings,
      a rapid multiplication technique for potatoes
      27 slides, 10 pp.
    • Esquejes de brote,
      una técnica de multiplicación rápida de papa
      27 diapositivas, 10 pp.
    • Bouturage de germes,
      une technique de multiplication rapide de la pomme de terre
      27 diapositives

see - ver - voir: 4) - p. 111-113

Seed Production	Producción de Semilla	Production de Semences

I-2 Bryan, J. E., Jackson, M. T., Quevedo, M., Meléndez, N. 1981
- Single-node cuttings,
  a rapid multiplication technique for potatoes
  20 slides, 8 pp.
- Esquejes de tallo juvenil,
  una técnica de multiplicación rápida de papa
  20 diapositivas, 8 pp.
- Bouturage de noeuds foliaires (cas d'une plante jeune),
  une technique de multiplication rapide de la pomme de
  terre
  20 diapositives

I-3 Bryan, J. E., Meléndez, N., Jackson, M. T. 1981
- Stem cuttings,
  a rapid multiplication technique for potatoes
  35 slides, 16 pp.
- Esquejes de tallo lateral,
  una técnica de multiplicación rápida de papa
  35 diapositivas, 16 pp.
- Bouturage de tiges latérales,
  une technique de multiplication rapide de la pomme de
  terre
  35 diapositives

I-4 Quevedo, M., Bryan, J. E., Jackson, M. T., Meléndez, N. 1981
- Leaf-bud cuttings,
  a rapid multiplication technique for potatoes
  25 slides, 10 pp.
- Esquejes de tallo adulto,
  una técnica de multiplicación rápida de papa
  25 diapositivas, 11 pp.
- Bouturage de noeuds foliaires (cas d'une plante d'age
  avancé),
  une technique de multiplication rapide de la pomme de
  terre
  25 diapositives

III-1 Malagamba, P. et al. 1983
- True potato seed:
  an alternative method for potato production
  42 slides, 12 pp.
- Semilla botánica:
  un método alterno para la producción de papa
  42 diapositivas, 12 pp.

see - ver - voir: 4) - p. 111-113

Seed Production	Producción de Semilla	Production de Semences

Training Charts - Carteles de Entrenamiento

Integrated system of rapid multiplication for potatoes International Potato Center, 1981 One chart, 60 × 85 cm	US$	1.-
Sistema integrado de multiplicación rápida de papa Centro Internacional de la Papa, 1981 Un cartel, 60 × 85 cm	US$	1.-
Rapid multiplication techniques for potatoes International Potato Center, 1981 Nine flip charts, 60 × 85 cm	US$	15.-
Técnicas de multiplicación rápida de papa Centro Internacional de la Papa, 1981 Guía ilustrada. 20 pp.	US$	1.-

Social Sciences	Ciencias Sociales	Sciences Sociales

General Publications - Publicaciones Generales

Social science research at the International Potato Center International Potato Center. 1982. 196 pp. Planning Conference Report 24 1)	US$	5.-
Partial budget analysis for on-farm potato research Horton, D. 1982. 16 pp. Technical Information Bulletin 16 3)	US$	1.-
Análisis de presupuesto parcial para investigación en papa a nivel de finca Horton, D. 1982. 16 pp. Boletín de Información Técnica 16 3)	US$	1.-

Social Sciences	Ciencias Sociales	Sciences Sociales

On-farm potato research in the Philippines
Potts, M. J. (ed.) 1983. 169 pp.
International Potato Center, Lima, Peru;
Ministry of Agriculture, Philippines; and
Philippine Council for Agriculture and
Resources Research and Development
    Ten papers presented by participants of a
    three-year program on farm-level research
    and technology transfer: Project approach
    agricultural systems, results, and farmer
    acceptance.
		US$ 5.-

Household gardens:
Theoretical considerations
on an old survival strategy
Niñez, V. K. 1984. 41 pp.
Potatoes in Food Systems Research Series No. 1
    Historical significance and function of
    home gardens. Definition and typology
    based on ecological and socioeconomic
    determinants. Cross-cultural comparative
    review of literature.
		US$ 1.-

World potato facts
International Potato Center. 1982. 54 pp.
    Statistical data on potato production and
    consumption for most countries of the world.
		US$ 1.-

Potato atlas
Atlas de la pomme de terre
Atlas de la papa
Horton, D. E., Fano, H. 1985. 136 pp.
    National and regional statistics on potato
    production and use in form of maps, figures,
    and tables.
    Statistiques au niveau national et régional
    sur la production et l'utilisation de la
    pomme de terre sous forme de cartes, figures
    et tableaux.
    Estadísticas a nivel nacional y regional de
    la producción y utilización de la papa,
    presentadas en forma de mapas, figuras y
    cuadros.
		US$ 5.-

Social Sciences	Ciencias Sociales	Sciences Sociales

Social scientists in agricultural research:
Lessons from the Mantaro Valley Project,
Peru
Horton, D. E. 1984. 67 pp.
International Development Research Centre
(IDRC)
Box 8500, Ottawa, Canada K1G 3H9
   Experiences and results of the project
   carried out in the highlands of Peru
   from 1977 to 1980.

Contact
IDRC

Los científicos sociales en la investigación
agrícola:
Lecciones del Proyecto del Valle del Mantaro,
Perú
Horton, D. E. 1984. 71 pp.
Centro Internacional de Investigaciones
para el Desarrollo (CIID)
Apartado Aéreo 53016, Bogotá, Colombia
   Experiencias y resultados del proyecto
   llevado a cabo en las tierras atlas del
   Perú de 1977 a 1980.

Contacte
CIID

Spécialistes des sciences sociales et recherche
agricole:
Enseignements du projet de la vallée du Mantaro,
Pérou.
Horton, D. E. 1984. 72 pp.
Centre de recherches pour le développement
international (CRDI)
C. P. 8500, Ottawa, Canada K1G 3H9
   Expériences et resultats du Projet de la
   Vallée de Mantaro exécuté dans les montagnes
   du Pérou de 1977 á 1980.

Contactez le
CRDI

Country Studies - Estudios de Paises

Land use in the Andes:
Ecology and agriculture in the Mantaro valley
of Peru with special reference to potatoes
Mayer, E. 1979. 115 pp.
   Anthropological study of agricultural practices
   and use of land. Description of methodologies
   used. Maps of growing areas, ecological
   characteristics and land tenure patterns.

US$ 3.-

Social Sciences	Ciencias Sociales	Sciences Sociales

Uso de la tierra en los Andes:
Ecología y agricultura en el valle del Mantaro
del Perú con referencia especial a la papa
Mayer, E. 1981. 127 pp.
   Estudio antropológico de prácticas agrícolas
   y utilización de tierras. Descripción de
   metodologías aplicadas. Mapas de áreas de
   producción, características ecológicas y
   patrones de tenencia de tierra.
		US$ 3.-

Potato production and utilization in Kenya
Duerr, G., Lorenzl, G. 1980. 133 pp.
Country study
International Potato Center, Lima, Peru;
Technical University, Berlin, West Germany;
and University of Nairobi, Kenya
   Historical and present role of the potato
   and potato industry in Kenya. Deficiencies
   and proposals for improvement.
		US$ 5.-

Potato storage systems in the Mantaro valley region of Peru Werge, R. W. 1980. 49 pp.		US$ 1.-

Marketing Bhutan's potatoes:
Present patterns and future prospects
Scott, G. J. 1984. 80 pp.
   Case study on marketing patterns for
   Bhutanese potatoes, marketing policies, and
   future prospects for potato marketing.
   Recommendations for development of foreign
   and domestic markets.
		US$ 2.-

Theses - Tesis

Los límites socioecológicos del crecimiento agrícola en la ceja de selva Recharte, J. 1981. 208 pp. Tesis Lic. Antropología Pontificia Universidad Católica del Perú, Lima, Perú		US$ 3.-
Introducción de semilla botánica de papa en un sistema de producción hortícola: Análisis prospectivo en la región de Tarma, Perú Alarcón, J. 1983. 238 pp. Tesis M. S. Universidad Nacional Agraria, La Molina, Lima, Perú		US$ 2.-

Social Sciences	Ciencias Sociales	Sciences Sociales

Análisis econométrico de funciones
de consumo de carne y papas en el Perú
Apaza, J. 1983. 194 pp.
Tesis Ing. Economista
Universidad Nacional Tecnica del Altiplano, Puno,
Perú — US$ 2.-

Cambio tecnológico y tendencias de la
producción de papa en la region central del Perú
Fano, H. R. 1983. 208 pp.
Tesis Economista
Universidad Nacional Agraria, La Molina, Lima,
Perú — US$ 2.-

Términos de intercambio de la agricultura
en la década del 70.
El caso de los productores de papa
en el Valle del Mantaro
Lafosse, R. 1983. 127 pp.
Tesis Bs. Economía
Universidad del Pacífico, Lima, Perú — US$ 2.-

*Training Documents - Documentos de
Entrenamiento* — US$ 1.-

The purpose of these publications is to make
available in preliminary form, materials in
preparation for future technical publications.

El propósito de estas publicaciones es darle
circulación en forma preliminar, a materiales que
están siendo preparados para futuras
publicaciones técnicas.

1980 - 4   Análisis económico de la producción y uso de
semilla de papa.
Monares, A. 10 pp.

1980 - 5   Investigation of farming in Peru by means of a
multiple visit survey.
Werge, R., Benavides, M. 12 pp.

1980 - 7   On-farm evaluation of seed stores
Cortbaoui, R., Booth, R. 10 pp.

1980 - 10   On-farm research
for optimizing potato productivity.
(A description of CIP's current approach)
Cortbaoui, R. T. 10 pp.

Social Sciences	Ciencias Sociales	Sciences Sociales

1981 - 1   Investigación a nivel de finca
para optimizar la productividad de la papa.
(Una descripción del enfoque actual del CIP)
Cortbaoui, R. 12 pp.

1982 - 1   Optimizing Potato Productivity:
Planning and implementing on-farm trials
Cortbaoui, R. 16 pp.

1982 - 2   The art of the informal agricultural survey.
Rhoades, R. E. 40 pp.

1982 - 3   Understanding small farmers:
Sociocultural perspectives
on experimental farm trials
Rhoades, R. E. 10 pp.

1982 - 5   Optimizing potato productivity.
Evaluating and utilizing results
of on-farm trials
Cortbaoui, R. T. 13 pp.

1982 - 6   Tips for planning formal farming surveys
in developing countries
Horton, D. 17 pp.

1982 - 8   Para comprender a los pequeños agricultores:
Perspectivas socioculturales
de la investigación agrícola
Rhoades, R. E. 9 pp.

Working Papers - Documentos de Trabajo   US$ 1.-

These publications deals with specific social
science topics to encourage debate, exchange of
ideas, and advancement of social science
knowledge.

En estas publicaciones se tratan temas específicos
de ciencias sociales, con el fin de estimular el
debate, el intercambio de ideas, y el avance del
conocimiento en ciencias sociales.

1979 - 4   The agricultural strategy of rural households in
three ecological zones of the central Andes.
Werge, R. 27 pp.

Social Sciences	Ciencias Sociales	Sciences Sociales

1980 - 1
Tecnología de la producción de papa
en el valle del Mantaro, Perú:
Resultados de una encuesta agroeconómica
de visita múltiple
Horton, D., Tardieu, F., Benavides, M.,
Tomassini, L., Accatino, P. 68 pp.

1980 - 4
Evaluación agroeconómica de ensayos conducidos
en campos de agricultores en el valle del Mantaro
(Perú).
Campaña 1978/1979
Franco E., et al.

1980 - 6
The potato as a food crop
for the developing world
Horton, D. 15 pp.

1982 - 1
Farmer-back-to-farmer:
a model for generating acceptable agricultural
technology
Rhoades, R. E., Booth, R. H. 15 pp.

1982 - 2
Un método estadístico de selección
y evaluación de muestras
para encuestas agrícolas
Alarcón J., Rubio, A. 23 pp.

1982 - 3
Potato preferences:
A preliminary examination
Poats, S. V. 18 pp.

1983 - 1
Potato production and utilization in Rwanda
Duerr, G. 85 pp.

1983 - 2
Producción y utilización de la papa en la región del
Cuzco
Franco, E., Moreno, C., Alarcón, J. 103 pp.

1984 - 1
Changing a post-harvest system: impact of diffused
light potato stores in Sri Lanka.
Rhoades, R. E. 22 pp.

1984 - 2
A small effective seed multiplication program:
Tunisia
Horton, D., Monares, A. 13 pp.

Training and Communications	Capicitación y Comunicaciones	Formation et Communications

General Publications - Publicaciones Generales

La papa en la bibliografía peruana de 1965 a 1980 Crosby, M. 1981. 172 pp.	US$	3.-

Vocabulario de enfermedades y plagas de la papa
Crosby, M. 1981. 22 pp.
  Guía de la terminología de enfermedades de la
  papa con nombres en latín (nombre científico),
  inglés y español.
	US$	1.-

Potato production course.
Siri, C., Zachmann, R. 1983. 65 pp.
  Guide for the organization of a 28-day course,
  and for the training materials to be used.
  May be adapted flexibly in duration and
  content.
	US$	2.-

Technical Information Bulletins (TIBs) *Boletines de Información Técnica*(TIBs)	US$	1.-

Specific topics of potato production and research
to support transfer of technology between science
and farmers.

Temas específicos de investigación y producción
de papa con el fin de apoyar la transferencia de
tecnología entre el nivel científico y los
agricultores.

1   Effect of stem density on potato production
     Wiersema, S. G. 1981. 15 pp.

     Effecto de la densidad de tallos en la producción de papa
     Wiersema, S. G. 1981. 15 pp.

2   Transmission of potato viruses by aphids
     Raman, K. V. 1985. 23 pp.

     Transmisión de virus de papa por áfidos
     Raman, K. V. 1985. 23 pp.

3   Potato tuber moth
     Raman, K. V. 1980. 14 pp.

     La polilla de la papa
     Raman, K. V. 1980. 14 pp.

4   Late blight of potato; Phytophthora infestans
     Henfling, J. W. 1979. 13 pp.

     El tizón tardío de la papa; Phytophora infestans
     Henfling, J. W. 1980. 16 pp.

Training and Communications	Capicitación y Comunicaciones	Formation et Communications

5
Roguing potatoes
Cortbaoui, R. 1984. 13 pp.

Descarte de plantas de papa
Cortbaoui, R. 1984. 13 pp.

6
Systematic botany and morphology of the potato
Huamán, Z. 1980. 19 pp.

Botánica sistemática y morfología de la papa
Huamán, Z. 1980. 20 pp.

7
On-farm seed improvement by potato seed plot technique
Bryan, J. E. 1983. 11 pp.

Técnica de parcelas de semilla de papa
a nivel de agricultor
Bryan, J. E. 1983. 11 pp.

8
Parasitic nematodes of potatoes
Jatala, P. 1981. 17 pp.

Nematodos parásitos de la papa
Jatala, P. 1981. 17 pp.

9
Potato cyst nematodes; Globodera spp.
Franco, J. 1981. 21 pp.

Nematodos del quiste de la papa; Globodera spp.
Franco, J. 1981. 21 pp.

10
Evaluating resistance to potato cyst nematodes
Scurrah, M. M. 1981. 16 pp.

Evaluación de la resistencia en papa
a los nematodos del quiste
Scurrah, M. M. 1981. 16 pp.

11
Planting potatoes
Cortbaoui, R. 1981. 17 pp.

Siembra de papa
Cortbaoui, R. 1981. 17 pp.

12
Clonal selection in potato seed production
Bryan, J. E. 1981. 15 pp.

Selección clonal en producción de semilla de papa
Bryan, J. E. 1981. 15 pp.

13
Bacterial wilt of potato; Pseudomonas solanacearum
Martin, C. 1981. 15 pp.

La marchitez bacteriana de la papa; Pseudomonas
solanacearum
Martin, C. 1981. 15 pp.

Training and Communications	Capicitación y Comunicaciones	Formation et Communications

14   Soil fertility requirements for potato production
Vander Zaag, P. 1981. 20 pp.

Necisidades de fertilidad de suelos
para la producción de papa
Vander Zaag, P. 1981. 20 pp.

15   Water management in potato production
Haverkort, A. J. 1982. 22 pp.

Manejo del agua en la producción de papa
Haverkort, A. J. 1982. 22 pp.

16   Partial budget analysis for on-farm potato research
Horton, D. 1982. 16 pp.

Análisis de presupuesto parcial
para investigación en papa a nivel de finca
Horton, D. 1982. 16 pp.

17   Early blight of potato; Alternaria solani
Zachmann, R. 1982. 14 pp.

El tizón temprano de la papa; Alternaria solani
Zachmann, R. 1982. 14 pp.

18   Virus detection in potato seed production
Salazar, L. F. 1982. 14 pp.

Detección de virus en la producción de semilla de papa
Salazar, L. F. 1982. 14 pp.

19   Virus diseases of potato
Hooker, W. J. 1982. 17 pp.

Enfermedades virosas de la papa
Hooker, W. J. 1982. 17 pp.

20   Physiological development of potato seed tubers
Wiersema, S. G. 1985. 16 pp.

Desarollo fisiologico de tubérculos-semillas
de papa
Wiersema, S. G. 1985. 16 pp.

Slide Training Series and Guidebooks
Series de Diapositivas Didácticas y Guías Didácticas
Jeux de Diapositives Didactiques

CIP's Slide Sets describe potato production and research technologies, as well as general information on potato.

Slide Set plus Guidebook   see below

Guidebook alone   US$ 1.-

Training and Communications	Capicitación y Comunicaciones	Formation et Communications

Los Juegos de Diapositivas de CIP describen tecnologías de producción o investigación, o bien presentan información general sobre la papa.

Juegos de Diapositivas con Guía   ver abajo
Guíasolamente   US$ 1.-

Les Jeux de Diapositives du CIP décrivent des technologies de production ou de recherche ou bien présentent des informations générales sur la
Pomme de terre.

Jeux de Diapositives   voir
(Il n'y a pas de guide en francais)   ci-dessous

Series I Rapid Multiplication Techniques
Técnicas de Multiplicacción Rápida
Techniques de Multiplication Rapide

I-1   Bryan, J. E., Meléndez, N., Jackson, M. T.   US$ 15.-
1981.

- Sprout cuttings,
  a rapid multiplication technique
  for potatoes
  27 slides, 10 pp.
- Esquejes de brote,
  una técnica de multiplicación rápida
  de papa
  27 diapositivas, 10 pp.
- Bouturage de germes,
  une technique de multiplication rapide
  de la pomme de terre
  27 diapositives

I 2   Bryan, J. E., Jackson, M. T., Quevedo. M.,   US$ 15.-
Meléndez, N. 1981

- Single-node cuttings,
  a rapid multiplication technique
  for potatoes
  20 slides, 8 pp.
- Esquejes de tallo juvenil,
  una técnica de multiplicación rápida
  de papa
  20 diapositivas, 8 pp.
- Bouturage de noeuds foliaires
  (cas d'une plante jeune),
  une technique de multiplication rapide
  de la pomme de terre
  20 diapositives

Training and Communications	Capicitación y Comunicaciones	Formation et Communications

I-3   Bryan, J. E., Meléndez, N., Jackson, M. T. 1981   US$ 15.-

- Stem cuttings,
  a rapid multiplication technique for potatoes
  35 slides, 16 pp.
- Esquejes de tallo lateral,
  une técnica de multiplicación rápida de papa
  35 diapositivas, 16 pp.
- Bouturage de tiges latérales,
  une technique de multiplication rapide de la
  pomme de terre
  35 diapositives

I-4   Quevedo, M., Bryan, J. E., Jackson, M. T.,   US$ 15.-
      Meléndez, N. 1981

- Leaf-bud cuttings,
  a rapid multiplication technique for potatoes
  25 slides, 10 pp.
- Esquejes de tallo adulto,
  una técnica de multiplicación rápida de papa
  25 diapositivas, 11 pp.
- Bouturage de noeuds foliaires (cas d'une plante
  d'age avancé),
  une technique de multiplication rapide de la
  pomme de terre
  25 diapositives

Series II   Potato Virus and Viroid Detection Methods
            Métodos de Detección de Virus y Viroides de Papa

II-1   Fribourg, C. and Nakashima, J. 1981   US$ 15.-

- Latex test for detecting potato viruses
  31 slides, 12 pp.
- Prueba de látex para detectar virus de papa
  31 diapositivas, 12 pp.

II-2   Salazar, L. F. 1981   US$ 25.-

- Detection of PSTV by gel eletrophoresis
  38 slides, 10 pp.

II-3   Salazar, L. F. 1983   US$ 15.-

- Detection with ELISA of potato viruses
  32 slides, 12 pp
- Detección con ELISA de virus de papa
  32 diapositivas, 12 pp.

Training and Communications	Capicitación y Comunicaciones	Formation et Communications

Series III — Potato Production from True Potato Seed
Producción de Papa a partir de la Semilla Botánica

III-1 — Malagamba, P. et al. 1983 — US$ 15.-

- True potato seed:
  an alternative method for potato production
  42 slides, 12 pp.
- Semilla botánica:
  un método alterno para la producción de papa
  42 diapositivas, 12 pp.

Series IV — Potato Diseases and Pests
Enfermedades y Plagas de la Papa

IV-1 — International Potato Center. 1983 — US$ 20.-
Centro Internacional de la Papa. 1983

- Major potato diseases, nematodes and insects
  100 slides, 39 pp.
- Principales enfermedades, nematodos, insectos y
  ácaros de la papa
  100 diapositivas, 44 pp.

IV-2 — Raman, K. V. 1984 — US$ 15.-

- Monitoring aphid populations
  35 slides, 12 pp.
- Estudio de poblaciones de áfidos
  35 diapositivas, 13 pp.

Training Charts - Cartales de Entranamiento

Integrated system of rapid multiplication for
potatoes
International Potato Center, 1981
One chart, 60 × 85 cm — US$ 1.-

Sistema integrado de multiplicación rápida de papa
Centro Internacional de la Papa, 1981
Un cartel, 60 × 85 cm — US$ 1.-

Rapid multiplication techniques for potatoes
International Potato Center, 1981
Nine flip charts, 60 × 85 cm — US$ 15.-

Técnicas de multiplicación rápida de papa
Centro Internacional de la Papa, 1981
Guía ilustrada. 20 pp. — US$ 1.-

Training and Communications	Capicitación y Comunicaciones	Formation et Communications	
The potato plant Zachmann, R. 1980. Technical Information Poster 1, 56 × 83 cm		US$	1.-
La planta de la papa Zachmann, R. 1980. Afiche de Información Técnica 1, 56 × 83 cm		US$	1.-
La plante de pomme de terre Zachmann, R. 1980. Information Technique Poster 1, 56 × 83 cm		US$	1.-

CIMMYT

Centro Internacional de Mejoramiento de Maiz y Trigo
(International Center for Maize and Wheat Improvement)

Londres 40 - Apdo. Postal 6-641 - 06600 Mexico D.F.
Cable address: CENCIMMYT - Telex address: 1772023 CIMTME
Telephone: 585-43-55

General Information

Maize and wheat constitute the principal sources of carbohydrates and protein for nearly half the people of the world. CIMMYT's primary objective is to develop superior germplasm that will provide higher and more stable yields, as well as better nutritional quality. Toward this objective, CIMMYT promotes and helps to implement research, training and information programs designed to improve maize and wheat production in the developing countries of the world. The Center currently conducts research on maize, bread wheat, durum wheat and triticale.

To be consistent with its mandate to serve the Third World, CIMMYT provides single copies of most of its publications *free of charge* to over 5,000 agricultural scientists and libraries in developing countries. Beyond this complementary distribution, CIMMYT employs a two-tiered pricing structure for its publications. Separate prices are listed in this catalog for Highly Developed Countries (HDCs) and Less Developed Countries (LDCs). All prices include air mail postage. (Note: CIMMYT reserves the right to adjust prices on an annual basis.)

In addition, all CIMMYT publications are available on microfishe. If interested in this delivery format, please write for cost information and other details.

Research Highlights

CIMMYT Staff

CIMMYT Research Highlights 1984

1985. 80 pages. 22 x 25 cm. Paperback. Perfect bound. Available in English. HDC $6.00. LDC $3.00.

A highlighted annual research report on selected CIMMYT research activities in maize, bread wheat, durum wheat and triticale crop improvement. Highlights of selected crop management and economics research activities are also provided.

Also available: Research Highlights for 1983.

Annual Reports

CIMMYT Staff

CIMMYT Annual Report 1984

1985. 68 pages. 21.5 × 28 cm. Paperback. Saddle wire stitched. Available in English and Spanish. HDC $8.00. LDC $4.00.

This non-technical annual report is intended primarily for policy makers and members of the donor community. It presents CIMMYT's research program activities in a highly summarized form (the "Management Report" and "The Year In Review"), and gives emphasis to the financial and special project aspects of the Center's operations (the "Independently Audited Financial Statement" and the "Extra-Core Grant Summaries"). This report is supplemented by the Research Highlights and two technical reports, one focusing on maize research and the other on wheat research.

CIMMYT Maize Program Staff

CIMMYT Biennial Report on Maize Improvement 1982-83

Available January, 1986. Paperback. Perfect bound. Available in English. HDC $8.00. LDC $4.00.

This technical report is produced every two years and is intended for informed, technically oriented individuals working on the various aspects of maize improvement. The report describes in detail the CIMMYT Maize Program's research activities in gene pool development, population improvement, quality protein maize improvement, special projects, wide crosses, training and regional programs.

Also available: CIMMYT Biennial Report on Maize Improvement 1980-81

CIMMYT Wheat Program Staff

CIMMYT Report on Wheat Improvement 1984

Available January, 1985. Paperback. Perfect bound. Available in English. HDC $8.00. LDC $4.00.

This technical report is produced annually and is intended for informed, technically oriented individuals working on the various aspects of wheat improvement. The report describes the CIMMYT Wheat Program's research on bread wheat, durum wheat and triticale, as well as the CIMMYT/ICARDA cooperative program for barley improvement. Also included are reports on the activities of the various research support programs, and those of the regional programs.

Also available: CIMMYT Report on Wheat Improvement 1983, 1982 and 1981

Monographs

H. Hanson, N.E. Borlaug and R.G. Anderson

Wheat in the Third World

1982. 174 pages. Paperback. Perfect bound. Available in English and Spanish. HDC $10.00. LDC $5.00.

This book addresses decision makers in developing countries and in international development agencies, providing essential information about the prospects for increasing the production of wheat in the Third World. The authors examine the characteristics of the wheat plant as a crop and as a source of food, explore recent scientific findings related to producing and handling the crop, and suggest important areas for future research. The specific wheat production problems and potentials in eight countries are also examined, and the authors describe what is required to organize and operate an effective national wheat research program. The book closes by forecasting the outlook for food production in general, wheat production in particular, and population growth to the end of the century.

Conference Proceedings

P.A. Burnett, Editor

Barley Yellow Dwarf: A Proceedings of the Workshop

1984. 209 pages. 17 × 24 cm. Paperback. Perfect bound. Available in English. HDC $12.00. LDC $6.00.

An international workshop on barley yellow dwarf (BYD) disease of wheat, barley and oats was held in Mexico City on December 6-8, 1983, and attended by selected professionals from developing countries, as well as from research institutions in the developed world. The proceedings of the workshop include a review of the global incidence of BYD, the yield losses it causes, and the current state of knowledge on BYD in the countries and regions represented in the workshop. Also included are technical papers by selected contributors on the following topics:
- Biology of the virus, the plant and the aphid
- Resistance to BYD and screening techniques for all the small grain cereals
- Control of BYD by chemical and/or cultural methods
- Methods for surveying for the incidence of BYD
- Purification of the BYD virus
- Specific BYD research programs
- *Diuraphis noxia,* an aphid recently recognized as causing losses in cereals, which also may be a vector of the BYD virus

R.L. Villareal and A. Klatt, Editors

Wheats for More Tropical Environments: Proceedings of an International Symposium

1985. 185 pages. 17 × 24 cm. Paperback. Perfect bound. Available in English. HDC $12.00. LDC $6.00.

An international symposium on wheats for more tropical environments was held in Mexico City on September 25-28, 1984, and was attended by representatives of over 50 developing countries located in the tropics. The proceedings contains "country reports" that summarize the status of wheat production in selected tropical countries, and a range of papers that present and discuss the various technical and economic constraints to producing wheat in more tropical environments.

B. Gelaw, Editor

Maize in Eastern, Central and Southern Africa: A Proceedings of the First Regional Workshop

Available January, 1986. 17 × 24 cm. Paperback. Perfect bound. Available in English. HDC $12.00. LDC $6.00.

A regional workshop on maize in Eastern, Central and Southern Africa was held in Lusaka, Zambia, on March 10-17, 1985, and was attended by maize breeders and agronomists from 17 African countries. The proceedings contains "country reports" that summarize the status of maize production in the 17 countries represented at the workshop, as well as technical papers by selected specialists. These technical papers present and discuss 1) various approaches to maize research, 2) the constraints to maize production in African environments, 3) the breeding efforts necessary to overcome these constraints, 4) seed production issues, and 5) questions regarding various economic aspects of maize production under the conditions of representative farmers.

Guidebooks and Manuals

F.J. Zillinsky

Common Diseases of Small Grain Cereals: A Guide to Identification

March 1983. 150 pages. 350 color plates. 21.5 × 28 cm. Paperback. Sewn and perfect bound. Available in English, Spanish and French. HDC $20.00. LDC $12.00

Practical, easy-to-use, full color laboratory and field guide to the identification of over 70 common diseases afflicting small grain cereals (wheat, barley, triticale, oats, rye). Diseases included are the rusts, helminthosporiums, septorias, smuts and bunts, fusarium diseases, weak pathogens and saprophytes, common bacterial diseases, viral and mycoplasmal diseases, and certain miscellaneous diseases affecting roots, crowns, leaves and heads. Injuries caused by parasites and pests, as well as physiological and environmental abnormalities are also described. Guidance in how to collect, preserve and use samples of diseased plant material are included.

CIMMYT Staff

Field Manual of Common Wheat Diseases and Pests

1985 (Third Edition). 120 pages. 110 color photographs. 10 × 18 cm. Saddle wire stitched. Available in English and Spanish (French version will be available in 1986). HDC $7.50. LDC $5.50.

An illustrated, compact (pocket-size) field guide to common wheat diseases and pests, designed for use in the field by all agricultural workers. A simple diagnostic key, numerous color photographs of diseased and pest-infested plants, and a helpful glossary of pathology-related terms are included as aids to identification.

C. De León

Maize Diseases: A Guide for Field Identification

1984 (Third Edition). 116 pages. 90 color photographs. 10 × 18 cm. Saddle wire stitched. Available in English and Spanish (French version will be available in 1986). HDC $7.50. LDC $5.50.

This booklet is intended to be used in the field. The text describes important maize diseases, their associated symptoms and causal organisms. A simple diagnostic key and color photographs of diseased plants are included as aids to identification.

A. Ortega

Maize Insects: A Guide for Field Identification

Available (in English) January, 1986. 10 × 18 cm. Saddle wire stitched. (Spanish and French versions will be available later in 1986). HDC $7.50. LDC $5.50.

Designed as a companion volume to "Maize Diseases: A Guide for Field Identification" this booklet is also intended to be used in the field, and includes numerous color photographs and a simple diagnostic key.

R.K. Perrin et al.

From Agronomic Data to Farmer Recommendations: An Economics Training Manual

First edition 1977. 5th printing 1983. 55 pages. 21 × 28 cm. Saddle wire stitched. Available in English, Spanish and French. HDC $7.50. LDC $3.00.

A manual intended for use by agronomists as they develop farmer recommendations on the basis of agronomic experiments. The manual describes a series of factors about farmers' circumstances that need to be taken into account as production recommendations are formulated. Descriptions and examples of economic analyses of agronomic data are included.

D. Byerlee, M. Collinson et al.

Planning Technologies Appropriate to Farmers: Concepts and Procedures

First edition 1980. 3rd printing 1984. 72 pages 21 × 28 cm. Saddle wire stitched. Available in English, Spanish and French. HDC $7.50. LDC $3.00.

This manual was prepared for professionals involved in on-farm research activities, particularly in survey research on farmers' circumstances. It describes a collaborative research process involving biological scientists and economists that is directed toward the development of production technologies that are both appropriate to farmers' circumstances and that help to meet the goals of national policy.

CIMMYT Staff

Development, Maintenance and Seed Multiplication of Open-Pollinated Maize Varieties

1984. 12 pages. 21.5 × 28 cm. Saddle wire stitched. Available in English, Spanish and French. HDC $3.00. LDC No charge.

An information bulletin describing the characteristics of open-pollinated maize varieties and the procedures recommended by CIMMYT for 1) maintaining the genetic integrity of these varieties, and 2) conducting seed multiplication activities to produce high-quality foundation seed for use in commercial seed production programs.

Periodicals

CIMMYT Economics Program Staff

1985 World Wheat Facts and Trends, Report Three: Prices, Marketing Margins and Pricing Policies for Wheat in Developing Countries

1985. 40 pages. 21.5 × 28 cm. Saddle wire stitched. Available in English. HDC $5.00. LDC $3.00.

Part 1 of this report presents and discusses in summary form how the international wheat market works, the pricing and market structures of selected developing countries, and some alternative pricing policies for Third World nations. Part 2 presents the current global wheat situation, including information about production, consumption, trade and prices. Part 3 presents (in tabular form) recent data related to the wheat economy in countries that either grow over 100,000 hectares of wheat or consume over 100,000 tons of the grain (or both).

Also available: Wheat Facts and Trends, Reports One (1981) and Two (1983)

CIMMYT Economics Program Staff

1984 World Maize Facts and Trends, Report Two: An Analysis of Changing Trends in Global Maize Consumption and Utilization

1984. 40 pages. 21.5 × 28 cm. Saddle wire stitched. Available in English. HDC $5.00. LDC $3.00.

This report details the major current maize consumption patterns of the world, considering maize as human food, animal feed, and import/export commodity. Part 1 of the report focuses on the changing trends in the utilization of maize and a human food and animal feed; Part 2 presents information on the current global maize situation, including information about production, consumption, trade and prices; Part 3 presents (in tabular form) recent data related to the maize economy in countries that either grow over 100,000 hectares of maize or consume over 100,000 tons of the grain (or both).

Also available: World Maize Facts and Trends, Report One (1982)

CIMMYT Today Series: Length varies from 12 to 16 pages. 21.5 × 28 cm. Saddle wire stitched. Full color presentation. Available in English and Spanish. HDC $3.00. LDC No charge.

(Note: CIMMYT Today Nos. 1, 2, 4, 7 and 13 are out of print)
Return of the Medic. S.A. Breth, 1975. (CT No. 3)
Wheat $\times$ Rye $=$ Triticale. A. Wolff, 1976. (CT No. 5)
Turkey's Wheat Research and Training Program. S.A. Breth, 1977. (CT No. 6)
Transforming Maize Farming in Zaire. S.A. Breth, 1978 (CT No. 8)
CIMMYT Training. A. Wolff, 1978. (CT No. 9)
CIMMYT's International Testing Program in Wheat, Triticale and Barley. 1979 (CT No. 10)
Speeding the Breeding. J.V. Mertin, 1979 (CT No. 11)
Probing the Gene Pools: Spring $\times$ *Winter Crosses in Bread Wheat.* K.R. Kern, 1980 (CT No. 12)
Maize Research and Production in Guatemala. 1981 (CT No. 14)
Wheat in Bangladesh. 1982 (CT No.15)
Patronato of Sonora. M.V. Lynch, 1985 (CT No. 16)
Advances in Bread Wheat: The Green Revolution Revisited. 1985 (CT No. 17)
CIMMYT Economics: Providing Information for Decision Makers. 1985 (CT No. 18)

IBPGR

International Board for Plant Genetic Resources

Crop Genetic Resources Centre Food and Agriculture Organization
of the United Nations, Via delle Terme de Caracalla, 00100
Rome, Italy
Cable: FOODGARI, ROME - Telex: 610181 FAO I

IBPGR Annual Report 1983

1984. 120 pages. 19 × 26 cm. Paper. ISBN 92-9043-109-1. Available free to developing countries but restricted distribution to developed countries.

The annual report provides a detailed overview of the IBPGR's activities during the year.

IBPGR Annual Report 1984

1985. 122 pages. 19 × 26 cm. Paper. ISBN 92-9043-117-2. Available free to developing countries but restricted distribution to developed countries.

The annual report provides a detailed overview of the IBPGR's activities during the year.

Coffee Genetic Resources

1980. 13 pages. 21.5 × 27.6 cm. Paper. Available free to developing countries but restricted distribution to developed countries.

This report followed an IBPGR Working Group on Coffee Genetic Resources. It examines existing collections and genetic erosion and recommends exploration and collecting priorities. It also discusses maintenance, exchange, quarantine and provides a minimum descriptor list.

F. B. Armitage, P. A. Joustra and B. Ben Salem

Genetic Resources of Tree Species in Arid and Semi-Arid Areas

1980. 118 pages. 21.5 × 27.5 cm. Paper. Available free to developing countries but restricted distribution to developed countries.

This is a report of the first phase of the FAO/IBPGR forestry project. Country reports from participating states (Chile, India, Israel, Mexico, Peru, Senegal, Sudan and PDR Yemen) are included.

G. J. H. Grubben and D. H. van Sloten

Genetic Resources of Amaranths

1981. 60 pages. 21.6 × 27.6 cm. Paper. Available free to developing countries but restricted distribution to developed countries.

This report outlines a global plan of action for *Amaranthus*.

Genetic Resources of Sweet Potato

1981. 30 pages. 21.4 × 27.7 cm. Paper. Available free to developing countries but restricted distribution to developed countries.

This report followed an *Ad Hoc* Working Group on Sweet Potato Genetic Resources. After reviewing the major existing collections and identifying gaps, storage and maintenance, evaluation and documentation and quarantine questions, it made a series of recommendations. A descriptor list is also included.

Genetic Resources of Cocoa

1981. 25 pages. 21.7 × 27.7 cm. Paper. Available free to developing countries but restricted distribution to developed countries.

This report followed a Working Group on Genetic Resources of Cocoa. It examines the existing collections and methods of conservation and sampling techniques, identifies priority areas for collecting, proposes collections for designation as major repositories, and a framework for action and discusses quarantine questions. It also includes passport descriptors.

D. Astley, N. L. Innes and Q. P. van der Meer

Genetic Resources of *Allium* Species

1982. 38 pages. 21.3 × 27.8 cm. Paper. Available free to developing countries but restricted distribution to developed countries.

This report provides considerable detailed information on the genetic resources of the genus *Allium*. It also includes a list of descriptors.

Genetic Resources of Sugarcane

1982. 19 pages. 21.5 × 28 cm. Paper. Available free to developing countries but restricted distribution to developed countries.

This report was written following an expert Working Group on Sugarcane. The report advises on action to be taken on the collection, conservation and documentation of sugarcane germplasm.

Genetic Resources of *Vigna* Species

1982. 18 pages. 21.7 × 27.6 cm. Paper. Free to developing countries but restricted distribution to developed countries.

This report was written after the convening of an expert Working Group on *Vigna*. It reviews major *Vigna* collections; defines crop and area priorities for collection; identifies base collection centres and centres to hold duplicates.

Genetic Resources of Citrus

1982. 13 pages. 21 × 29.7 cm. Paper. Free to developing countries but restricted distribution to developed countries.

This report was written following an expert Working Group on Citrus. The report reviews the citrus genepool and advises on action to be taken on the collection, conservation and documentation of citrus germplasm.

Genetic Resources of *Vitis* Species

1983. 11 pages. 21 × 29.7 cm. Paper. Free to developing countries but restricted distribution to developed countries.

This report follows a Working Group on *Vitis* Genetic Resources and contains reviews of existing collections in Europe and of collecting activities by country. It discusses additional collecting activities necessary, methods of conservation, characterization, evaluation and documentation and quarantine problems.

P. J. Gulick, C. Hershey and J. T. Esquinas Alcazar

Genetic Resources of Cassava and Wild Relatives

1983. 56 pages. 21 × 29.5 cm. Paper. Free to developing countries but restricted distribution to developed countries.

This report was written after meetings of two working groups and a consultant report. The publication reviews major existing collections of cassava, advises on known genetic erosion, identifies collection priorities, outlines collection procedures, identifies potential depositories for international designation, advises on storage and maintenance procedures, identifies quarantine problems affecting transfer of material and provides a list of descriptors.

Genetic Resources of *Capsicum*

1983. 49 pages. 29.5 × 21 cm. Paper. Free to developing countries but restricted distribution to developed countries.

This report was written after preliminary studies undertaken by CATIE and an expert consultation. This is a comprehensive report on the genetic resources of *Capsicum* in which a global plan of action is set out. A full descriptor list is also included.

Genetic Resources of Soyabean

1983. 19 pages. 29.5 × 21 cm. Paper. Free to developing countries but restricted distribution to developed countries.

This report was written after an *ad hoc* Working Group on the Genetic Resources of *Glycine* species was held. It deals with the genetic resources of Soyabean. A minimum list of descriptors is also included.

J. T. Esquinas-Alcazar and P. J. Gulick

Genetic Resources of Cucurbitaceae

1983. 101 pages. 29.5 × 21 cm. Paper. ISBN 92-9043-112-1. Available free to developing countries but restricted distribution to developed countries.

This report provides a comprehensive overview of the genetic resources of Cucurbitaceae. It also contains descriptor lists for *Cucurbita* spp., *Cucumis melo,* and *C. sativus.*

E. Acheampong, N. Murthi Anishetty and J. T. Williams

A World Survey of Sorghum and Millets Germplasm

1984. 45 pages. 29.5 × 21 cm. Paper. Available free to developing countries but restricted distribution to developed countries.

This report surveys the existing collections of sorghum and millets in order to identify gaps and inadequacies which exist in the collections and to provide a basis for the formulation of recommendations for further collecting.

W. Ellis Davies

A Plan of Action for Forage Genetic Resources

1984. 30 pages. 21 × 29.7 cm. Paper. Available free to developing countries but restricted distribution to developed countries.

This booklet describes the present state of forage genetic resources, lists the main repositories of germplasm and designates 3 areas for priority and operational purposes (Mediterranean and adjacent semi-arid/arid areas; and temperate areas).

R. K. Arora and E. R. Nayar

Wild Relatives of Crop Plants of India

1984. 106 pages. 16.1 × 24.1 cm. Paper. Available free to developing countries but restricted distribution to developed countries.

This book describes wild relatives and related types of Indian crop species by category: cereals and millets, legumes, vegetables and fruits, oil seeds and fibre and spice plants.

Genetic Resources of *Hevea*

1984. 14 pages. 21.5 × 28 cm. Paper. Available free to developing countries but restricted distribution to developed countries.

This report assesses the availability for applied research of germplasm of *Hevea brasiliensis* and other *Hevea* species, considers whether material is being lost or under threat of loss in its centre of origin and diversity and considers steps to ensure that representative samples of genetic variability are conserved for present and future use.

Forage and Browse Plants for Arid and Semi-Arid Africa

1984. 293 pages. 21 × 29.7 cm. Paper. Available free to developing countries but restricted distribution to developed countries.

This book contains information on 100 trees and shrubs, grasses, herbaceous legumes and forbs. The information is laid out in standard format by categories, treating each species separately. Some have accompanying maps and illustrations. Includes references, glossary and index.

Genetic Resources of *Abelmoschus* (Okra)

1984. 61 pages. 21 × 29.7 cm. Paper. Available free to developing countries but restricted distribution to developed countries.

This is an English translation of a report originally published in French by IBPGR. It includes a bibliographical review of cultivated okra and related species and presents an overview of taxonomy, distribution, cytogenetics, evolution of cultivated and wild forms and varietal improvement. It also presents information on collecting missions, on existing collections and on their evaluation, and concludes with suggestions to enrich and maintain collections.

L. A. Withers

Institutes Working on Tissue Culture for Genetic Conservation

1982. 104 pages. 22 × 28 cm. Paper. Available free to developing countries but restricted distribution to developed countries.

In vitro culture techniques have been proposed for the conservation of genetic resources of vegetatively propagated crops. This report is a survey of institutes involved with *in vitro* culture techniques.

A. S. Cromarty, R. H. Ellis and E. H. Roberts

The Design of Seed Storage Facilities for Genetic Conservation

1982. 96 pages. 21.3 × 27.5 cm. Paper. Available free to developing countries but restricted distribution to developed countries.

This documents provides an in-depth report of the factors to be taken into consideration when designing seed storage facilities for genetic conservation.

R. H. Ellis and E. H. Roberts

Use of Deep-Freeze Chests for Medium and Long-Term Storage of Small Seed Collections

1983. 9 pages. 17 × 24.5 cm. Paper. Available free to developing countries but restricted distribution to developed countries.

This booklet provides, in summary form, all information necessary for the setting up of deep-freeze chests for medium and long-term storage of small seed collections.

J. B. Dickie, S. Linington and J. T. Williams (Editors)

Seed Management Techniques for Genebanks

1984. 298 pages. 21 × 29.5 cm. Paper. ISBN 92-9043-111-1. Available free to developing countries but restricted distribution to developed countries.

This publication provides an exposition of the proceedings of a Workshop held at the Royal Botanic Gardens, Kew, 6-9 July 1982.

Institutes Conserving Crop Germplasm: the IBPGR Global Network of Genebanks

1984. 25 pages. 21 × 29.5 cm. Paper. Available free to developing countries but restricted distribution to developed countries.

This book describes "base" and "active" collections and the IBPGR global network of base collections, lists institutes conserving crop germplasm, notes minimal standards for genebanks in addition to practical constraints to efficient conservation.

W. R. Scowcroft

Genetic Variability in Tissue Culture: Impact on Germplasm Conservation and Utilization

1984. 41 pages. 21 × 29.7 cm. Paper. Available free to developing countries but restricted distribution to developed countries.

This status report itemizes variability arising during *in vitro* culture, particularly that which might affect the genetic integrity of a germplasm resource.

Descriptors for Mungbean

1980. 18 pages. 16.8 × 24.4 cm. Paper. Available free to developing countries but restricted distribution to developed countries.

In order for the global network of crop genetic resources centres to readily exchange data about samples along with the plant materials, it is necessary for each centre to develop a data bank with a certain degree of standardization. This standardization is provided by the descriptors for each crop agreed internationally. The use of these descriptors permits scientists of different nationalities to readily communicate among themselves.

Published in co-operation with International Rice Research Institute (IRRI)

Descriptors for Rice (*Oryza sativa*)

1980. 21 pages. 21.5 × 27.5 cm. Paper. Free to developing countries but restricted distribution to developed countries.

In order for the global network of crop genetic resources centres to readily exchange data about samples along with the plant materials, it is necessary for each centre to develop a data bank with a certain degree of standardization. This standardization is provided by the descriptors for each crop agreed internationally. The use of these descriptors permits scientists of different nationalities to readily communicate among themselves.

Revised Winged Bean Descriptors

1981. 18 pages. 17 × 24 cm. Paper. Available free to developing countries but restricted distribution to developed countries.

In order for the global network of crop genetic resources centres to readily exchange data about samples along with the plant materials, it is necessary for each centre to develop a data bank with a certain degree of standardization. This standardization is provided by the descriptors for each crop agreed internationally. The use of these descriptors permits scientists of different nationalities to readily communicate among themselves.

Published in cooperation with International Crops Research Institute for the Semi-Arid Tropics (ICRISAT)

Descriptors for Pearl Millet

1981. 34 pages. 17 × 24 cm. Paper. Available free to developing countries but restricted distribution to developed countries.

In order for the global network of crop genetic resources centres to readily exchange data about samples along with the plant materials, it is necessary for each centre to develop a data bank with a certain degree of standardization. This standardization is provided by the descriptors for each crop agreed internationally. The use of these descriptors permits scientists of different nationalities to readily communicate among themselves.

Descriptores de Lupinos/Lupin Descriptors

1981. 68 pages. 17 × 24 cm. Paper. Free to developing countries but restricted distribution to developed countries.

In order for the global network of crop genetic resources centres to readily exchange data about samples along with the plant materials, it is necessary for each centre to develop a data bank with a certain degree of standardization. This standardization is provided by the descriptors for each crop agreed internationally. The use of these descriptors permits scientists of different nationalities to readily communicate among themselves.

In English and Spanish.

Barley Descriptors

1982. 14 pages. 17 × 24.5 cm. Paper. Available free to developing countries but restricted distribution to developed countries.

In order for the global network of crop genetic resources centres to readily exchange data about samples along with the plant materials, it is necessary for each centre to develop a data bank with a certain degree of standardization. This standardization is provided by the descriptors for each crop agreed internationally. The use of these descriptors permits scientists of different nationalities to readily communicate among themselves.

Published in cooperation with Centro Internacional de Agricultura Tropical (CIAT)

Lima Bean Descriptors

1982. 36 pages. 17 × 24.5 cm. Paper. Available free to developing countries but restricted distribution to developed countries.

In order for the global network of crop genetic resources centres to readily exchange data about samples along with the plant materials, it is necessary for each centre to develop a data bank with a certain degree of standardization. This standardization is provided by the descriptors for each crop agreed internationally. The use of these descriptors permits scientists of different nationalities to readily communicate among themselves.

R. Watkins and R. A. Smith (editors); published in cooperation with Commission of European Communities (CEC) and United States Department of Agriculture (USDA)

Apple Descriptors

1982. 46 pages. 21.2 × 29.5 cm. Paper. Available free to developing countries but restricted distribution to developed countries.

In order for the global network of crop genetic resources centres to readily exchange data about samples along with the plant materials, it is necessary for each centre to develop a data bank with a certain degree of standardization. This standardization is provided by the descriptors for each crop agreed internationally. The use of these descriptors permits scientists of different nationalities to readily communicate among themselves.

B. Thibault, R. Watkins and R. A. Smith (editors); published in cooperation with Commission of European Communities (CEC) and United States Department of Agriculture (USDA)

Pear Descriptors

1983. 39 pages. 21.2 × 29 cm. Paper. Available free to developing countries but restricted distribution to developed countries.

In order for the global network of crop genetic resources centres to readily exchange data about samples along with the plant materials, it is necessary for each centre to develop a data bank with a certain degree of standardization. This standardization is provided by the descriptors for each crop agreed internationally. The use of these descriptors permits scientists of different nationalities to readily communicate among themselves.

Descriptors for Cowpea

1983. 29 pages. 16.8 × 24.6 cm. Paper. Available free to developing countries but restricted distribution to developed countries.

In order for the global network of crop genetic resources centres to readily exchange data about samples along with the plant materials, it is necessary for each centre to develop a data bank with a certain degree of standardization. This standardization is provided by the descriptors for each crop agreed internationally. The use of these descriptors permits scientists of different nationalities to readily communicate among themselves.

Safflower Descriptors

1983. 15 pages. 17 × 24 cm. Paper. Available free to developing countries but restricted distribution to developed countries.

In order for the global network of crop genetic resources centres to readily exchange data about samples along with the plant materials, it is necessary for each centre to develop a data bank with a certain degree of standardization. This standardization is provided by the descriptors for each crop agreed internationally. The use of these descriptors permits scientists of different nationalities to readily communicate among themselves.

Kodo Millet Descriptors

1983. 27 pages. 16.8 × 24.6 cm. Paper. Available free to developing countries but restricted distribution to developed countries.

In order for the global network of crop genetic resources centres to readily exchange data about samples along with the plant materials, it is necessary for each centre to develop a data bank with a certain degree of standardization. This standardization is provided by the descriptors for each crop agreed internationally. The use of these descriptors permits scientists of different nationalities to readily communicate among themselves.

Echinochloa Millet Descriptors

1983. 17 pages. 16.8 × 24.6 cm. Paper. Available free to developing countries but restricted distribution to developed countries.

In order for the global network of crop genetic resources centres to readily exchange data about samples along with the plant materials, it is necessary for each centre to develop a data bank with a certain degree of standardization. This standardization is provided by the descriptors for each crop agreed internationally. The use these descriptors permits scientists of different nationalities to readily communicate among themselves.

Published in cooperation with Centro Internacional de Agricultura Tropical (CIAT)

Phaseolus coccineus Descriptors

1983. 34 pages. 16.8 × 24.6 cm. Paper. Available free to developing countries but restricted distribution to developed countries.

In order for the global network of crop genetic resources centres to readily exchange data about samples along with the plant materials, it is necessary for each centre to develop a data bank with a certain degree of standardization. This standardization is provided by the descriptors for each crop agreed internationally. The use of these descriptors permits scientists of different nationalities to readily communicate among themselves.

Published in cooperation with Office International de la Vigne et du Vin (OIV) and International Union for the Protection of New Varieties of Plants (UPOV)

Grape Descriptors

1983. 97 pages. 16.8 × 24.6 cm. Paper. Available free to developing countries but restricted distribution to developed countries.

In order for the global network of crop genetic resources centres to readily exchange data about samples along with the plant materials, it is necessary for each centre to develop a data bank with a certain degree of standardization. This standardization is provided by the descriptors for each crop agreed internationally. The use of these descriptors permits scientists of different nationalities to readily communicate among themselves.

Banana Descriptors (Revised)

1984. 35 pages. 16.8 × 24.6 cm. Paper. Available free to developing countries but restricted distribution to developed countries.

In order for the global network of crop genetic resources centres to readily exchange data about samples along with the plant materials, it is necessary for each centre to develop a data bank with a certain degree of standardization. This standardization is provided by the descriptors for each crop agreed internationally. The use of these descriptors permits scientists of different nationalities to readily communicate among themselves.

Published in cooperation with International Crops Research Institute for the Semi-Arid Tropics (ICRISAT)

Sorghum Descriptors (Revised)

1984. 36 pages. 17 × 24.5 cm. Paper. Available free to developing countries but restricted distribution to developed countries.

In order for the global network of crop genetic resources centres to readily exchange data about samples along with the plant materials, it is necessary for each centre to develop a data bank with a certain degree of standardization. This standardization is provided by the descriptors for each crop agreed internationally. The use of these descriptors permits scientists of different nationalities to readily communicate among themselves.

Soyabean Descriptors

1984. ca. 50. 16.8 × 24.6 cm. Paper. Available free to developing countries but restricted distribution to developed countries.

In order for the global network of crop genetic resources centres to readily exchange data about samples along with the plant materials, it is necessary for each centre to develop a data bank with a certain degree of standardization. This standardization is provided by the descriptors for each crop agreed internationally. The use of these descriptors permits scientists of different nationalities to readily communicate among themselves.

In Chinese and English.

Report of IBPGR Workshop on South Asian Plant Genetic Resources

1978. 56 pages. 21.1 × 28 cm. Paper. Available free to developing countries but restricted distribution to developed countries.

The IBPGR develops its activities in association with national governments and those in adjacent countries which share common cultural and agricultural features, thereby benefiting from mutual cooperation. The IBPGR attempts to organize such regional cooperation in countries located within regions of crop diversity.

(1) Report of the Fourth Regional Committee Meeting for Southeast Asia; (2) Report of the Fifth Regional Committee Meeting for Southeast Asia

1982, 1983. 65 and 91 pages. 21.5 × 28 cm. Paper. Available free to developing countries but restricted distribution to developed countries.

The IBPGR develops its activities in association with national governments and those in adjacent countries which share common cultural and agricultural features, thereby benefiting from mutual cooperation. The IBPGR attempts to organize such regional cooperation in countries located within regions of crop diversity.

IBPGR Advisory Committee on Seed Storage: Report of the First Meeting

1982. 13 pages. 21.5 × 27.5 cm. Paper. Available free to developing countries but restricted distribution to developed countries.

This document reports the findings of the first meeting of one of the IBPGR's two Standing Committees on Conservation.

Report of South Asia Liaison Officers Meeting

1982. 39 pages. 21.5 × 27.5 cm. Paper. Available free to developing countries but restricted distribution to developed countries.

The IBPGR develops its activities in association with national governments and those in adjacent countries which share common cultural and agricultural features, thereby benefiting from mutual cooperation. The IBPGR attempts to organize such regional cooperation in countries located within regions of crop diversity.

Report of Regional Meeting of Liaison Officers for the Mediterranean Programme

1982. 23 pages. 21.7 × 28 cm. Paper. Available free to developing countries but restricted distribution to developed countries.

The IBPGR develops its activities in association with national governments and those in adjacent countries which share common cultural and agricultural features, thereby benefiting from mutual cooperation. The IBPGR attempts to organize such regional cooperation in countries located within regions of crop diversity.

S. Blixt and J. T. Williams (Editors)

Documentation of Genetic Resources: a Model

1982. 84 pages. 17 × 24.5 cm. Paper. Available free to developing countries but restricted distribution to developed countries.

This report provides the full proceedings of a small group of experts who met to discuss all aspects of documentation of pea genetics and germplasm conservation.

IBPGR Advisory Committee on *In Vitro* Storage

1983. 11 pages. 21.5 × 27.5 cm. Paper. Available free to developing countries but restricted distribution to developed countries.

This document reports the findings of the first meeting of one of the IBPGR's two Standing Committees on Conservation.

Published in cooperation with International Rice Research Institute (IRRI)

1983 Rice Germplasm Conservation Workshop

1983. 109 pages. 15.2 × 23 cm. Paper. ISBN 971-104-115-4. Available free to developing countries but restricted distribution to developed countries.

This report provides the lectures given at the Rice Germplasm Conservation Workshop.

IBPGR Cocoa Working Group: Report of the Second Meeting

1984. 34 pages. 21 × 29.5 cm. Paper. Available free to developing countries but restricted distribution to developed countries.

This report summarizes the second meeting of an IBPGR Working Group, which considered a genetic resources action plan for the Western Hemisphere. The Working Group met to review information gathered to date on the activities of individual national programmes, the need for further collecting of germplasm and its maintenance in collections and to review field work carried out.

IBPGR Advisory Committee on Seed Storage: Report of the Second Meeting

1984. 27 pages. 21 × 29.7 cm. Paper. Available free to developing countries but restricted distribution to developed countries.

This report summarizes a meeting of an IBPGR Advisory Committee which reviewed current problems affecting seed stores, recalcitrant seeds,training, development of handbooks, data handling, equipment and standards in genebanks.

A Cooperative Regional Programme in Southeast Asia: Five-Year Plan of Action (1985-1989)

1984. 50 pages. 21.5 × 28 cm. Paper. Available free to developing countries but restricted distribution to developed countries.

This book provides details of the IBPGR programme in Southeast Asia including priorities for action on collecting, conservation, characterization and documentation of crop germplasm, and training of scientists in the region.

Tropical and Sub-Tropical Forages: Report of a Working Group

1984. 29 pages. 21.5 × 28 cm. Paper. Available free to developing countries but restricted distribution to developed countries.

This report summarizes a meeting of an IBPGR Working Group, which reviewed the current state of tropical forage genetic resources, listed legume and grass taxa meriting action, identified areas and countries where there is severe genetic erosion and recommended future work on documentation and basic research on the biology of important taxa.

IBPGR Advisory Committee on Seed Storage: Report of the Third Meeting

1985. 22 pages. 21 × 29.7 cm. Paper. Available free to developing countries but restricted distribution to developed countries.

This report identifies minimal standards necessary to ensure the safety of seeds of genetic resources samples and also the safety of personnel working in genebanks. Standards for both equipment and procedure are outlined.

J. G. Hawkes

Crop Genetic Resources Field Collection Manual

1980. 37 pages. 13.5 × 19 cm. Paper. Available free to developing countries but restricted distribution to developed countries.

This manual has been written as a general guide for the collection of genetic resources materials of seed crops, root and tuber crops and tree fruit crops, and their related wild species.

J. T. Esquinas Alcazar

Los Recursos Fitogeneticos: Una Inversion Segura para el Futuro

1981. 44 pages. 16.9 × 23.8 cm. Paper. ISBN 84-7498-082-8. Available free to developing countries but restricted distribution to developed countries.

This booklet provides a general account of crop genetic resources.

In Spanish.

Revised Priorities Among Crops and Regions

1981. 18 pages. 17 × 24.5 cm. Paper. Available free to developing countries but restricted distribution to developed countries.

This booklet presents the criteria used by the IBPGR for selecting priorities among crops and regions and provides detailed lists of these priorities.

Crop Genetic Resources: an Introduction to the IBPGR

1981. 14 pages. 17 × 24.5 cm. Paper. Available free to developing countries but restricted distribution to developed countries.

This booklet provides an overview of the work of the IBPGR including: priority crops and priority regions; national, regional and international centres; collection and conservation procedures; information, documentation and publications; training; and staff and funding.

IBPGR Brochure

1982. 10 × 21.3 cm. Paper. Available free to developing countries but restricted distribution to developed countries.

This brochure provides a general overview in outline form of the IBPGR.

Plant Varieties Rights and Genetic Resources

1983. 8 pages. 17 × 24.5 cm. Paper. Available free to developing countries but restricted distribution to developed countries.

This booklet presents a policy statement of the IBPGR on plant varieties rights and touches on the topics of their effect on the availability of genetic resources, on genetic erosion, and on genebanks and the future of variety improvement.

Practical Constraints Affecting the Collection and Exchange of Samples of Wild Species and Primitive Cultivars

1983. 11 pages. 17 × 24.5 cm. Paper. Available free to developing countries but restricted distribution to developed countries.

This report was commissioned by the IBPGR to identify any practical constraints, other than quarantine, affecting the collection and exchange of samples of wild species and primitive cultivars and to suggest improvements.

Facts about the IBPGR/Qu'est-ce que le CIRP/Datos sobre el CIRP

1983. 8 pages. 17.3 × 24.4 cm. Paper. Available free to developing countries but restricted distribution to developed countries.

This brochure uses the question and answer approach to provide basic information concerning the origin, purpose, function, funding, working methodology and future plans of the IBPGR.

Available in English, French or Spanish.

A Global Network of Genebanks/Un Reseau Mondial de Banques de Genes/Red Mundial de Bancos de Genes

1983. 8 pages. 17.3 × 24.4 cm. Paper. Available free to developing countries but restricted distribution to developed countries.

This brochure uses the question and answer approach to provide basic information concerning the global network of genebanks for conservation.

Available in English, French or Spanish.

J. Toll, N. Murthi Anishetty and G. Ayad

Directory of Germplasm Collections; 3. Cereals: III Rice

1980. 20 pages. 21.7 × 27.6 cm. Paper. Available free to developing countries but restricted distribution to developed countries.

This directory lists the germplasm holdings in institutes around the world. The information contained therein has been provided by the curators. These data are a preliminary attempt to assess total holdings and the information aids scientists in making contact with other scientists working on the same crop.

N. Murthi Anishetty, W. G. Ayad and J. Toll

Directory of Germplasm Collections; 3. Cereals: IV Sorghum and Millets

1981. 37 pages. 21.6 × 28 cm. Paper. Available free to developing countries but restricted distribution to developed countries.

This directory lists the germplasm holdings in institutes around the world. The information contained therein has been provided by the curators. These data are a preliminary attempt to assess total holdings and the information aids scientists in making contact with other scientists working on the same crop.

J. T. Williams and A. B. Damania

Directory of Germplasm Collections; 5. Industrial Crops: I Cacao, Coconut, Pepper, Sugarcane and Tea

1981. 50 pages. 21.5 × 27.5 cm. Paper. Available free to developing countries but restricted distribution to developed countries.

This directory lists the germplasm holdings in institutes around the world. The information contained therein has been provided by the curators. These data are a preliminary attempt to assess total holdings and the information aids scientists in making contact with other scientists working on the same crop.

J. Toll and D. H. van Sloten

Directory of Germplasm Collections; 4. Vegetables

1982. 187 pages. 21.5 × 27.5 cm. Paper. Available free to developing countries but restricted distribution to developed countries.

This directory lists the germplasm holdings in institutes around the world. The information contained therein has been provided by the curators. These data are a preliminary attempt to assess total holdings and the information aids scientists in making contact with other scientists working on the same crop.

N. Murthi Anishetty, J. Toll, W. G. Ayad and J. R. Witcombe

Directory of Germplasm Collections; 3. Cereals: V Barley

1982. 26 pages. 21.4 × 27.7 cm. Paper. Available free to developing countries but restricted distribution to developed countries.

This directory lists the germplasm holdings in institutes around the world. The information contained therein has been provided by the curators. These data are a preliminary attempt to assess total holdings and the information aids scientists in making contact with other scientists working on the same crop.

P. J. Gulick and D. H. van Sloten

Directory of Germplasm Collections; 6.: I Tropical Fruits

1984. 190 pages. 21 × 29.5 cm. Paper. ISBN 92-9043-113-X. Available free to developing countries but restricted distribution to developed countries.

This directory lists the germplasm holdings in institutes around the world. The information contained therein has been provided by the curators. These data are a preliminary attempt to assess total holdings and the information aids scientists in making contact with other scientists working on the same crop.

W. Ellis Davies and B. T. McLean

Directory of Germplasm Collections; 7. Forages (Legumes, Grasses, etc.)

1984. 46 pages. 21 × 29.5 cm. Paper. ISBN 92-9043-110-5. Available free to developing countries but restricted distribution to developed countries.

This directory lists the germplasm holdings in institutes around the world. The information contained therein has been provided by the curators. These data are a preliminary attempt to assess total holdings and the information aids scientists in making contact with other scientists working on the same crop.

Collection of Crop Germplasm: the First Ten Years

1984. 119 pages. 17 × 24.5 cm. Paper. Available free to developing countries but restricted distribution to developed countries.

This report summarizes widespread and generalized collecting in which IBPGR has had sponsoring and organizing involvement from 1974 to 1984. Country, year, number of samples collected and deposition of samples are listed for each collecting mission.

Catalog of *Arachis* Germplasm Collections in South America

1984. 79 pages. 28 × 21.5 cm. Paper. Available free to developing countries but restricted distribution to developed countries.

This book lists material gathered during collecting missions of *Arachis* in South America 1976-1983 and includes data about the samples and the collecting sites.

J. G. Hawkes, J. T. Williams and R. P. Croston

A Bibliography of Crop Genetic Resources

1983. 442 pages. 21.5 × 27.5 cm. Paper. ISBN 92-9043-108-3. Available free to developing countries but restricted distribution to developed countries.

This document provides an extensive bibliography of the literature in the field of crop genetic resources.

FAO/IBPGR Plant Genetic Resources Newsletter (through No. 61)

Quarterly. Approx. 40 pages. 21 × 29.7 cm. Paper. Available free to developing countries but restricted distribution to developed countries.

This newsletter describes the diverse activities of the global programme developed by the IBPGR and organized co-jointly with FAO.

IBPGR Regional Committee for Southeast Asia Newsletter

Quarterly. Approx. 16 pages. 21.2 × 28.5 cm. Paper. Available free to developing countries but restricted distribution to developed countries.

This Newsletter reports on crop genetic resources activities in the Southeast Asian Region (Available from IBPGR/SEAP Regional Office for Asia and the Pacific, Maliwan Mansion, Phra Atit Road, Bangkok - 10200, Thailand).

Crop Genetic Resources: a slidepack with text

1983. 68 pages. 19 × 13 cm. Paper. ISBN 92-9043-106-7. Available free to developing countries but restricted distribution to developed countries.

This booklet provides the text and colour photos of the IBPGR slidepack. A general introduction to crop genetic resources is given and a description of the role of the IBPGR.

Available in four different volumes of three languages each: Chinese, Japanese, Korean; English, Indonesian, Thai; English, Portuguese, Spanish; Arabic, English, French.

ICARDA

The International Center for Agricultural Research in the Dry Areas

P.O. Box 5466, Aleppo, Syria
Cable: ICARDA - Aleppo
Telex: 331206 SY; 331208 SY; 331263 SY
Telephone: 550465; 551280; 213433 Tel Hadya; 213477 Tel Hadya

General Information

Objectives:
To improve the agricultural systems and major food crops of the drier regions of Western Asia and North Africa. Mediterranean-type of climate of cool, moist winter and hot, dry summers, and the high elevation plateaux, with extremes of winter cold and summer heat, and snow cover for up to five months a year.

Research Highlights

RH-1   ICARDA Research Highlights 1981 (En - Ar), 55 pp.
RH-2   ICARDA Research Highlights 1982 (En - Ar), 64 pp.
RH-3   ICARDA Research Highlights 1983 (En - Ar), 84 pp.
RH-4   ICARDA Research Highlights 1984 (En - Ar), (in press)

Annual Reports

AR-1   Report on Research Program 1978/79 (En - Ar), 43 pp.*
AR-2   Report on Research Program 1979/80 (En -Ar), 48 pp.
AR-3   ICARDA Annual Report 1980/81 (En), 51 pp.*
AR-4   ICARDA Annual Report 1981 (En), 164 pp.
AR-5   ICARDA Annual Report 1982 (En), 215 pp.
AR-6   ICARDA Annual Report 1983 (En, Ar), 280 pp.
AR-7   ICARDA Annual Report 1984 (En, Ar), (in press)

Information Brochures

IB-1   ICARDA: Aims and Programs (Ar), 1982, 41 pp.
IB-2   ICARDA's Pasture and Forage Improvement Program (En, Ar), 1982, 19 pp.
IB-3   ICARDA's Food Legume Improvement Program (En, Ar) 1982, 24 pp.
IB-4   Opportunities for Training and Post-Graduate Research at ICARDA (En, Ar,) 1983, 20 pp.
IB-5   ICARDA: Objectives and Endeavors (En, Ar), 1983, 16 pp.
IB-6   ICARDA in Syria (En, Ar), 1982, 30 pp.*
IB-7   ICARDA's Guide to Aleppo, 1983, 25 pp.
IB-8   New Dimensions for Faba Bean Research and Production in Egypt and Sudan: A Profile of the ICARDA/IFAD Nile Valley Project (En, Ar), 1982, 24 pp.

IB-9 Better Harvests in Dry Areas (En), ICARDA's Cereal Improvement Program, 1983, 24 pp.
IB-10 A Precious Heritage: Genetic Resources at ICARDA (En), 1985, 13 pp.
IB-11 Harvest of Research: Highlights of the IFAD/ICARDA Nile Valley Project 1979-1985 (En), 1985, 48 pp.

Information Bulletins

IBU-1 Field Manual to Major Pests and Diseases of Wheat and Barley, by A. H. Kamel (Ar), 1984, 93 pp.
IBU-2 Field Guide to Major Insect Pests of Faba Bean in the Nile Valley by C. Cardona, E. Z. Farm, S. I. Beshara, A. G. Bushara (En), 1984, 60 pp.
IBU-3 Field Manual of Common Faba Bean Diseases in the Nile Valley by C. C. Bernier, S. B. Hanuonik, M. M. Hussein, H. A. Mohamed (En), 1984, 41 pp.

Discussion Papers

DP-1 Preliminary Agronomic Studies on Wheat and Barley in the 1978/79 season, 14 pp.
DP-2 Experimental Designs for Off-Station Agronomy Trials, by Roger J. Petersen, 1980, 28 pp.
DP-3 Augmented Designs for International Yield Trials, by Roger J. Petersen, 1980, 9 pp.
DP-4 Post-Harvest Processing of Winter Crops in N. W. Syria, Farming Systems Research Program, 1980, 15 pp.
DP-5 Summer Crops in Syria, by J. A. Harvey, 1980, 36 pp.
DP-6 Planting Methods for Winter Crops in N. W. Syria, by J. A. Harvey, 1980, 29 pp.
DP-7 Balanced Incomplete Block Designs, by P. Walker and S. Ceccarelli, 1982, 14 pp.
DP-8 Wheat Production with Supplementary Irrigation in Two Hama Villages, by E. Bailey, 1982, 48 pp.
DP-9 Livestock-Crop Interactions: The Case of Green Stage Barley Grazing, by T. L. Nordblom, March 1983, 37 pp.*
DP-10 Livestock-Crop Interactions: The Decision to Harvest or to Graze Mature Grain Crops, by T. L. Nordblom, May 1983, 21 pp.*
DP-11 Nutrition-Related Dimensions of Agricultural Research at ICARDA, By Kutlu Somel, 1983, 17 pp.
DP-12 Labour-Related Issues: An Assessment with Respect to Agricultural Research in the Middle East and North Africa, 1983, 15 pp.
DP-13 An Economic Guideline for Applied Agricultural Research (Minimum Yield Increase Requirements (MYIR), by Kutlu Somel, 1983, 9 pp.

Technical/Educational Manuals

No. 1   An Introduction to Food Legumes (En, Ar), 1978, 86 pp.*
No. 2   An Introduction to Breeding Food Legumes (En, Ar), 1978, 109 pp.
No. 3   An Introduction to Genetics (En, Ar), 1978, 12 pp.
No. 4   An Introduction to Major Pests of Food Legume Crops in West Asia (En, Ar), 1979, 93 pp.
No. 5   Important Legume Diseases of West Asia and North Africa (En, Ar), 1978, 82 pp.
No. 6   An Introduction to Agriculture in West Asia and North Africa (En, Ar), 1979, 88 pp.
No. 7   An Introduction to Statistics and Experimental Design, 1979, 198 pp.
No. 8   An Introduction to Wheat and Barley in the Near East and North Africa (En, Ar), 1979, 88 pp.
No. 9   Introduction to the Major Insect Pests of Wheat and Barley in the Middle East and North Africa (En, Ar), 1982, 112 pp.
No. 10   Introduction to Seed Science and Technology, 1982, 112 pp.
No. 11   Statistics and Experimental Design, Roger Petersen, 1982, 81 pp (revised 1985, 96 pp).
No. 12   Legume Crop Physiology, 1982, 83 pp.
No. 13   Seed Production Technology, 1982, 92 pp.

Symposia, Books, Proceedings and Conferences

B-1   Fourth Regional Winter Cereal Workshop on Barley, Amman, Jordan, 1977. Vol. 1, 273 pp; Vol. 2, 420 pp.
B-2   Food Legume Improvement and Development: Proceedings of a Workshop held at the University of Aleppo, Syria, 2-7 May 1978. G. C. Hawtin and G. C. Chancellor, Editors. Published jointly by ICARDA and IDRC, 1979, 216 pp.

This volume comprises material presented at the First Regional Food Legume Improvement Workshop that was held at the University of Aleppo, Aleppo, Syria, on 2-7 May 1978. It constitutes a synopsis of the current status of food legume production in the countries of Western Asia and North Africa.

The workshop was sponsored jointly by the International Center for Agricultural Research in the Dry Areas (ICARDA) and the University of Aleppo, and had as its primary aims the bringing together of research workers and other legume specialists from all parts of the region served by ICARDA, and the identification of priorities for research attention. More than 15 countries were represented, together with several international and regional research institutions concerned with food legume development.

The workshop and this resultant text are the first steps toward achieving an increasing awareness of the common problems facing food legume production and improvement in the region. They provide a solid base from which to develop appropriate cooperative strategies planned to minimize the constraints that at present limit the production of leguminous crops that are so important both to human nutrition and to the agricultural systems of Western Asia and North Africa.

B-3   Regional Workshop on Cereal Diseases Methodology. Published jointly by the Government of The Netherlands, ICARDA, and CIMMYT, 1979, 150 pp.

B-4   Recommendations, Summaries of Discussion, Program and Participants from the ICARDA/UNDP Workshop "Increasing the Effectiveness of Wheat and Nitrogen In Rainfed Farming Systems in Mediterranean-type Environments," 1980, 48 pp.

B-5   Soil-Water and Nitrogen in Mediterranean-type Environments. (Review papers). Colin Webb and John Monteith, Editors. Published for ICARDA by Martinus Nijhoff Publishers, The Hague, The Netherlands, 1981, 352 pp.

Based on a workshop cosponsored by the International Center for Agricultural Research in the Dry Areas and the United Nations Development Programme.

Of the international research centers, ICARDA in particular has an important role to play in economic well-being in the rainfed agricultural areas of the Middle East and North Africa. One important focus is on the development of more productive and stable dry farming systems.

There has been a special concern for the cereal and legume crops which constitute the basis of the crop rotation in the region. A special effort is being made to devise ways of integrating legume-based ley pastures into the cropping rotation to improve fertility and to provide additional feed for the sheep and goat population — an important source of farmer cash flow.

In these rainfed systems, soil and water resources, as a function of the environment, largely determine the level of production and ultimately the economic well being of the region. Better use of soil water and nitrogen is a primary focus of the book.

B-6   Lentils. By Colin Webb and Geoffrey Hawtin, Editors. Published jointly by ICARDA and CAB, 1981, 216 pp.

Although lentils are a relatively minor crop on a world scale, in certain regions they assume considerable local importance. This is particularly true in countries surrounding the Mediterranean, in the northern parts of the Indian sub-continent and in several countries of Latin America. In these regions lentils are not only a valuable food item but may also form an important component of local or international trade. In recent years there has been an increasing interest in the crop both as a protein source and as a means of improving soil fertility — properties lentils have in common with other food legume species.
This book is the outcome of that seminar. While it is not a summary of proceedings, it is an attempt by several of the scientists who attended the seminar to summarize and review the latest research findings. Lentils is the first comprehensive book to cover all main aspects of the crop. It should be of interest not only to students and scientists already working with lentils or other food legumes, but also to anyone concerned with investigating or promoting minor crops in general.

B-7   Crop Production in Lebanon: Proceedings of a Research Conference held at Chtaura, Lebanon, 1981, 1982, 74 pp.

B-8   Proceedings of the Seed Production Symposium, sponsored by The Netherlands Government and ICARDA, 23-26 March 1981. Published jointly by the Government of The Netherlands and ICARDA, 1982, 124 pp.

B-9   Proceedings of the International Faba Bean Conference, ICARDA/IFAD Nile Valley Project in Cooperation with the Government of Egypt, 7-11 March 1981, Cairo, Egypt, 1982, 147 pp.

B-10 Faba Bean Improvement: Proceedings of the Faba Bean Conference, Cairo, 7-11 March 1981, by G. C. Hawtin and Collin Webb, Editors. Published for ICARDA by Martinus Nijhoff Publishers, The Hague, The Netherlands, 1983, 348 pp.

This authoritative reference book is based on papers presented at the international conference on faba beans sponsored by the ICARDA/IFAD Nile Valley Project in Cairo.

Compiled by 60 accepted world authorities, faba bean improvement focuses on genetic improvement, the development of cultural practices; control of pests, diseases, and weeds; nitrogen fixation; and the use of faba beans as a human food.

This book will be of wide interest and relevance to scientists, students, nutritionists, extension workers, and others who are interested in this important crop.

B-11 Proceedings of the International Workshop on Ascochyta Blight and Winter Sowing of Chickpeas, ICARDA. M. C. Saxena and K. B. Singh, Editors. Published for ICARDA by Martinus Nijhoff Publishers, The Hague, The Netherlands, 1983.

B-12 Faba Bean, Kabuli Chickpeas, and Lentils in the 1980s: Proceedings of an International Workshop ICARDA. M. C. Saxena and S. Varma, Editors. 1983.
B-13 Genetic Resources and Their Exploitation — Chickpeas, Faba Beans and Lentils, by J. R. Witcombe, and W. Erskine, Editors. Published for ICARDA by Martinus Nijhoff/Dr. W. Junk, The Hague, The Netherlands, 1984, 256 pp.*
B-14 Proceedings of the Soils Directorate — Syrian Ministry of Agriculture and Agrarian Reform/ICARDA Workshop on Fertilizer Use in the Dry Areas held at ICARDA, 26-29 March 1984. 1984, 214 pp.
B-15 Fourth Annual Farming Systems Research and Extension Symposium — Kansas University/ICARDA. The Nile Valley Project: A Model of Cooperation between International and National Programs in Research and Extension, 8-10 October 1984, 14 pp.
B-16 Proceedings of the Third International Symposium on Parasitic Weeds, ICARDA/International Parasitic Seed Plant Research Group, 7-9 May 1984, Aleppo, Syria. By C. Musselman, L. J. Polhill, R. M., and Wilson, A. K. Editors. 265 pp.
B-17 Report on Recommendations of the Rainfed Agricultural Information Network (RAIN) Workshop, 17-20 March 1985, Amman, Jordan. Organized by ICARDA/USAID. Sponsored by Ministry of Agriculture, Jordan, 1985, 32 pp.
B-18 The Arab Conference for Agricultural Research on Basic Food Crops, 31 March - 5 April 1985, Arab Fund for Economic and Social Development (AFESD)/ICARDA, 1985. (5 papers presented).

Periodicals

SCIENTIFIC NEWS SERVICE

A)
FABIS: (Faba Bean Information Service), Supported by The International Development Research Centre (IDRC), Canada.

FABIS contains scientific articles mainly, book reviews and news items about training and conferences on faba beans. Free for developing countries, US$10 per year for others. Annual-Nos 1-3, 1979-1981. Biannual-Nos 5-7, 1982-83 average 40 pp. Nos 8-10, 1984. (Nos 5-6 NA). No. 11, 1985.

Genetic Variation Within *Vicia faba* by G. P. Chapman, 1981, 12 pp.

— Directory of World Faba Bean Research, by J. Issa, 1981, 67 pp.
— Faba Bean Abstracts. Quarterly, CAB/ICARDA; free supplement to FABIS Newsletter.

— Nile Valley Faba Bean Abstracts. CAB/ICARDA/IFAD. 568 abstracts from Egypt and Sudan, 1982. Supplement to Faba Bean Abstracts.

B)
LENS (Lentil Experimental News Services), supported by IDRC. LENS Newsletter from Vol. 9 (1982) is published by ICARDA in cooperation with the University of Saskatchewan, Canada, publishers of the first eight annual issues.

Vol. 10 (1) published in August 1983; biannual from then on. Contains mainly scientific articles but includes also a bibliography, key abstracts, book reviews and news items on lentils, 36 pp. (NA)

LENS Vol. 11 No. 1) 1984
LENS Vol. 11 No. 2)

— Lentil Abstracts. Annual, CAB/ICARDA; free supplement to LENS Newsletter.

C)
RACHIS (Barley, Wheat and Triticale Newsletter) supported by ICARDA.

RACHIS Newsletter. Biannual. Contains mainly scientific articles but also includes book reviews, and news items about training, conferences and scientists in wheat, barley, and triticale. (English). Arabic Version was started from Vol. 3(1) January 1984, supported by IDRC.

No. 1	November 1982 (E)
No. 2	May 1983 (E)
Vol. 3 (1)	January 1984 (E) (A)
Vol. 3 (2)	June 1984 (E) (A = in press)
Vol. 4 (1)	January 1985 (E)

D)
FARMING SYSTEMS RESEARCH NEWS for the Middle East and North Africa

Contains short articles on any aspect related to Farming Systems Research. Also book reviews, bibliographies, abstracts of recent research papers, and meeting announcements (New Publications).

Vol. 1	October 1984
Vol. 2 No. 1.	June 1984
Vol. 3 No. 1	April 1985

E)
CURRENT AWARENESS
Contains complete listing of books, relevant journal articles, reprints, bulletins and reports received in ICARDA's Library. It is a Selective Dissemination of Information (SDI) service produced by the Library.

Numbers published: from 1-21.

LIBRARY BULLETIN
No. 1-2, 1985

ICARDA IN NEWS
This is a collection of news stories and press releases made from the
sources available and is not intended to be a complete record of all
media coverage of ICARDA's work. January 1985.

GERMPLASM CATALOGS

GC-1   Kabuli Chickpea Germplasm Catalog, by K. S. Singh, R. S.
       Malhotra, and J. R. Witcombe, 1983, 284 pp.
GC-2   Lentil Germplasm Catalog by W. Erskine, J. R. Witcombe,
       1984, 363 pp.

Bibliographies

Bib-1   An Anotated Bibliography of Chickpea Genetics and Breeding
        — 1915-1983 by K. B. Singh, R. S. Malhotra, and F. J.
        Muehlbauer, October 1984, 195 pp.

INTERNATIONAL NURSERY REPORTS

Food Legumes
NRF-1   Report on the International Cooperative Program on Food
        Legume Improvement 1977/78, 63 pp.
NRF-2   International Nursery Report 1978/79, No. 3 - 115 pp.
NRF-3   International Nursery Report 1979/80, No. 4 - 190 pp.
NRF-4   International Nursery Report 1980/81, No. 5 - 258 pp.
NRF-5   International Nursery Report 1981/82, No. 6 - 282 pp.
NRF-6   International Nursery Report 1982/83, No. 7 - (in press)

Cereals
NRC-1   Cereals International Nurseries Report 1977/78, 115 pp.
NRC-2   Regional Wheat and Barley Nurseries Preliminary Report
        1978/79, 88 pp.
NRC-3   Cereals International Nurseries Report 1979/80, 151 pp.
NRC-4   Cereals International Nurseries Report 1980/81, 228 pp.
NRC-5   Regional Wheat and Barley Nurseries Preliminary Report,
        1980, 47 pp.
NRC-6   Regional Yield Trials and Preliminary Observation Nurseries
        1980/81, Final Report, 1983, 228 pp.
NRC-7   Regional Yield Trials and Preliminary Observation Nurseries:
        Preliminary Report 1980/81, 58 pp.
NRC-8   Cereals International Nurseries Report 1981/82, 215 pp.
NRC-9   Regional Yield Trials and Preliminary Report 1983/84,
        198 pp.

ICRISAT

International Crops Research Institute for the Semi-Arid Tropics

Patancheru P.O. Andhra Pradesh 502324, India
Cable: ICRISAT, Hyderabad - Telex: 0152-203

General Information

The International Crops Research Institute for the Semi-Arid Tropics (ICRISAT) was founded in 1972 and is one of 13 centers in an international research network operating under the aegis of the Consultative Group on International Agricultural Research (CGIAR). Its headquarters and main research farm, ICRISAT Center, is near the village of Patancheru, 25 km northwest of Hyderabad, India. Other ICRISAT scientists are posted in nine countries of Africa, in Mexico, and in Syria.

The mandate of ICRISAT is to:

1. Serve as a world center for the improvement of grain yield and quality of sorghum, millet, chickpea, pigeonpea, and groundnut, and to act as a world repository for the genetic resources of these crops.

2. Develop improved farming systems that will help to increase and stabilize agricultural production through more effective use of natural and human resources in the seasonally dry semi-arid tropics.

3. Identify constraints to agricultural development in the semi-arid tropics and evaluate means of alleviating them through technological and institutional changes.

4. Assist in the development and transfer of technology to the farmer through cooperation with national and regional programs, and by sponsoring workshops and conferences, operating training programs, and assisting extension activities.

Since its foundation ICRISAT has realized the importance of communicating its research findings and production technology via a wide range of publications. The ones listed here are bound in paper covers unless otherwise stated.

Ordering Instructions

The per-copy prices are listed separately for: highly-developed countries (HDCs) and less-developed countries (LDCs)—a price discounted by 67%.

HDCs include Australia, Canada, European countries, Iran, Iraq, Japan, Kuwait, Libya, New Zealand, Saudi Arabia, South Africa, and USA.

On purchases of five or more copies of any publication, a discount of 30% on the per-copy price will be allowed. Book dealers will be given the 30% discount regardless of the number of copies ordered.

Prepayment is required, i.e., purchase orders are not acceptable without payment. Enclose checks expressed in U.S. dollars, made payable to ICRISAT, or UNESCO coupons, with your order.

Prices, and cost of postage and packing (P&P), are quoted for all priced publications and must be remitted in full. Please use ICRISAT codes when ordering publications. They are given in parentheses after the title of each catalog entry.

Organizations that formally exchange publications with ICRISAT library, journal editors who review ICRISAT publications, and national program staff who collaborate with ICRISAT research programs are entitled to single free copies.

Publications will be despatched via air bookpost because experience shows that surface mail packages may not always reach their destinations, they may be damaged, and delayed for months.

Send orders to:

Information Services
ICRISAT
Patancheru P.O.
Andhra Pradesh 502 324
India

Research Highlights.

(RHE 001 to 006) Published annually since 1979 approx.50 pages. 17 x 23.5 cm. Saddle stitched.
Free

Each calendar-year issue describes major research achievements by ICRISAT scientists, illustrated in color, and written to inform a wide audience, of advances in the Institute's work throughout the semi-arid tropics. Information on each of the Institutes programs is presented together with concise summaries on the transfer of technology and cooperative programs at national and international levels.

Progrès de la recherche.

(RHF 002 to 006) Série de rapports annuels de synthèse d'environ 50 pages publiés depuis 1979. 17 x 23.5 cm.
Gratuit

Cette publication annuelle porte sur les faits saillants de la recherche effectuée par l'ICRISAT. Elle vise à informer le grand public sur les progrès réalisés par l'Institut dans les zones tropicales semi-arides du monde. Elle contient une information générale sur les programmes de recherche et de coopération internationale de l'ICRISAT, ainsi qu'un bref résumé sur le transfert de technologie. Le texte est illustré par des photographies en couleurs.

Annual Reports

ICRISAT Annual Reports.

(ARE 001 to 011) Published annually since 1973/74, approx.350 pages. 18.5 x 24 cm.
HDC $15.0, LDC $5.0, P&P $12.60.

These reports give a comprehensive coverage of ICRISAT's major research programs: sorghum, pearl millet, chickpea, pigeonpea, groundnut, farming systems, and economics. Work at ICRISAT Center is described along with results from collaborative programs at various locations in India, and at national and international research facilities throughout the semi-arid tropics.

Monographs

Books

L.R.House.

A guide to sorghum breeding (2nd edition).

(BOE 002) 1985. 216 pages. 21.5 x 28 cm.
HDC $25.00, LDC $7.00, P&P $9.00.

The author has drawn on knowledge and experience gained from 25 years' work on sorghum. Illustrated with photographs and figures, the first two sections describe the crop and its taxonomy, the third provides genetic background, the fourth considers applied breeding aspects, and the fifth discusses problems of seed production. This second fully- indexed, extensively—revised edition contains additional information on procedures to screen for resistance to disease, insect pests, environmental stresses, and the parasitic witchweed *Striga*. A sorghum discriptor list and a glossary are appended.

L.R. House.

Manuel pour la sélection du sorgho (2ème édition).

(BOF 002) 1985. 216 pages. 21.5 x 28 cm.
HDC $25.00, LDC $7.00, P&P $9.00.

L'auteur a puisé pour ce livre, dans sa grande expérience et compétence acquéries au cours de 25 ans de travail sur le sorgho. Avec de nombreuses photos et figures, les sujets sont abordés en cinq parties : présentation descriptive de la culture, caractéristiques essentielles et taxonomie; croissance et mor-phologie; génétique et hérédité; méthodes et procédés d'amélioration, techniques de croisements et de criblage; organisation et développement de la production de semences. La nouvelle édition (2e) de l'ouvrage, considérablement révisée, comporte des informations supplémentaires sur les procédures de sélection pour la résistance aux maladies, aux insectes, aux pressions de l'environnement et au *Striga*, ainsi qu'une liste des descripteurs du sorgho et un glossaire.

L.R.House.

El sorgo, guía para su mejoramiento genético.
Solamente primero edición.

(BOS 001) 1982. 427 páginas. 14.5 x 21 cm.
Universidad Autonoma Chapingo, Mexico.

Este libro fue escrito para ayudar a los ingenieros agrónomos y a los técnicos. Las dos primeras partes describen el cultivo y su taxonomia, la tercera parte explica el fondo genético, la cuarta considera aspectos de mejoramiento y la quinta explica los problemas en la produccion de semillas. También contiene un glosario y una lista describiendo los diferentes tipos de sorgo.

Obtengase en la casa de imprenta:
Universidad Autonoma Chapingo,
Patronato Universitario,
Chapingo, Mexico.

K.O.Rachie and J.V.Majumdar.

Pearl Millet

(BOE 003) 1980. 317 pages. 18.5 x 26 cm. Hardcover. ISBN 0-271-00234-4.
Pennsylvania State University Press

This book covers pearl millet's origin and history, cytology and cytogenetics, breeding and growing, and storage. It is a comprehensive review of the world research literature on the subject, and incorporates reports of research by ICRISAT scientists.

Order from publisher:
The Pennsylvania State University Press,
215 Wagner Bldg., University Park,
Pa. 16802, USA.

A.H.Kassam.

Crops of the West African semi-arid tropics.

(OPE 001) 1976. 160 pages. 21 x 26.5 cm.
(Microfiche only)

Written to help technicians and planners gain a better understanding of the production requirements of the region, covers the ecology, cultivation, and diseases and pests of 23 major crops grown there. There are six crop groups: cereals, legumes, roots and tubers, vegetables, fibers, and others.

K.O.Rachie and L.V.Peters.

The Eleusines: a review of the world literature.

(OPE 002) 1977. 179 pages. 14 x 21.5 cm.
(Microfiche only)

H.P.Binswanger.

The economics of tractors in South Asia: an analytical review.

(OPE 003) 1978. 104 pages. 15 x 23 cm.
HDC $4.80, LDC $1.60, P&P $3.25.

This analysis of farm-level tractor surveys in South Asia does not find convincing evidence that tractors are responsible for substantial increases in cropping intensity, yields, timeliness, and gross returns in India, Pakistan, and Nepal. Tractors appear to provide little economic advantage as substitutes for labor and bullocks.

Research Bulletins

A series of occasional bulletins by which Institute staff and consultants report their research findings, including the results of long-term research projects, that do not readily lend themselves to publication in a journal or a similar scientific literature outlets. The target audience is peer scientists worldwide.

No.1

S.M.Virmani, M.V.K.Sivakumar, and S.J.Reddy.

Rainfall probability estimates for selected locations of semi-arid India (second enlarged edition).

(RBE 000) 1982. 180 pages. 21.5 x 28 cm.
HDC $8.10, LDC $2.70, P&P $6.60.

Quantification of rainfall in agronomically relevant terms is an important step in the development of improved farming systems for the semi-arid tropics. The authors of this monograph present analyzed rainfall probabilities for 56 semi-arid locations in the Indian subcontinent to arrive at the dependable rainfall periods for selected agricultural operations.

No.2

H.P.Binswanger, S.M.Virmani, and J.Kampen.

Farming systems components for selected areas in India: evidence from ICRISAT.

(RBE 001) 1980. 44 pages. 21.5 x 28 cm. Saddle stitched.
HDC $2.10, LDC $0.70, P&P $2.40.

Conclusions that can be drawn from 7 years of soil and water management research in the Farming Systems Program of ICRISAT are summarized, together with selected conclusions from other ICRISAT programs and subprograms. Findings that need to be evaluated on farmers' fields are clearly stated. Areas of further research that need continuing intensive investigations and evaluation are also indicated.

No.3

H.P.Binswanger and B.C.Barah.

Yield risk, risk aversion, and genotype selection: conceptual issues and approaches.

(RBE 002) 1980. 29 pages. 21.5 x 28 cm. Saddle stitched.
HDC $1.50, LDC $0.50, P&P $2.25.

Several methods of stability and adaptability analysis are discussed. A practical way of measuring risk with several years of coordinated yield trial data is proposed. Region specificity is examined by exploring regression approaches on plant independent variables. The relationship between stability and adaptability in the context of the regression framework is discussed.

No.4

D.W.Norman, M.D.Newman, and I.Ouedraogo.

Vol.1 Farm and village production systems in the semi-arid tropics of West Africa: an interpretive review of research.

(RBE 003) 1981. 100 pages. 21.5 x 28 cm.
HDC $4.50, LDC $1.50 P&P $5.0.

This report is an analytical review of socioeconomic research on farm and village production systems in the semi-arid tropics (SAT) of West Africa. It uses a farming systems framework for the development of guidelines on crop technologies in the region. The authors hypothesize that heterogeneity in both the technical and human environments at the farm level is increasing in many areas, resulting in increasing distributional inequalities. Specific socioeconomic research areas important to the West African SAT are suggested.

No.4

I.Ouedraogo, M.D.Newman, and D.W.Norman.

Vol.2 The farmer in the semi-arid tropics of West Africa: a partially annotated bibliography.

(RBE 004) 1982. 132 pages. 21.5 x 28 cm.
HDC $6.0, LDC $2.0, P&P $5.20.

The 1045 references included were used to prepare ICRISAT Research Bulletin No. 4(Vol.1). Most of them date from 1960 to 1980. The topics covered include methodology, external institutions, community structure, norms and beliefs, land, labor, capital and cash flows, goals and management, crops, livestock, level of living, village-level studies, and bibliographic studies. Author and subject indexes are provided.

No.5

S.S.Bisen and A.R.Sheldrake.

The anatomy of the pigeonpea.

(RBE 005) 1981. 28 pages. 21 x 27.5 cm. Saddle stitched.
HDC $1.20, LDC $0.40, P&P $2.0.

Reports a 3-year study at ICRISAT on the anatomy of the tissues and organs of the pigeonpea plant. An attempt has been made to emphasize the unique features of this plant. The text is illustrated with a collection of 52 photographs of the many pertinent microscope slides made during the study.

No.6

R.K.Maiti and F.R.Bidinger.

Growth and development of the pearl millet plant.

(RBE 006) 1981. 16 pages. 21.5 x 28 cm. Saddle stitched.
HDC $0.90, LDC $0.30, P&P $1.80.

Describes the phenological and morphological development of the pearl millet plant in terms of nine early identifiable growth stages. It also provides quantitative and qualitative information on the development and growth of the major plant organs.

No.7

J.G.Ryan, P.D.Bidinger, N.Prahlad Rao, and P.Pushpamma.

The determinants of individual diets and nutritional status in six villages of Southern India.

(RBE 007) 1984. 140 pages. 21.5 x 28 cm. Saddle stitched.
HDC $9.00, LDC $3.00, P&P $5.00.

Limiting nutrients in diets and the factors that influence nutrient intakes of 1200 people are reported. Major deficiencies were energy, calcium, vitamins A, B complex, and C. Important findings are that village children need more vitamin A and calcium, and that crop improvement programs should emphasize yields and stability rather than protein content and quality.

No.8

H.L.S.Tandon and J.S.Kanwar.

A review of fertilizer use research on sorghum in India.

(RBE 008) 1984. 68 pages. 21.5 x 28 cm. Saddle stitched.
HDC $4.20, LDC $1.40, P&P $3.70.

This literature review covers the period 1960-83 and is concerned with the response of grain sorghum to all soil nutrients, related soil and climatic conditions, and the seasons in which the crop is grown. The bulletin attempts to quantify the responses and their relationship to different environments, and shows that fertilizer use is the key for increasing production.

No.9

D.Jha and R.Sarin.

Fertilizer use in semi-arid tropical India.

(RBE 009) 1984. 80 pages. 21.5 x 28 cm. Saddle stitched.
HDC $5.00, LDC $2.00, P&P $4.00.

This report focuses attention on the amount of fertilizer consumed in both irrigated and non-irrigated areas, growth patterns in consumption, the allocations of fertilizer between crops made by farmers, the agronomic practices of fertilizer application, and the major determinants in fertilizer use. It ends with a summary and itemized implications for research, extension, and development programs.

Information Bulletins

This occasional series of short, well-illustrated bulletins has become the main vehicle for information from ICRISAT that, derived from research findings, has immediate application. The target audience is national agricultural program scientists and extension staff in countries of the semi-arid tropics.

No.1

Y.L.Nene, A.Mengistu, J.B.Sinclair, and D.J.Royse.

An annotated bibliography of chickpea diseases, 1915-1976.

(IBE 001) 1978. 49 pages. 21.5 x 28 cm. Saddle stitched. P&P $7.0.

Includes 331 citations and is intended to provide a working file of the literature that will assist in the identification of fungal and bacterial microorganisms associated with chickpea seeds. The index is divided into five sections according to the causal agent of the disease.

No.2

R.J.Williams, R.A.Frederiksen, and J-C.Girard.

Sorghum and pearl millet disease identification handbook.

(IBE 002) 1978. 88 pages. 10 x 18 cm. Saddle stitched. HDC $5.70, LDC $1.90, P&P $2.0.

A pocket-size field guide produced to help agricultural scientists identify the common diseases and parasitic weeds of sorghum and pearl millet. The symptoms of 32 diseases are discussed and illustrated with 59 color photographs. A disease identification key is included, enabling field diagnosis on the visual symptoms alone.

No.2

R.J.Williams, R.A.Frederiksen et J-C.Girard.

Manuel d'identification des maladies du sorgho et du mil.

(IBF 002) 1978. 88 pages. 10 x 18 cm.
HDC $5.70, LDC $1.90, P&P $2.0.

Ce manuel permet aux chercheurs agricoles d'identifier les maladies courantes du sorgho et du petit mil. Les symptômes de 32 maladies sont décrits et illustrés grâce à 59 photos en couleurs. Ce manuel contient également une clef d'identification des maladies de ces cultures.

No.2

R.J.Williams, R.A.Frederiksen, y J-C.Girard.

Manual para la identificación de las enfermedades del sorgo y mijo.

(IBS 002) 1978. 88 páginas. 10 x 18 cm.
HDC $5.70, LDC $1.90, P&P $2.0.

Este manual ha sido producido para ayudar a los científicos agrícolas a identificar las enfermedades y parásitos más comunes del sorgo y mijo. Los síntomas de 32 enfermedades son explicados e ilustrados con 59 fotos a color. Este manual contiene también una clave para identificar las enfermedades de esos cultivos en el campo.

No.3

Y.L.Nene, M.P.Haware, and M.V.Reddy.

Diagnosis of some wilt-like disorders of chickpea (*Cicer arietinum* L.).

(IBE 003) 1978. 44 pages. 10 x 18 cm. Saddle stitched.
HDC $3.0, LDC $1.0, P&P $1.40.

Characteristic symptoms of some important diseases of chickpea such as wilt, collar rot, root rot, dry root rot, stem rot, stunt, mosaic, and root-knot are described. Other chickpea disorders, such as iron chlorosis, the effects of salinity, and frost injury are also discussed. A list of chickpea diseases observed in India and a few other countries is provided. The text is illustrated by 37 color photographs.

No.3

Y.L.Nene, M.P.Haware, and M.V.Reddy.

Diagnosis de algunos marchitamientos del garbanzo (*Cicer arletinum* L.).

(IBS 003) 1978. 44 páginas. 10 x 18 cm.
HDC $2.10, LDC $0.70, P&P $1.40.

Síntomas de algunos marchitamientos y de otras enfermedades del garbanzo estan descritos. También contiene una lista de varias enfermedades que fueron observadas en la India y en otros países. El texto esta ilustrado con 37 fotos en blanco y negro.

No.4

Shree Ram Yadav, S.Prasannalakshmi, and P.S.Jadhav.

Indian theses on groundnut: a bibliography 1948-1977.

(IBE 004) 1978. 54 pages. 21.5 x 28 cm.
(Microfiche only)

No.5

M.V.K.Sivakumar, S.M.Virmani, and S.J.Reddy.

Rainfall climatology of West Africa: Niger.

(IBE 005) 1980. 72 pages. 21.5 x 28 cm.
(Microfiche only)

No.6

R.K.Maiti, and S.S.Bisen.

Pearl millet anatomy.

(IBE 006) 1980. 24 pages. 21.5 x 28 cm. Saddle stitched.
HDC $1.50, LDC $0.50, P&P $2.40.

During the past several years the authors have studied in detail the structure and the development of various tissues and organs of a number of pearl millet (*Pennisetum americanum* (L.) Leeke) cultivars. This publication presents a summary of that work. It is an attempt to present a complete, simplified, illustrated description of the anatomy of pearl millet.

No.7

S.M.Virmani, S.J.Reddy, and M.N.S.Bose.

A handbook on the rainfall climatology of West Africa: data for selected locations.

(IBE 007) 1980. 56 pages. 21 x 13.5 cm.
(Microfiche only)

The authors assembled rainfall and evapotranspiration data for Chad, Mali, Niger, Senegal, and Burkina Faso. The data are presented here to help readers interpret the agricultural climate of the West Africa region.

No.7

S.M.Virmani, S.J.Reddy et M.N.S.Bose.

Manuel de climatologie pluviale de l'Afrique occidentale: Données pour des stations sélectionnées.

(IBF 007) 1980. 56 pages. 21 x 13.5 cm.
(Seulement en version microfiche)

Les auteurs ont rassemblé des données sur les pluies et l'évapotranspiration de la Burkina Faso, du Mali, du Niger, du Sénégal et du Tchad. Ce manuel a pour but d'informer les lecteurs sur les caractéristiques climatiques de cette région d'Afrique.

No.8

G.E.Thierstein.

The animal-drawn wheeled tool carrier.

(IBE 008) 1983. 12 pages. 21 x 13.5 cm. Saddle stitched.
Free

A multipurpose animal-drawn wheeled tool carrier that permits farmers to carry out their basic operations of tillage, planting, fertilization, and weeding quickly and precisely is a key instrument in ICRISAT's farming systems technology. It can also be used as a cart. This bulletin describes the carrier. It is illustrated by 13 photographs of the carrier and its attachments performing various farming operations.

No.8

G.E. Thierstein.

Le polyculteur : porte-outils à traction animale à roues

(IBF 008) 1985. 12 pages 21 x 13.5 cm.
Gratuit

Le polyculteur est une machine polyvalente qui permet au paysan de réaliser, rapidement et avec une grande précision, les opérations essentielles de culture, notamment le labour, les semis, l'épandage d'engrais et le sarclage; il constitue un élément clé de la technologie des systèmes de production de l'ICRISAT. La machine peut également être transformée en charrette. Ce bulletin donne une description du polyculteur et en présente les diverses utilisations à l'aide de 13 photos de la machine avec les outils en position de travail.

No.8

G.E. Thierstein.

El portaherramienta de ruedas tirado por animales

(IBS 008) 1985. 12 páginas. 21 x 13.5 cm.
Gratis

El portamienta de ruedas tirado por animales es una máquina de uso multiple que le permite al agricultor realizar sus operaciones básicas de labranza, siembra, fertilización, y deshierbe con rapidez y precisión, es una máquina importante en los sistemas tecnologías agrícolas de ICRISAT. Además, puede ser usado como carreta. Este boletín describe dicho portaherramienta y es ilustrado con 13 fotos del mismo con sus accesorios realizando varias operaciones agrícolas.

No.9

Y.L.Nene, J.Kannaiyan, and M.V.Reddy.

Pigeonpea diseases: resistance screening techniques.

(IBE 009) 1981. 16 pages. 18.5 x 24 cm. Saddle stitched.
HDC $1.50, LDC $0.50, P&P $1.55.

Describes easy and effective techniques for screening germplasm and breeding material for resistance to some of the economically important diseases of pigeonpea: wilt, phytophthora blight, and sterility mosaic. A combination of techniques is also described for screening for multiple disease resistance to all three diseases. The text is illustrated by 23 color photographs, and a nine-point scale is given for easy identification of resistant breeding material.

No.10

Y.L.Nene, M.P.Haware, and M.V.Reddy.

Chickpea diseases: resistance-screening techniques.

(IBE 010) 1981. 12 pages. 18.5 x 24 cm. Saddle stitched.
HDC $1.20, LDC $0.40, P&P $1.40.

Describes effective field and laboratory techniques for screening chickpea germplasm and breeding material for resistance to important diseases: wilt, dry root rot, and ascochyta blight. A nine-point scale is given to facilitate identification of resistant material. The text is illustrated by 10 color photographs.

No.11

J.Kampen.

An approach to improve productivity on deep Vertisols.

(IBE 011) 1982. 16 pages. 21.5 x 28 cm. Saddle stitched.
HDC $0.60, LDC $0.20, P&P $1.55.

The traditional rainfed farming systems on deep Vertisols in India are often characterized by low productivity and high soil and water losses. Research at ICRISAT has shown that there is considerable scope for increasing production on these soils, through a small-watershed management technology that includes improved soil-, water-, and crop-management practices. A description of this technology is presented in this bulletin.

No.12

G.L.Teetes, K.V.Seshu Reddy, K.Leuschner and L.R.House.

Sorghum insect identification handbook.

(IBE 012) 1983. 124 pages. 10 x 18 cm. Saddle stitched.
HDC $8.10, LDC $2.70, P&P $2.70.

Helps readers to identify common insect and mite pests at all stages of the crop's growth and in storage. There are six sections: soil insects, foliage, stem and head feeders, predators and parasites, and storage insects. The text, describing the distribution, attack symptoms, identification characteristics, biology, and the methods for controlling each pest is printed on pages facing the 59 color illustrations.

No.12

G.L.Teetes, K.V.Seshu Reddy, K.Leuschner, et L.R.House.
Manuel d'identification des insectes nuisibles au sorgho.
(IBF 012) 1983. 124 pages. 10 x 18 cm.
HDC $8.10, LDC $2.70, P&P $2.70.

Le but de cet ouvrage est d'aider les lecteurs à identifier les insectes et les acariens les plus connus du sorgho à tous les stades de développement et au cours du stockage. Il est réparti en six thèmes relatifs aux insectes du sol, des feuilles, des panicules, aux prédateurs/parasites et aux insectes des denrées. Des descriptions sur la répartition, les symptômes d'attaque, les caractéristiques, la biologie et les méthodes de lutte pour chaque ravageur sont présentées à l'aide de 59 illustrations en couleur.

No.12

G.L.Teetes, K.V.Seshu Reddy, K.Leuschner, y L.R.House.
Manual para la identificación de las plagas del sorgo.
(IBS 012) 1983. 124 páginas. 10 x 18 cm.
HDC $8.10, LDC $2.70, P&P $2.70.

Ayuda a identificar las plagas de insectos y ácaros más comunes mundiales, los cuales atacan al cultivo durante las varias épocas del crecemiento y del grano almacenado. El libro se divide en seis partes: insectos del suelo, plagas del follaje, del tallo y de la inflorescencia, depredadores y parásitos y plagas del grano almacenado. El texto describe la distribución, los síntomas del dano, las características específicas la biología y los métodos de combate para cada plaga. Estos detalles se ubican en las páginas opuestas a cada una de las 59 illustraciónes a color.

No.13

P.Subrahmanyam and D.McDonald.

Rust disease of groundnut.

(IBE 013) 1983. 15 pages. 18.5 x 24 cm. Saddle stitched.
HDC $1.50, LDC $0.50, P&P $1.75.

Describes disease symptoms, the morphology and taxonomy of
the causal fungus, *Puccinia arachidis*, and outlines the disease
cycle. It stresses the need for an integrated approach to control,
and consitituent cultural, chemical, and cultural measures are
described. Sufficient advice and data are given, with color illustrations, for readers to evolve locally suitable control policies.

No.14

L.J.G.van der Maesen.

World distribution of pigeonpea.

(IBE 014) 1983. 44 pages. 21.5 x 28 cm. Saddle stitched.
HDC $2.10, LDC $0.70, P&P $2.40.

Based on herbarium records compared with available statistics,
this bulletin describes and maps the world distribution of
pigeonpea. It emphasizes the importance of this crop which is
cultivated in many small areas, often unreported in national
statistics. Pigeonpea is present in virtually every tropical country, growing in gardens if not on a field scale, and making a
larger contribution to the protein diet than is often thought.

No.15

K.V.Ramaiah, C.Parker, M.J.Vasudeva Rao, and L.J.Musselman.

Striga identification and control handbook.

(IBE 015) 1983. 56 pages. 10 x 18 cm. Saddle stitched.
HDC $4.20, LDC $1.40, P&P $1.70.

Creates increased awareness of the need to reduce crop losses
associated with *Striga* infestations in the semi-arid tropics. The
handbook describes the most important of the 25 species of this
parasitic weed occurring in the world, their biology, and symptoms of attack. It also provides concise information about the
options for control. A key is presented to assist in the identification of the seven most damaging species, and the text is supported by 34 illustrations in color.

No.15

K.V.Ramaiah, C.Parker, M.J.Vasudeva Rao, et L.J.Musselman.

Manuel d'identification et de lutte contre le *Striga*.

(IBF 015) 1983. 56 pages. 10 x 18 cm.
HDC $4.20, LDC $1.40, P&P $1.70.

Cet ouvrage a pour objet de sensibiliser les lecteurs au besoin de réduire les pertes importantes de récolte dues au *Striga* dans les zones tropicales semi-arides. Il traite des plus importantes parmi les 25 espèces de cette plante parasite, leur biologie, les symptômes d'attaque et présente des recommandations succinctes sur les méthodes de lutte. Une clé de détermination est donnée afin de faciliter l'identification des sept espèces les plus nuisibles de *Striga*. 34 illustrations en couleur complètent le texte.

No.16

D.J.Andrews and G.Harinarayana.

Procedures for the seed production of pearl millet varieties.

(IBE 016) 1984. 8 pages. 18.5 x 24 cm. Saddle stitched.
HDC $0.60, LDC $0.20, P&P $1.20.

A brief guide to plant breeders on the techniques required to produce as pure and vigorous seed as possible from varieties of this cross-pollinated crop. Guidance on how to multiply improved seed is also given for extension staff to pass on to farmers.

No.17

J.A.Thompson.

Production and quality control of carrier-based legume inoculants.

(IBE 017) 1984. 40 pages. 18 x 24 cm. Saddle stitched.
HDC $2.20, LDC $0.70, P&P $1.80.

The bulletin outlines the use of small fermentation vessels for production of *Rhizobium* inoculant. Production is explained, with emphasis on quality control measures. Types and suitabilities of various worldwide inoculant carriers are discussed along with treatment and packaging techniques to ensure maximum effectiveness and sterility. Sources and selection of inoculant strains, and transportation of inoculants are also considered. Examples of national and private bodies governing the manufacture of inoculants are reviewed, and standards for production are suggested.

No.18

O.P.Rupela, P.J.Dart, B.Toomsan, D.V.Singh, and B.K.Sharma.
Facility for growing plants in test tubes at ICRISAT.
(IBE 018) 1984. 12 pages. 18.5 x 24 cm. Saddle stitched.
HDC $3.00, LDC $1.00, P&P $1.30.

This short, illustrated bulletin gives details of construction, operation, and costs of two plant growth units manufactured at ICRISAT Center and used for the growing of plants under sterile conditions in microbiological *Rhizobium* research.

No.19

M.V.K.Sivakumar, M.Konate, and S.M.Virmani.
Agroclimatology of West Africa: Mali.
(IBE 019) 1984. 300 pages. 21.5 x 28 cm.
HDC $18.30, LDC $6.30, P&P $8.70.

In semi-arid tropical countries such as Mali, a landlocked country without access to the sea, variable and erratic rainfall during the growing season influences crop potential. In this study a description of the rainfall variability is attempted through the analysis of annual, monthly, seasonal, and weekly totals. Through probability analysis of rainfall carried out weekly, its seasonality and dependability are described, as is the application of this analysis to crop planning.

No.19

M.V.K.Sivakumar, M.Konate et S.M.Virmani.
Agroclimatologie de l'Afrique de l'Ouest: le Mali.
(IBF 019) 1984. 300 pages. 21.5 x 28 cm.
HDC $18.30, LDC $6.30, P&P $8.70.

Dans les pays des zones tropicales semi-arides tels que le Mali qui est sans accès à la mer le potentiel de la production agricole est conditionné par une pluviométrie aléatoire. Cet ouvrage a pour but de décrire la fluctuation pluviométrique à l'aide de l'analyse de la pluviométrie annuelle, mensuelle, saisonnière et hebdomadaire. Une analyse probabiliste de la pluviométrie hebdomadaire permet à décrire son occurence dans la saison et sa stabilité ainsi que l'application de cette analyse pour la planification agricole.

No.20

M.J. Vasudeva Rao.

Techniques for screening sorghums for resistance to *Striga*.

(IBE 020) 1985. 20 pages. 18 x 24 cm. Saddle stitched.
HDC $3.00, LDC $1.00, P&P $2.00.

Striga is a root parasite of cereals and legumes which causes serious losses to host crops. Breeding genotypes of host crops with resistance to this parasite is recognized as the most economic way to combat *Striga*, however, breeding progress has been slowed by the absence of valid screening techniques. All laboratory and field methodologies currently used for screening *Striga* resistance are described in detail, as well as new screening techniques such as the 3-stage methodology and checkerboard layout. suggestions are offered to assist the development and management of *Striga* -sick fields.

No.21

D. McDonald, P. Subrahmanyam, R.W. Gibbons and D.H. Smith.

Leaf spots of groundnut

(IBE 021) 1985. (In press).

Provides extension workers with sufficient information to evolve measures of control of leaf spots well suited to local disease situations. The color-illustrated text describes disease symptoms, the morphology and taxonomy of the causal fungus, and outlines the disease cycle. It advocates an integrated approach to disease management, and the effectiveness of chemical and cultural control measures are assessed.

Workshop and Symposia Proceedings

The Institute's sponsorship of a conference or workshop, especially at the international level, includes publication of the proceedings by ICRISAT.

International workshop on farming systems, Nov 18-21 1974.

(CPE 001) 1975. 548 pages. 21.5 x 28 cm.
(Microfiche only)

International workshop on grain legumes, 13-16 Jan 1975.

(CPE 002) 1976. 360 pages. 21.5 x 28 cm.
HDC $16.20, LDC $5.40, P&P $13.60.

These proceedings deal almost exclusively with chickpea and pigeonpea, ICRISAT's two mandate pulse crops. The 31 reports cover aspects of breeding, diseases, insects, physiology, and general research in many of the countries where these crops are grown.

Consultants' group meeting on downy mildew and ergot of pearl millet, 1-3 Oct 1975.

(CPE 003) 1976. 152 pages. 20.5 x 26 cm. Saddle stitched.
HDC $6.90, LDC $2.30, P&P $5.75.

In order to review the state of knowledge on the downy mildew and ergot diseases of pearl millet, and to establish a priority of activities likely to lead to their effective control, 14 international scientists met in Hyderabad for in-depth discussions with ICRISAT staff on the topics of biology, epidemiology and control. This volume contains the papers presented and summaries of the discussion sessions at the meeting.

International workshop on agroclimatological research needs of the semi-arid tropics, 22-24 Nov 1978.

(CPE 004) 1980. 236 pages. 17 x 23.5 cm.
HDC $10.80, LDC $3.60, P&P $.6.30.

Workshop participants discussed: climatologic features of the semi-arid tropics, including the agrometeorologic network and data collection in semi-arid areas of India, Africa, and Brazil; water budgeting models and their application; crop-environment interaction; and interdisciplinary research needs of agroclimatologic studies. Recommendations for future action were made for each topic.

Sorghum diseases, a world review: proceedings of the international workshop, 11-15 Dec 1978.

(CPE 005) 1980. 477 pages. 17 x 23.5 cm.
HDC $21.60, LDC $7.20, P&P $11.65.

Contents include: review papers on sorghum grain molds, sorghum downy mildew, leaf diseases caused by fungi, smuts, ergot, stalk rots, viral and bacterial diseases; research papers on these diseases; transcripts of the discussions; workshop recommendations; and addresses of the 45 workshop participants. The workshop was cosponsored by Texas A&M University and ICRISAT.

Consultants' group discussion on the resistance to soil-borne disease of legumes, 8-11 Jan 1979.

(CPE 006) 1980. 175 pages. 21.5 x 28 cm.
HDC $8.10, LDC $2.70, P&P $.650.

A comprehensive review of ICRISAT's work on soil-borne diseases of pigeonpea and chickpea, particularly the work on resistance to fusarium wilts and phytophthora blight. Also included are papers presented by 10 consultants from five countries based on their experiences with soil-borne diseases of beans and peas.

International workshop on intercropping, 10-13 Jan 1979.

(CPE 007) 1981. 409 pages. 17 x 23.5 cm.
HDC $18.60, LDC $6.20, P&P $10.0.

Thirty-five papers on intercropping from 13 countries are presented. The major subjects covered are: plant population; the use of light, nutrients, and water; genotypes; weeds, insects, and diseases; statistics; economics; and operational management.

International workshop on socioeconomic constraints to development of semi-arid tropical agriculture, 19-23 Feb 1979.

(CPE 008) 1980. 447 pages. 17 x 23.5 cm.
HDC $20.40, LDC $6.80, P&P $10.65.

A collection of 25 papers and discussion summaries. Topics include: the analysis of existing farming systems and practices, socioeconomics of prospective technologies, field assessment of prospective technologies, issues in food grain marketing, nature and significance of risk, rural labor markets, and the economics of improved animal-drawn implements and mechanization.

International workshop on chickpea improvement, 28 Feb-2 Mar 1979.

(CPE 009) 1980. 306 pages. 17 x 23.5 cm.
HDC $14.10, LDC $4.70, P&P $7.60.

These proceedings contain 19 papers, 12 country reports, and discussions and recommendations. They cover all aspects of current chickpea research, including breeding, agronomy, physiology, plant protection, and microbiology.

International symposium on development and transfer of technology for rainfed agriculture and the SAT farmer, 28 Aug-1 Sep 1979.

(CPE 010) 1980. 330 pages. 17 x 23.5 cm.
(Microfiche only)

Participants from 28 countries and four international organizations reviewed field experience in the semi-arid tropics (SAT) and discussed basic problems of technology transfer. The role of international institutes and national programs is discussed and the need to establish effective linkages at all levels of the technology transfer process is emphasized. The symposium was held in connection with the inauguration of ICRISAT Center.

Climatic classification: a consultants' meeting, 14-16 Apr 1980.

(CPE 011) 1980. 171 pages. 21.5 x 27.5 cm.
(Microfiche only)

Characterization of crop climate is an important prerequisite to developing improved farming systems in the semi-arid tropics, where rainfall is unpredictable, evaporation high, and soil poor. This volume briefly reviews current concepts, identifies gaps in research, and suggests some approaches to a practical agronomically relevant classification system.

International workshop on groundnuts, 13-17 Oct 1980.

(CPE 012) 1980. 333 pages. 17 x 23.5 cm.
(Microfiche only)

Approximately 100 researchers from 19 countries presented papers on research organization and development, genetics and breeding, cytogenetics and utilization of wild species crop rotation and agronomy, entomology, and pathology. The proceedings also includes 17 country reports on groundnut production, utilization, research problems, further research needs, and discussion summaries.

Abstracts of the international workshop on groundnuts, 13-17 Oct 1980.
Résumés des communications. Colloque international sur l'arachide, 13-17 October 1980.

(CPE 013) 1981. 100 pages. 17 x 23.5 cm.
HDC $4.50, LDC $1.50, P&P $3.10.

International workshop on pigeonpeas, 15-19 Dec 1980

(CPE 014) 1981. Vol. 1. 516 pages. 17 x 23.5 cm.
HDC $23.40, LDC $7.80, P&P $12.20.
(CPE 015) 1981. Vol. 2. 459 pages. 17 x 23.5 cm.
HDC $21.0, LDC $7.0, P&P $11.10.

Volume 1 contains 48 invited papers by acknowledged authorities on pigeonpea (*Cajanus cajan*) research. These papers summarize research on all aspects of pigeonpea improvement, production, and utilization, including cropping systems, physiology, pathology, entomology, microbiology, and breeding methodology.
Volume 2 contains 63 specialized volunteer papers on pigeonpea research.

Improving the management of India's deep black soils; report of a seminar, 21 May 1981.

(CPE 016) 1981. 112 pages. 21.5 x 28 cm.
HDC $5.10, LDC $1.70, P&P $4.50.

Discusses the distribution and properties of deep black soils, ICRISAT's contribution to research on their management, the economies of technology options in their use for cropping, and various aspects of their improved management for production of cereals, pulses, and oilseeds. A proposed action plan is largely endorsed in the final recommendations of this high-level meeting with Indian agriculture officials in New Delhi; sponsored by the Indian Ministry of Agriculture, the Indian Council of Agricultural Research (ICAR), and ICRISAT.

Second international *Striga* workshop, 5-8 Oct 1981.

(CPE 017) 1983. 138 pages. 17 x 23.5 cm.
HDC $6.30, LDC $2.10, P&P $5.75.

The root parasitic *Striga* species can cause heavy yield losses, especially in poor soils. Control is difficult because seed persists in the soil for years and, because of strict quarantine regulations, research is limited to the areas where *Striga* occurs. These proceedings summarize worldwide *Striga* research, describe the known *Striga* species, and discuss control methods such as cultural practices; use of herbicides and germination stimulants; and breeding resistant cultivars. Abstracts, summaries of discussions, and recommendations are presented in English and French.

International symposium on sorghum grain quality, 28-31 Oct 1981.

(CPE 018) 1982. 415 pages. 17 x 23.5 cm.
HDC $19.80, LDC $6.60, P&P $10.25.

A valuable information source on: traditional sorghum foods; quality differences among cultivars; laboratory methods for quality evaluation; genetics; structure of kernel; physicochemical properties; dehulling techniques; polyphenols; consumer acceptance; human nutrition; and potential quality standards. The proceedings contains 40 presented papers, discussions, and final recommendations. The symposium was cosponsored by INTSORMIL, ICRISAT, and ICAR.

Sorghum in the eighties: proceedings of the international symposium, 1-7 Nov 1981.

(CPE 019) 1982. In two volumes: 473 and 326 pages. 17 x 23.5 cm.
HDC $36.30, LDC $12.10, P&P $19.0.

Contains reviews in important areas of sorghum research and projections for research and production for the 1980s. Over 40 presented papers covered such topics as: environmental effects; physical and biotic stress factors; agronomy and production technology; germplasm; food quality and utilization; breeding; and socioeconomic considerations. Recommendations and discussions by approximately 220 participants from 37 countries are also included. The symposium was cosponsored by INTSORMIL, ICRISAT, and ICAR.

International workshop on *Heliothis* management, 15-20 Nov 1981.

(CPE 020) 1982. 428 pages. 17 x 23.5 cm.
HDC $19.50, LDC $6.50, P&P $10.10.

This workshop brought together many of the world's most active researches on the management of these important insect pests. The proceedings record the 35 presented papers, discussions, and proposals for future work. Subjects covered include biology, ecology, populations, migration, biological control, pesticide and pheromone use, and plant resistance.

Agrometeorology of sorghum and millet in the semi-arid tropics; a symposium/planning meeting, 15-19 Nov 1982.

(CPE 021) 1984. 332 pages. 17 x 23.5 cm.
HDC $15.00, LDC $5.00, P&P $8.45.

Scientists from 18 countries reviewed the present state of knowledge on agroclimatic factors that influence the growth and development of sorghum and millet. Topics covered include global production of these crops; climatic requirements, agroclimatological studies on crop environments, and modeling of climatic responses. A planning session identified research needs, stressing the practical application of climatic knowledge to increase sorghum and millet production. The proceedings include abstracts in English and French.

Minimum data sets for agrotechnology transfer: ICRISAT-IBSNAT-SMSS symposium, 21-26 March 1983.

(CPE 022) 1984. 220 pages. 17 x 23.5 cm.
HDC $9.90, LDC $3.30, P&P $5.25.

ICRISAT hosted this symposium for the International Benchmark Sites Network for Agrotechnology Transfer (IBSNAT) and Soil Management Support Services (SMSS). Scientists from 24 countries met to identify the number and nature of research stations in the network; agree on the number of crops to be researched and on the design of experiments; identify the minimum data set to collect from each experiment; formulate plans for data base management and analysis; assign responsibilities for data collection and analysis, and data base management. These recommendations are described.

Watershed-based dryland farming in black and red soils of peninsular India, 3-4 Oct 1983.

(CPE 023) 1984. 96 pages. 21.5 x 27.5 cm. Saddle stitched.
HDC $4.50, LDC $1.50, P&P $4.10.

This meeting was the first on the subject in which banking and financial institutions participated, and the report is intended to further the expressed urgencies of improving dryland agriculture in India on a watershed basis.

International workshop on agricultural markets in the semi-arid tropics, 24-28 Oct 1983.

(CPE 024) (In press).

The objective of the workshop was to deepen the understanding of the role that agricultural marketing plays in the process of agricultural development in the semi-arid tropics, taking into account economic efficiency as well as social equity as important goals of development. The workshop was designed to bring out controversial issues and to also open interdisciplinary dialogues at an international level. The workshop was cosponsored by ICRISAT, International Geographical Union (IGU), and Osmania University.

International workshop on cytogenetics of *Arachis*, 31 Oct-2 Nov 1983.

(CPE 025) 1985. 200 pages. 17 x 23.5 cm.
HDC $12.00, LDC $4.00, P&P $5.00

This workshop reviewed the cytogenetics of *Arachis*, with emphasis on the utilization of wild species in the improvement of groundnut, especially in transferring pest resistance. A number of invited review papers, and a series of reports from scientists engaged on research involving wild *Arachis* species are presented. Topics covered include extending collection of the 70 known wild species, techniques for hybridization, use of plant growth hormones, and tissue culture by which desirable genes can be transferred to cultivated groundnut.

Sorghum root and stalk rots, a critical review: proceedings of the consultative group discussion, 27 Nov-2 Dec 1983.

(CPE 026) 1984. 278 pages. 17 x 23.5 cm.
HDC $15.0, LDC $5.0, P&P $6.40.

The objectives of this meeeting, held under the joint sponsorship of ICRISAT and INTSORMIL, were to assess current knowledge, available control measures, and constraints to research and to determine future research emphasis. Using the group nominal technique for selecting interdisciplinary research priorities recommendations were made for fusarium root and stalk disease, charcoal rot, and anthracnose stalk rot. Reports of these discussions together with 20 major papers on various diseases are presented. (Includes 13 color plates).

Dryland management alternatives and research needs for SAT Alfisols: summary report of a consultants' workshop on the state of the art and management alternatives for optimizing the productivity of SAT Alfisols and related soils, 1-3 Dec 1983.

(CPE 027) 1984. 44 pages. 17 x 23.5 cm. Saddle stitched.
HDC $3.00, LDC $1.00, P&P $1.30.

Discusses distribution and properties of Alfisols and related soils in the SAT, their agroclimatic characteristics, their management and conservation requirements, and cropping alternatives. Priority areas of research and systematic development and testing of promising management components and combinations at different locations are emphasized, and a cooperative network research approach recommended.

Consultants' workshop on the state of the art and management alternatives for optimizing the productivity of SAT Alfisols and related soils, 1-3 Dec 1983.

(In preparation).

Scientists from 9 countries joined 25 from ICRISAT to discuss constraints to increased productivity of these tropical red soils under rainfed conditions and assess the current state of knowledge on their effective management. Major subjects covered are: inventory and properties of Alfisols and related soils in the SAT, their agroclimatic characteristics, their management and conservation requirements, and cropping alternatives.

Grain legumes in Asia: summary proceedings of the consultative group meeting for Asian regional research on grain legumes, 11-15 Dec 1983.

(CPE 028) 1984. 104 pages. I7 x 23.5 cm. Saddle stitched.
HDC $4.80, LDC $1.60, P&P $3.0.

The objectives of the meeting were to develop an understanding of ongoing national and regional grain legume research programs in Asia, and to formulate plans for ICRISAT's participation in supporting and strengthening research programs on groundnut, chickpea, and pigeonpea. The recommendations and a keynote address are supported by summaries of papers from 5 members of ICRISAT staff, 10 country representatives, and 9 members of international organizations.

Regional groundnut workshop for southern Africa, 26-29 March 1984.

(CPE 029) 1985. 300 pages 17 x 23.5 cm.
(In press)

Twenty-five papers plus discussions, review the current state of groundnut research and production in six southern African countries. Regional work by ICRISAT, SADCC, and IDRC is also discussed.

Miscellaneous Reports and Brochures

General Audience Publications

Single copies issued free.

About ICRISAT.

(GAE 001) 1984. 8 pages. 10 x 23 cm. Saddle stitched.

A concise leaflet presenting basic facts on ICRISAT's constitution, area of work, and training programs, mandate crops, client groups, and donors. Also available in French (GAF 001); Hindi (GAH 001); Marathi (GAM 001) Tamil (GAT 001) and Telugu (GAL 001).

Training at ICRISAT.

(GAE 004) 1983. 2 pages. 23.5 x 12.5 cm.

An illustrated leaflet that outlines the programs available at ICRISAT for international interns, in-service and research fellows, research scholars, in-service trainees, and apprentices.

Stages de formation à l'ICRISAT

(GAF 004) 1983. 2 pages. 23.5 x 12.5 cm.

Ce dépliant donne un aperçu général des programmes de formation offerts à l'ICRISAT, à savoir le perfectionnement des chercheurs, la formation post-universitaire, la formation universitaire et la formation professionnelle, ainsi que des stages de courte durée.

ICRISAT Publications catalog.

(GAE 009) 1985. 24 pages. 18.5 x 24 cm. Saddle stitched.

A full listing of all ICRISAT publications together with descriptions of series, and information on the contents of each individual item. Includes ordering instructions, prices and order form.

Plant Material Descriptions

These free leaflets provide brief descriptions of crop genotypes identified or developed by ICRISAT, to facilitate the identification of cultivars and lines for plant breeding purposes and promote their wide utilization.

No.1

Pearl millet variety ICMV 1.

(PME 001) 1985. 4 pages. 14 x 21.5 cm.

No.2

Pearl millet variety ICMV 2.

(PME 002) 1985. 4 pages. 14 x 21.5 cm.

No.3

Pearl millet variety ICMV 3.

(PME 003) 1985. 4 pages. 14 x 21.5 cm.

No.4

Pearl millet male-sterile line ICMA 1 and maintainer line ICMB 1.

(PME 004) 1985. 4 pages. 14 x 21.5 cm.

No.5

Pearl millet male-sterile line ICMA 2 and maintainer line ICMB 2.

(PME 005) 1985. 4 pages. 14 x 21.5 cm.

No.6

Pearl millet male-sterile line ICMA 3 and maintainer line ICMB 3.

(PME 006) 1985. 4 pages. 14 x 21.5 cm.

No.7

Pearl millet male-sterile line ICMA 4 and maintainer line ICMB 4.

(PME 007) 1985. 4 pages. 14 x 21.5 cm.

No.8

Sorghum variety ICSV 1.

(PME 008) 1985. 4 pages. 14 x 21.5 cm.

No.9

Chickpea cultivar ICCV I.

(PME 009) 1984. 4 pages. 14 x 21.5 cm.

ICRISAT R&D Leaflets

This unnumbered occasional series issued free for limited distribution comprises of brief statements of potentially adoptable ICRISAT research findings, or executive summaries of forthcoming ICRISAT publications.

D.G.Faris and J.F.Scheuring.

Pigeonpea: a drought-tolerant pulse crop for West Africa.

(RLE 001) 1984. 2 pages. 21.5 x 28 cm.

Improved seeds from ICRISAT: plant material descriptions nos. 1-9.

(RLE 002) 1985. 2 pages. 21.5 x 28 cm.

Other Publications

ICRISAT farming systems research: a special report.

(OPE 005) 1983. 68 pages. 21.5 x 28 cm. Saddle stitched. HDC $3.20, LDC $1.10, P&P $3.25.

In the agricultural development of the semi-arid tropics, ICRISAT has recognized the need to supplement research for the improvement of individual crops with research into the integration of the various components of farm productivity, in terms of stable and socioeconomically viable farming systems. This report evaluates current farming systems research at ICRISAT in depth and makes recommendations concerning its future development, with particular regard to the organization of multidisciplinary research, the use of well-distributed benchmark sites across the semi-arid tropics, and the strengthening of collaboration between international and national research staff.

The Tropicultor operators' manual.

(OPE 006) 1985. 64 pages. 21.5 x 28 cm. Saddle stitched.
HDC $4.00, LDC $1.50, P&P $3.50.

This step-by-step instructional manual, illustrated by 194 photographs and 12 line diagrams, shows how to assemble the animal-drawn wheeled tool carrier on delivery, and how to use it in the field. One part deals with operations on flat land and the other on broadbeds. Both show how plowing and cultivation work is done, and the latter additionally describes how the hand-fed and mechanical planters are used for sowing seeds and applying fertilizer. Also available in Telugu (OPL 006).

Pigeonpea for West Africa

(OPE 007) 1984. 8 pages. 10 x 21.5 cm. Saddle stitched.
Free

This brief color-illustrated guide for research and extension workers describes how to obtain information and planting material, how to grow the crop in this region of Africa, and how to dehull the grain for domestic use.

Pois d'Angole : Légumineuse à grand potentiel en Afrique de l'Ouest

(OPF 007) 1984. 8 pages 10 x 21.5 cm.
Gratuit

Ce petit guide illustré en couleurs, rédigé à l'intention de chercheurs et de vulgarisateurs, désigne des sources pouvant fournir de plus amples informations sur le pois d'Angole ainsi que le matériel d'essai, explique comment la plante peut être cultivée dans cette partie de l'Afrique, et présente une méthode de décorticage des graines pour la consommation domestique.

Directory of sorghum and millets research workers.

(OPE 004) 1982. 184 pages. 21.5 x 28 cm.
HDC $8.40, LDC $2.80, P&P $7.0.

Includes 2223 entries relating to research workers in 88 countries. Entries are organized alphabetically by country and, within countries, by organizations. Name and organization indexes are included.

Bibliographies

Copies available from Library, ICRISAT, Patancheru P.O., Andhra Pradesh 502 324, India.

Single copies issued free, postage and packing must be prepaid.

Sorghum and Millets Information Center (SMIC)

An unnumbered series of bibliographies, with sufficiently complete citations to permit the reader to identify the original document. Subjects covered include: botany, physiology, biochemistry, breeding, cytology, genetics, varieties, agronomy, mechanization, seeds, soil microbiology, pathology, entomology, postharvest operations, crop products, and economics. Author, subject, and geographic indexes are provided.

Sorgho: bibliographie annotée de la documentation internationale en français, 1900-1976.

(BBF 001) 1980. 165 pages. 21.5 x 28 cm. Relie.
P&P $11.00.

Sorghum bibliography, 1970-73.

(BBE 002) 1982. 138 pages. 21.5 x 28 cm. Hardcover.
P&P $11.00.

Sorghum bibliography, 1974-76.

(BBE 003) 1983. 200 pages. 21.5 x 28 cm.
P&P $7.50.

Sorghum bibliography 1977-80.

(BBE 004) 1984. 540 Pages. 21.5 x 28 cm.
P&P $7.50.

Sorghum bibliography, 1981.

(BBE 005) 1984. 304 pages. 21.5 x 28 cm.
P&P $7.50.

Millets bibliography, 1970-76.

(BBE 006) 1983. 160 Pages. 21.5 x 28 cm.
P&P $7.50.

Millets bibliography, 1977-80.

(BBE 007) 1983. 226 pages. 21.5 x 28 cm.
P&P $7.50.

Millets bibliography, 1981.

(BBE 008) 1985. 212 pages. 21.5 x 28 cm.
P&P $7.50.

Y.L.Nene, A.Mengistu, J.B.Sinclair and D.J.Royse.

An annotated bibliography of chickpea diseases, 1915-1976.

(IBE 001) 1978. 49 pages. 21.5 x 28 cm. Saddle stitched. Information Bulletin No.1.
P&P $7.50.

Includes 331 citations and is intended to provide a working file of the literature that will assist in the identification of fungal and bacterial microorganisms associated with chickpea seeds. The index is divided into five sections according to the causal agent of the disease.

B.S.Dahiya.

An annotated bibliography of pigeonpea, 1900-1977.

(BBE 009) 1980. 195 pages. 21.5 x 28 cm. Hardcover.
P&P $11.00.

Contains 1275 citations including all traceable references prior to 1900, a comprehensive coverage 1900-1977, and some from 1978. References are grouped under broad subject categories. Author, subject, and word indexes are provided.

Shree Ram Yadav, S.Prasannalakshmi, and T.C.Jain.

Bibliography of theses on ICRISAT specialities.

(BBE 010) 1977. 232 pages. 13.5 x 20 cm.
(Microfiche only)

K.B.Singh and L.J.G.van der Maesen.

Chickpea bibliography, 1930-1974.

(BBE 011) 1977. 229 pages. 21.5 x 28 cm. Hardcover.
(Microfiche only)

Shree Ram Yadav, S.Prasannalakshmi, and P.S.Jadhav.

Indian theses on groundnut: a bibliography 1948-1977

(IBE 004) 1978. 54 pages. 21.5 x 28 cm. Information Bulletin
No.4.
(Microfiche only)

P.S.Jadhav, T.C.Jain, and S.Prasannalakshmi.

Sorghum, millets, peas: a bibliography of the Indian literature, 1969-73.

(BBE 012) 1975. 128 pages. 14 x 21.5 cm.
(Microfiche only)

Y.L.Nene, W.J.Kaiser, J.S.Grewal, J.Kannaiyan, and
S.P.S.Beniwal.

An annotated bibliography of pigeonpea diseases, 1906-81.

(BBE 013) (In press).

Contains 594 citations, most of them with short abstracts, supported by an index to fungal, bacterial, viral, and nematode-associated diseases. Cross-references are given to methods of fungicidal and chemical control.

Newsletters

Single copies free of cost and postal charges.

At ICRISAT.

(NAE) Quarterly. 21.5 x 28 cm.

Reports recent ICRISAT activities and research development to a broad audience of agriculturalists, donor agencies, government officials, and the general public. Lists latest publications and articles by ICRISAT scientists accepted by professional journals in the current quarter.

Nouvelles de l'ICRISAT.

(NAF) Bulletin trimestriel. 21.5 x 28 cm.

Bulletin d'information portant sur la recherche et les activités de l'ICRISAT; y figure également la liste des nouvelles publications de l'Institut et des articles de ses chercheurs acceptès par diverses revues scientifiques. Ce bulletin est destiné aux autorités agricoles, aux agronomes, aux institutions donatrices et au grand public.

International chickpea newsletter.

(NCE) Semiannual. 21.5 x 28 cm. Saddle stitched.

Published in June and December of each year, as a means of rapid communication between scientists working on chickpeas in all parts of the world. The newsletter features short communications that inform colleagues of new developments, progress, or problems in their work, and contains information on publications and meetings of interest to chickpea workers.

International pigeonpea newsletter.

(NPE) Semiannual. 21.5 x 28 cm. Saddle stitched.

Serves as a communication link for all those interested in the research and development of pigeonpeas throughout the world. Contains news, views, requests, and research notes (including previously collected data of importance that has not been published elsewhere), an abstract section, and information regarding publications and meetings of interest to pigeonpea workers.

SMIC newsletter.

(NSE) Three times a year. 21.5 x 28 cm. Side stitched.

Produced by the Sorghum and Millets Information Center (SMIC), this newsletter serves as a vehicle for research and extension workers in sorghum and millets to exchange ideas and information on research efforts to improve production and utilization of these crops. It includes worldwide information and selected bibliography on current sorghum and millets literature.

La Lettre du SMIC

(NSF) Trois fois par an. 21.5 x 28 cm.

Bulletin périodique du Centre d'information sur le sorgho et les mils (SMIC) visant à stimuler l'échange d'informations et d'idées entre les chercheurs du sorgho et du mil et les encadreurs chargés de la vulgarisation de ces cultures. Les sujets portent sur différents domaines de recherche et régions du monde; chaque numéro contient une bibliographie thématique et une revue de la documentation la plus récente sur le sorgho et les mils.

Audiovisual Materials

ICRISAT at a glance

(SSE 001) 1984. 70 color transparencies, script, 20 minutes. HDC and LDC $30. (includes P&P).

A general audiovisual presentation on ICRISAT's work and its place in the CGIAR system.

IFPRI

International Food Policy Research Institute

1776 Massachusetts Avenue, N.W., Washington, D. C. 20036, USA
Cable: IFPRI - Telex: 440954
Telephone: (202) 862-5600

General Information

Objectives

Research at IFPRI is aimed at contributing to the reduction of hunger and malnutrition in low-income countries through the analysis of underlying processes that extend beyond the narrowly defined food sector. The results of IFPRI's research and analysis are intended to assist policymakers in developing countries make informed decisions on food policy issues. IFPRI works closely wih other international and developing nation institutions to understand food production, consumption, and trade processes; to develop new options for policymakers; and to assess the efficacy of existing policies. Individual copies of most IFPRI publications are free but multiple copies will carry a fee plus shipping costs. The fee varies with size and weight of publications.

Annual Report

International Food Policy Research Institute

IFPRI Report 1983

April 1985. 67 + pages. $3 air postage. Free, except for multiple copies to booksellers.

The annual report summarizes the year's work in each of the Institute's four main programs: food trends analysis, food production and development strategy, food consumption and nutrition, and food trade and security, as well as in collaborative and regional projects.

Monographs

International Food Policy Research Institute

Research Reports

Number of pages varies. 17.8 × 25.4 cm. Saddle stitched.

The main vehicle for reporting IFPRI research results is its Research Report series. The reports, which range from 30 to 180 pages, present data, analysis, and findings of completed research projects. Reports are reviewed by experts from outside the Institute. Since 1976 IFPRI has published 40 research reports.

These are free, except for multiple copies and to booksellers, when they cost $3.00 for publications with 48 or fewer pages, and $4.00 for publications with 49 or more pages. Surface mail is free; air postage costs $3.00.

No. 1

Meeting Food Needs in the Developing World: Location and Magnitude of the Task in the Next Decade

(Out of print)

No. 2

Recent and Prospective Developments in Food Consumption: Some Policy Issues

July 1977. 61 pages. ISBN 0-89629-002-6. (Out of print)

No. 3

Food Needs of Developing Countries: Projections of Production and Consumption to 1990

December 1977. 157 pages. ISBN 0-89629-004-2.

The data in Research Report 1 are updated and widened in scope, and the projections are extended to 1990.

No. 4

Panos Konandreas, Barbara Huddleston, and Virabongsa Ramangkura

Food Security: An Insurance Approach

September 1978. 96 pages. ISBN 0-89629-005-2.

Two insurance schemes designed to assist food-deficit developing countries in stabilizing cereal consumption are evaluated.

No. 5

Shubh K. Kumar

Impact of Subsidized Rice on Food Consumption and Nutrition in Kerala

January 1979. 45 pages. ISBN 0-89629-006-2.

The contribution of subsidized "ration" rice to the nutrition of a sample of low-income families of rural Kerala state in India is analyzed.

No. 6

Yair Mundlak

Insectional Factor Mobility and Agricultural Growth

February 1979. 137 pages. ISBN 0-89629-007-7.

An econometric approach is applied to a study of Japanese agriculture that takes account of the interdependence of agriculture and the rest of the economy. Mundlak's work is based on the assumption that factor returns are not equal across sectors, which leads to a flow of factors from one sector to another. This is the basis of resource allocation.

No. 7

P. S. George

Public Distribution of Foodgrain in Kerala — Income Distribution Implications and Effectiveness

March 1979. 67 pages. ISBN 0-89629-008-5.

This study examines the operation and cost of the ration and procurement systems for rice in the Indian state of Kerala during the past 25 years.

No. 8

Raisuddin Ahmed

Foodgrain Supply, Distribution, and Consumption Policies Within a Dual Pricing Mechanism: A Case Study of Bangladesh

May 1979. 81 pages. ISBN 0-89629-009-3.

This report provides a framework for analysis of food policy issues as demonstrated by the foodgrain distribution system of Bangladesh. It shows that public foodgrain distribution is primarily urban oriented, although the urban poor are often better fed than their rural counterparts.

No. 9

Roger Fox

Brazil's Minimum Price Policy and the Agricultural Sector of Northeast Brazil

June 1979. 117 pages. ISBN 0-89629-010-7.

Brazil's minimum price and storage loan programs for corn, rice, beans, and cotton are analyzed.

No. 10

Peter Oram, Juan Zapata, George Alibaruho, and Shyamal Roy

Investment and Input Requirements for Accelerating Food Production in Low-Income Countries, 1990

September 1979. 178 pages. ISBN 0-89629-011-5.

This study, begun in 1977, estimates the level and comparison of investment required during the next 15 years to close specified food gaps in 36 low-income, food-deficit developing market economy countries.

No. 11

Kenneth L. Bachmann and Leonardo A. Paulino

Rapid Food Production Growth in Selected Developing Countries: A Comparative Analysis of Underlying Trends, 1961-1976

October 1979. 97 pages. ISBN 0-89629-012-3.

This analysis, based on published data, compares the major components of increased food intake in 16 countries in an attempt to shed light on the causes of their relativey high growth rates in agricultural production.

No. 12

J. S. Sarma and Shyamal Roy, and P. S. George

Two Analyses of Indian Foodgrain Production and Consumption Data

November 1979. 81 pages. ISBN 0-89629-013-1.

The two studies in this report, "Foodgrain Production and Consumption Behavior in India, 1960-67" by Sarma and Roy, and "Aspects of the Structure of Consumer Foodgrain Demand in India, 1961/62 to 1973/74" by George, explain why in a time of record production, per capita consumption in India was declining.

No. 13

James D. Gavan and Indrani Sri Chandrasekera

The Impact of Public Foodgrain Distribution on Food Consumption and Welfare in Sri Lanka

December 1979. 54 pages. ISBN 0-89629-014-X.

This report analyzes Sri Lanka's comprehensive public food distribution scheme and its contribution to the comparatively satisfactory living standards achieved in that country.

No. 14

Timothy Josling

Developed-Country Agricultural Policies and Developing-Country Supplies: The Case of Wheat

March 1980. 66 pages. ISBN 0-89629-015-8.

This study makes a quantitative assessment of the effect of developed-country policies on the world wheat market and their contribution to the instability of trade and prices.

No. 15

Anthony M. Tang and Bruce Stone

Food Production in the People's Republic of China

May 1980. 178 pages. ISBN 0-89629-016-6.

Two studies are included in this report: "Food and Agriculture in China: Trends and Projections, 1952-77 and 2000" by Tang, and

"China's 1985 Foodgrain Production Target: Issues and Prospects" by Stone. The former reviews development strategy, analyzes the historical experience of agricultural growth in the People's Republic of China, and projects the aggregate food supply/demand balance for the year 2000. The latter analyzes China's foodgrain production target in the light of past performance, production and input growth, and current policies.

No. 16

Bruce Stone

A Review of Chinese Agricultural Statistics, 1949-79

ISBN 0-89629-017-4. (Not yet available)

This compilation of updated historical data on the population and foodgrain economy of the People's Republic of China brings together and compares a number of scattered statistical series generated by China analysts from partial official series and fragmentary information contained in official pronouncements in new media.

No. 17

Francis Sulemanu Idachaba

Agricultural Research Policy in Nigeria

August 1980. 69 pages. ISBN 0-89629-018-2.

This study reviews the evolution of agricultural research in Nigeria and examines the relative emphasis research efforts in export and import crops; livestock, forestry, and fisheries; rainfed and irrigated agriculture; and agricultural production and inputs. It identifies political and economic policies that affect the efficiency of the agricultural research system and makes recommendations.

No. 18

Daniel T. Morrow

The Economics of the International Stockholding of Wheat

September 1980. 45 pages. ISBN 0-89629-019-0.

This report describes the behavior of stockholding in the world wheat economy since 1960, predicts stockholding behavior for the near future, and considers possible benefits from an international agreement to increase stockholding above the predicted level.

No. 19

Leonardo A. Paulino and Shen Sheng Tseng

A Comparative Study of FAO and USDA Data on Production, Area, and Trade of Major Food Staples

October 1980. 77 pages. ISBN 0-89629-020-4.

The differences between FAO and USDA statistics on production and area of the major staple food crops and those on cereal trade are examined. The study identifies the commodities and major countries for which wide data differences exist and measures the differences among countries, regions, economic groups, and world totals.

No. 20

Dharm Narain and Shyamal Roy

Impact of Irrigation and Labor Availability on Multiple Cropping: A Case Study of India

November 1980. 34 pages. ISBN 0-89629-021-2.

This report examines the differences in multiple cropping within and between states in India and indicates the extent to which irrigation can have an impact on agricultural growth by expanding multiple cropping.

No. 21

Alberto Valdes and Joachim Zietz

Agricultural Protection in OECD Countries: Its Cost to Less-Developed Countries

December 1980. 57 pages. ISBN 0-89629-022-0.

This study examines the costs of agricultural protection to developing countries based on a hypothetical 50 percent reduction for 100 agricultural commodities in 19 Organization for Economic Cooperation and Development (OECD) countries.

No. 22

Padma Desai

Estimates of Soviet Grain Imports in 1980-85: Alternative Approaches

February 1981. 47 pages. ISBN 0-89629-023-9.

The Soviet Union's massive imports of grain beginning in the 1970s contribute to instability in the international market, thus making it important to foretell their size. This study uses three different methods to predict the differences between Soviet supplies and requirements; estimates of domestic production based on past trends, estimates based on production functions, and regression estimates of import demand functions.

No. 23

Victor J. Elias

Government Expenditures on Agriculture in Latin America

May 1981. 67 pages. ISBN 0-89629-024-7.

This is a descriptive report analyzing data assembled on government spending on the rural sectors of nine Latin American countries from 1950 to 1978. It identifies expenditures in the budgets of central and state governments and decentralized government agencies in addition to departments of agriculture.

No. 24

Jorge Garcia Garcia

The Effects of Exchange Rates and Commercial Policy on Agricultural Incentives in Colombia: 1953-1978

June 1981. 88 pages. ISBN 0-89629-025-5.

This report traces the effects on Colombian agriculture and trade of tariffs, severe import restrictions, overvaluation of the peso, and export subsidies.

No. 25

Shakuntla Mehra

Instability in Indian Agriculture in the Context of the New Technology

July 1981. 55 pages. ISBN 0-89629-026-3.

Although the use of new-fertilizer technology in India has led to unprecedented production growth, yield variability has also increased. This report examines the causes of yield fluctuations and the possible role of new technology in accentuating them.

No. 26

John McIntire

Food Security in the Sahel: Variable Import Levy, Grain Reserves and Foreign Exchange Assistance

September 1981. 70 pages. ISBN 0-89629-027-1.

This report studies the costs and benefits of various proposed trade storage policies for the Sahel. It argues that increased storage of grain reserves would be expensive and, in most cases, not as effective as measures to liberalize trade or the establishment of a food insurance or compensatory financing scheme.

No. 27

Raisuddin Ahmed

Agricultural Price Policies Under Complex Socioeconomical and Natural Constraints: The Case of Bangladesh

October 1981. 78 pages. ISBN 0-89629-028-X.

This report examines the production response of small family farms in Bangladesh to price incentives, as well as the issue underlying the impact of prices on land resources, labor, modern inputs, consumption, export crop production, demand linkages, and welfare.

No. 28

J. S. Sarma

Growth and Equity: Policies and Implementation in Indian Agriculture

November 1981. 76 pages. ISBN 0-89629-029-8.

The agricultural policies and strategies that evolved in India after Independence are examined and their effects on agricultural growth and on interpersonal and interregional disparities are analyzed. This report also includes commentaries on the growth and equity experiences of Europe by Ester Boserup, Japan by S. Hirashima, and the United States by Olaf F. Larson.

No. 29

Grant M. Scobie

Government Policy and Food Imports: The Case of Wheat in Egypt

December 1981. 88 pages. ISBN 0-89629-030-1.

This report analyzes the relationship between food imports and government subsidies in Egypt, which has a longstanding and extensive policy of providing subsidies to the people, particularly of wheat.

No. 30

Peter B. R. Hazell

Instability in Indian Foodgrain Production

May 1982. 60 pages. ISBN 0-89629-031-X.

In this report, statistical decomposition analysis is applied to determine how much of the increase in yield variability accompanying the rise in India's foodgrain production is the result of new technologies.

No. 31

Guvant M. Desai

Sustaining Rapid Growth in India's Fertilizer Consumption: A Perspective Based on Composition of Use

August 1982. 72 pages. ISBN 0-89629-032-8.

Utilizing large sample surveys, this report identifies crops, varieties, and irrigated and unirrigated areas associated with India's growth in fertilizer use since the 1950s and discusses policies to sustain growth in fertilizer use during the 1980s.

No. 32

Cheryl Williamson Gray

Food Consumption Parameters for Brazil and their Application to Food Policy

September 1982. 78 pages. ISBN 0-89629-033-6.

Income and price elasticities of cereals, other foods, and total caloric intake are measured for both the malnourished and the adequately nourished and for different income groups in Brazil. The report also shows how consumption parameters can be applied to two policy problems: policies to increase caloric consumption through subsidies and the government's program to produce alcohol from crops.

No. 33

C. Rangarajan

Agricultural Growth and Industrial Performance in India: A Study of Interdependence

October 1982. 40 pages. ISBN 0-89629-034-4.

This report examines the production, demand, and savings and investment linkages between agriculture and industry in India, using a microeconomic model to determine the effects of agricultural growth on industry.

No. 34

Harold Alderman, Joachim von Braun, and Sakr Ahmed Sakr

Egypt's Food Subsidy and Rationing System: A Description

October 1982. 80 pages. ISBN 0-89629-035-2.

The institutional arrangements and regulations that make up Egypt's extensive food distribution network are described in detail.

No. 35

Ulrich Koester

Policy Options for the Grain Economy of the European Community: Implications for Developing Countries

November 1982. 90 pages. ISBN 0-89629-036-0.

This report analyzes how four policy options of the European Community (EC) would affect world grain markets, developing countries, and the EC itself. They include the continuation of past grain policy, a policy to eliminate EC grain tariffs, and two options aimed at reducing instability in the world grain market.

No. 36

Domingo Cavalio and Yair Mundlak

Agriculture and Economic Growth in an Open Economy: The Case of Argentina

December 1982. 162 pages. ISBN 0-89629-037-9.

A model consisting of a number of behavioral equations is used to explain the pattern of Argentine sectoral growth during 1940-72. The differential growth in sectoral inputs and productivities is related to differences in factor renumeration, which reflect the taxation of agricultural exports and protection of the nonagricultural sector.

No. 37

Sudhir Wanmali

Service Provision and Rural Development in India: A Study of Miryalguda Taluda

February 1983. 62 pages. ISBN 0-89629-038-7.

This report documents the development of rural services since the introduction of irrigation in Miryalguda, a small section of Andhra Pradesh, in 1968. It attempts to measure catalytic effects of government location policies on the growth of private enterprise in both the dry and irrigated portions of the study area.

No. 38

Raj Khrishna and Ajay Chhibber

Policy Modeling of a Dual Grain Market: The Case of Wheat in India

May 1983. 74 pages. ISBN 0-89629-039-5.

This report stresses the interaction of prices and quantities in the commercial and concessional wheat markets of India. It projects purchases, sales, imports, and stocks for 1979-92 and determines their least-cost values.

No. 39

Ammar Siamwalla and Stephen Haykin

The World Rice Market: Structure, Conduct, and Performance

June 1983. ISBN 0-89629-040-9.

According to this report, technological changes in rice production have favored importing countries more than exporting countries, and policies of individual countries have contributed to the market's thinness and volatility. The workings of the rice market are examined with an eye to reducing its inefficiencies.

No. 40

Grant M. Scobie

Food Subsidies: Their Impact on Foreign Exchange and Trade in Egypt

August 1983. ISBN 0-89629-041-7.

Using historical evidence, this report examines the impact of Egyptian subsidy expenditures on domestic inflation, the balance of payments, and foreign exchange.

No. 41

Peter B. R. Hazell and Ailsa Roell

Rural Growth Linkages: Household Expenditure Patterns in Malaysia and Nigeria

September 1983. 64 pages. ISBN 0-89629-042-5.

Data from two rural communities are examined to determine which kinds of farm households will spend the largest share of incremental income on locally produced, labor-intensive goods and services.

No. 42

Joachim von Braun and Hartwig de Haen

The Effects of Food Price and Subsidy Policies on Egyptian Agriculture

November 1983. 94 pages. ISBN 0-89629-043-3.

This report analyzes how the policies of the Egyptian government have affected agriculture and questions whether increases in food subsidies have reduced government support for agriculture.

No. 43

Barbara Huddleston

Closing the Cereals Gap with Trade and Food Aid

January 1984. 107 pages. ISBN 0-89629-044-1.

This report presents a comprehensive series of food aid data, which is then used to analyze past food aid trends and to project the requirements for the future. Major issues concerning food aid are also discussed.

No. 44

Michael Schluter

Constraints on Kenya's Food and Beverage Exports

April 1984. 118 pages. ISBN 0-89629-045-X.

The potential for improving Kenya's economy by increasing its agricultural exports, particularly to a group of oil-exporting countries, is examined. Schluter analyzes the potential of eight possible exports and the economic and political factors that constrain their development.

No. 45

Harold Alderman and Joachim von Braun

The Effects on Welfare and Distribution of the Egyptian Food Ration and Subsidy System

June 1984. 127 pages. ISBN 0-89629-046-8.

Determining which groups of the overall population — urban, rural, low-income, or high — benefit more from Egypt's extensive subsidy system is the objective of this study.

No. 46

Prasarn Trairatvorakul

The Effects on Income Distribution and Nutrition of Alternative Rice Price Policies in Thailand

November 1984. 102 pages. ISBN 0-89629-047-6.

This study argues that the net gains to the rural poor would be minimal if the Thai government were to change its rice policy to allow the domestic price of rice to reflect export prices.

No. 47

Nabil Khaldi

Evolving Food Gaps in the Middle East/North Africa: Prospects and Policy Implications

December 1984. 74 pages. ISBN 0-89629-048-4.

In this report Khaldi examines the widening deficit between production and consumption in many countries of the Middle East/North Africa, as rising incomes foster increased demand, especially for livestock products.

No. 48

Sudhir Wanmali

Rural Household Use of Services: A Study of Miryalguda Taluka, India

March 1985. 56 pages. ISBN 0-89629-049-2.

Effective planning for service provision requires an understanding of the patterns of service use. Here data are examined for 10 villages to identify differences in the use of public and retail services among inhabitants of dry versus irrigated tracts and among households of different economic status.

No. 49

J. S. Sarma and Patrick Yeung

Livestock Products in the Third World: Past Trends and Projections to 1990 and 2000

April 1985. ____ pages. ISBN 0-89629-050-6.

In this report past trends in production, consumption, and trade in meat, milk, and eggs are analyzed for 104 developing countries. These trends are then projected to 1990 and 2000 to give a broad indication of supply-demand balances.

Conference Proceedings

International Food Policy Research Institute

Food Policy Issues and Concerns in Sub-Saharan Africa

February 1981. 175 pages. 21.59 × 27.035 cm. Paper, perfect.

Papers presented by researchers at IFPRI and discussed with colleagues in Ibadan, Nigeria, 9-11 February 1981. Topics include an assessment of the food situation, production potentials, nutritional concerns, pricing policy, fertilizer use, expansion of food consumption, and food security.

International Food Policy Research Institute and UN Administrative Committee on Co-ordination/Sub-Committee on Nutrition
Edited by Per Pinstrup-Andersen, Alan Berg, and Martin Forman

International Agricultural Research and Human Nutrition

1984. 326 pages. ISBN 0-89629-303-3.

This collection of papers reviews the nutrition-related activities undertaken by the international agricultural research centers of the Consultative Group on International Agricultural Research and suggests ways research institutes can incorporate goals for improving human nutrition into research design and planning.

Miscellaneous Reports and Brochures

International Food Policy Research Institute

Some Commentaries on Food

October 1981. 20 pages. 20.32 × 20.32 cm. Paper, saddle stitched.

Prepared by IFPRI researchers for World Food Day, 16 October 1981. Commentaries deal with food security, food subsidies, accelerating production, food problems in Africa, fertilizer supplies, an international food financing facility, energy cropping, and crop insurance.

P. A. Oram and V. Bindish

Resource Allocations to National Agricultural Research. Trends in the 1976s (A Review of Third World Systems)

November 1981. 104 pages. 21.59 × 27.305 cm. Paper, perfect.

This study assesses recent progress on the development of the agricultural research systems in the Third World and identifies issues for further study. Published jointly with the International Service for National Agricultural Research (ISNAR).

International Food Policy Research Institute

Looking Ahead: The Development Plan for the International Food Policy Research Institute

June 1982. 20 pages. 21.59 × 27.035 cm. Saddle stitched.

This piece describes IFPRI'S growth and its future research and financial considerations. It identifies six major food policy questions for the 1980s.

Richard H. Adams

IFPRI Research and the Creation of the IMF Cereal Import Facility

August 1982. 10 pages. 3.51 × 1.56 cm. Saddle stitched.

Using the role IFPRI's research played in the creation of the IMF Cereal Facility as an example, this piece highlights the important contribution careful and timely research can make to public policy-making.

John W. Mellor

Food and the Structure of Economic Growth: Its Relevance to North-South Relations

October 1982. 12 pages. 3.02 × 1.56 cm. Saddle stitched.

Paper presented at the Symposium on the World Food Problem and Japan, sponsored by the Japan FAO Association on the occasion of World Food Day, October 16, 1982.

Eileen T. Kennedy and Per Pinstrup-Andersen

Nutrition-Related Policies and Programs: Past Performance and Research Needs

February 1983. 104 pages. 21.59 × 27.30 cm. Perfect.

This paper reviews the success of various government interventions aimed at improving human nutrition and proposes nutrition-related research to help policymakers plan and implement effective programs for reducing malnutrition.

J. S. Sarma

Contingency Planning for Famines and Other Acute Food Shortages: A Brief Review

April 1983. 28 pages. 21.59 × 27.30 cm. Saddle stitched.

This paper reviews the roles and functions of international agencies dealing with acute food shortages caused by droughts, cyclones, floods, etc. It suggests areas of research and various national and international agencies might undertake.

Abstracts and Reprints

Various Authors, IFPRI

IFPRI Abstract

About eight/year since August 1980. 4 pages. 21.59 × 27.94 cm.

A series of publications that summarize and highlight the policy implications of each research report published by IFPRI, beginning with Research Report 16.

International Food Policy Research Institute

IFPRI Reprints

IFPRI publishes a number of articles by IFPRI researchers reprinted from journals and other publications.

Periodicals

International Food Policy Research Institute

IFPRI Report

Each January, May and September since May 1979. 4 pages. 21.59 × 27.94 cm.

IFPRI's newsletter. Each issue contains a commentary on a specific food policy issue written by a member (or members) of IFPRI's senior research staff, and reports on completed and soon-to-be-completed research.

IITA

International Institute of Tropical Agriculture

P.M.B. 5320, Ibadan, Nigeria
Cable: TROFOUND, IJEKA, NIGERIA — Telex: TROPIB NG 31417

Publications of the International Institute of Tropical Agriculture 1984

About IITA

The International Institute of Tropical Agriculture is a non-profit research and training institute established in 1967 under the Laws of the Federal Republic of Nigeria. Like its other 12 sister institutions, IITA is principally financed through the Consultative Group on International Agricultural Research which is an informal group of donor countries, development banks, foundations, and agencies.

The mandate

IITA is responsible for developing viable alternatives to shifting cultivation in the humid and sub-humid tropics. Also, the Institute has a global mandate for research to improve cowpeas, yams, and sweet potatoes as well as carrying out research in Africa on maize, rice, cassava, and soybean. It organizes training, conferences and work-shops to increase the number of well qualified persons to carry out research and development.

Books published by commercial companies should be obtained through the book stores.

Research Highlights

IITA Research Highlights

Published yearly since 1976. Pages vary. 17 × 23 cm. Saddle stitched. Free.

These reports highlight some of the most significant research results from experiments conducted by the International Institute of Tropical Agriculture (IITA).

In English and French.

Annual Reports

International Institute of Tropical Agriculture, Annual Report

Published yearly since 1971. Pages vary. 21 × 28 cm. Wrap around. Free.

An account of the research IITA scientists carry out each year.

Monographs

Leaky, C.L.A. and J.B. Wills (Editors)

Food Crops of the Lowland Tropics

1977. 345 pages. 19 × 25 cm. Hard cover. US$25.00. Academic Press.

Based on 15 seminars, this book is for agricultural researchers, educators, and administrators. It covers food crops of tropical Africa including agricultural systems, mechanization, irrigation, land tenure, storage, research strategies, insect and mite pests and their control as well as fungi, bacteria, viruses, nematodes and their control. (Reprints of each of the 15 chapters available on request.)

Buddenhagen, I.W. and G.J. Persley (Editors)

Rice in Africa

1978. 365 pages. 16 × 23 cm. Hard cover. US$30.00

The book presents proceedings of a rice conference. There are 30 papers and 18 country statements from more than 50 authors covering the pathology, entomology, soil science, plant breeding and the economics of rice production in Africa. This volume highlights aspects of rice production unique to Africa.

Lal, R., J.C. Moomaw, and A. Fagbamiye

Soil-Water Relationships of Rice

1981. 355 pages. 21 × 28 cm. Wrap around. US$4.00

Drought in rice is a complex problem to be investigated for agronomic, genetic, engineering, soil water management as well as irrigation and plant physiological aspects. This report presents the scanty information on these aspects and reviews the results of experiments conducted at the International Institute of Tropical Agriculture.

Singh, S.R. and D.J. Allen

Cowpea Pests and Diseases

1979. 113 pages. 10 × 18 cm. Saddle stitched. US$4.00.

This is a field guide to recognizing cowpea pests and diseases for agricultural research workers, extension officers and cowpea growers. It includes laboratory diagnostic features, notes on collecting and preparing disease and pest specimens, non-parasitic diseases and current chemicals. It covers the world emphasizing tropical Africa.

In English, French and Spanish.

Rachie, K.O., K. Rawal, and J.D. Franchkowiak

A Rapid Method for Head Crossing Cowpea

Published in 1975. Reprinted in 1978. 5 pages. 15 × 23 cm. Saddle stitched. Free.

Cowpeas are generally easier than other grain legumes, but conventional methods have a number of limitations. This booklet describes a rapid and effective method for hand emasculating and crossing cowpeas.

Rachie, K.O. and K.M. Rawal

Integrated Approaches to Improving Cowpeas (*Vigna unguiculata* [L.] Walp.)

1976. 36 pages. 16 × 23 cm. Saddle stitched. Free.

This booklet describes the research methodology used to improve cowpea at the International Institute of Tropical Agriculture. It describes the origin, taxonomy, morphology, genetics, adaptation and production of cowpea. It also discusses the problems and desirable attributes of the crop, defines research goals, shows recent progress and breeding methodology.

Grain Legume Entomology

First published in 1977. Reprinted in 1982. 48 pages. 17 × 23 cm. Saddle stitched. Free.

Written primarily for IITA's legume training courses. It covers general classification and insect control with details on calculating insecticide dosages. The emphasis is on cowpea pests, pest status, identifying damage, biology, assessment methods and control measures.

In English and French.

Grain Legume Pathology

First published in I976. Reprinted in 1982. 91 pages. 17 × 21 cm. Saddle stitched. Free.

Written for IITA's grain legume courses, this book concentrates on the diseases of cowpea, soybean, lima bean and the French bean. It also discusses screening for bacterial, fungal and viral diseases of cowpea.

In English and French.

Wilson, Jill E.

Cocoyam Breeding by Flower Induction, Pollination and Seed Germination

1980. 15 pages. 10 × 18 cm. Saddle stitched. Free.

Traditionally, the cocoyams *Xanthosoma* spp. and *Celocsia* spp. are vegetatively propagated using corms. This booklet discusses techniques for promoting flowering, producing seed using hand pollination and growing seedlings.

Frison, E.A.

Recommendations for Transporting and Handling Tissue Culture Material — Sweet Potato and Cassava

1981. 20 pages. 10 × 18 cm. Saddle stitched. Free.

This is a step-by-step guide of how to handle tissue culture material from the time it is prepared for transportation in the laboratory to the time it is old enough to be transplanted in the field in the country where tissue culture has been sent.

In English and French.

Sadik Sidki, O.O. Okereke, and S.K. Hahn

Screening for Acyanogenis in Cassava

4 pages. 15 × 22 cm. Saddle stiched. Free.

The sodium picrate test for hydrocyanic acid was used to evaluate 88,510 cassava plants for cyanogenic content.

Acyanogenic plants were not found inspite of the large number of plants tested. The method is simple, sensitive and lends itself to large-scale screening.

Moormann, F.R., R. Lal, and A.S.R. Juo

The Soils of IITA

First printed in 1975. Reprinted in 1978. 48 pages. 15 × 23 cm. Saddle stitched. Free.

Describes the soils of the International Institute of Tropical Agriculture at Ibadan, Nigeria. It gives their distribution, morphology, classification as well as their physical, chemical, and mineralogical characteristics.

Knipscheer, H.C., K.M. Menz, and F.H. Khadr (Editors)

Benchmark Surveys of Three Crops in Nigeria — Wheat, Millet and Sorghum

1980. 78 pages. 18 × 25 cm. Saddle stitched. Free.

Six benchmark surveys were conducted on cassava, maize, rice, wheat, millet and sorghum when implementing the Nigerian National Accelerated Food Production Project in 1975 and 1976. Presented here are three surveys. The study covers factors such as sampling, farmer and farm characteristics as well as problems, labor and inputs.

Atayi, E.A. and H.C. Knipscheer

Survey of Food Crop Farming Systems in the "Zapi Est" East Cameroon, IITA/ONAREST

1980. 94 pages. 18 × 25 cm. Saddle stitched. Free.

Describes crop production in ZAPI, Cameroon, under a rural development project jointly lauched by the Cameroon Government, the World Bank and the International Institute of Tropical Agriculture. The survey covers methodology, farmer characteristics, land type, cropping systems, crop yields, labor, inputs, credit, revenue, expenditure, decision making, marketing and pricing.

In English and French.

Lal, R.

Soil Erosion Problems on an Alfisol in Western Nigeria and their Control

Published in 1976. Reprinted in 1979. 208 pages. 18 × 25 cm. Wrap around . US$25.00.

Presents a detailed study of soil erosion in relation to food crop production carried out at the International Institute of Tropical Agriculture in 1970. It discusses tropics such as design and construction of runoff plots, mulching effects, nutrient loss, properties of eroded sediments and changes in soil physical characteristics.

Ayanaba, A. and P.J. Darts (Editors)

Biological Nitrogen Fixation in Farming Systems of the Tropics

1977. 377 pages. 16 × 23 cm. Hard cover. US$25.00. John Wiley & Sons.

Based on a symposium held at the International Institute of Tropical Agriculture, this book presents 30 papers from 39 contributors. It covers legumes in tropical farming systems, rhizobia, ecology, physiology, nonlegume sources of biological nitrogen in nature and measurement of nitrogen gains and losses in farming systems.

Greenland, D.J. and R. Lal (Editors)

Soil Conservation and Management in the Humid Tropics

1977. 283 pages. 16 × 24 cm. Hard cover. US$35.00. John Wiley & Sons.

This book deals with soil characteristics, climate in the humid tropics, soil erosion under existing systems and new systems needed to control it. It is based on the "Proceedings of International Conference on Soil Conservation and Management in the Humid Tropics" held at Ibadan, Nigeria, in 1975.

Lal, R.

Role of Mulching Techniques in Tropical Soil and Water Management

First printed in 1975. Reprinted in 1983. 38 pages. 15 × 23 cm. Saddle stitched. Free.

This book details the potential of mulch farming techniques as a system of maintaining favorable soil physical conditions in the humid tropics. Broadly, it covers topics such as soil physical properties, effect of mulching on soils, crop yields and mulching.

Akobundu, I. Okezie, John A. Poku, and C.W. Agyakwa

Field Guide to Common Weeds of IITA and their Control

1983. 50 pages. 14 × 22 cm. Saddle stitched. Free.

Using photographs, it gives botanical descriptions of 35 common weeds at IITA and how to control them. It also shows how to carry out herbicide dosage calculations.

Lal, R.

No Till Farming — Soil and Water Conservation and Management in the Humid Tropics

1983. 64 pages. 17 × 26 cm. Saddle stitched. Free.

Soil degradation and the consequent decline in productivity in the tropics can be controlled by adopting conservation effective farming systems such as no-till. The data presented here covers 12 years of research done at IITA on no-till farming and deforestation.

Balasubramanian, V., J.O. Braide, and E.A. Atayi

An Appraisal of the Present Farming Systems of the Atebubu District of Ghana

1982. 88 pages. 18 × 24 cm. Saddle stitched. Free.

The Managed Inputs and Delivery of Agricultural Services (MIDAS) project is sponsored by the United States Agency for International Development (USAID) to organize and distribute all the inputs necessary for food production by small farmers in Brong-Ahafo Region of Ghana. This report appraises the present farming systems of the Atebubu District and identifies the constraints to increased production under the present system.

Tuber and Root Crops Production Manual

1982. 244 pages. 22 × 28 cm. Wrap around. Free.

Written for the tuber and root crops course at the International Institute of Tropical Agriculture, this manual gives information on various aspects of growing cassava, sweet potato, yam and cocoyams in the humid and sub-humid tropics of Africa. It points out the importance of tuber and root crops in Africa. For each crop it covers the origin, botany, soils, agronomy, pathology, entomology, breeding, harvesting, and postharvesting technology.

In English and French.

Cowpea Production Training Manual

1982. 198 pages. 22 × 28 cm. Wrap around. Free.

This is a basic reference manual for participants in the cowpea training course at the International Institute of Tropical Agriculture covering the humid and sub-humid tropics. It covers history, origin, importance, distribution, botany, physiology, agronomy, pathology, entomology, breeding as well as seed production and distribution.

In English and French.

Soybean Production Training Manual

A basic reference manual for the soybean production course at the International Institute of Tropical Agriculture for the humid and sub-humid tropics. It covers nutritive quality, botany, physiology, agronomy, pathology, entomology, breeding, nematology, harvesting, and storage.

In English and French.

Maize Production Manual

1982. Vol. 1, chapt. 1-9, 218 pages; Vol. 2, chapt 10-19. 201 pages. 22 × 28 cm. Wrap around. Free.

A maize training manual for the course at the International Institute of Tropical Agriculture for the humid and sub-humid tropics of Africa. Volume 1 covers history, importance in Africa, botany, agronomy, weed biology, mechanization, and breeding. Volume 2 is on seed production, pathology, entomology, nematology, harvesting, and storage.

In English and French.

Selected Methods for Soil and Plant Analysis

1979. 70 pages. 21 × 28 cm. Saddle stitched. US$4.00.

This is a working manual for research assistants, technicians and students in the soil laboratory at the International Institute of Tropical Agriculture.

In English and French.

Automated and Semi-Automated Methods for Plant Analysis

1982. 33 pages. 21 × 28 cm. Saddle stitched. Free.

This is a supplement to "Selected Methods for Soil and Plant Analysis." It describes the standard procedures used in IITA's Analytical Service Laboratory.

Tel, Dirk A. (Editor)

Soil and Plant Analysis

1984. 278 pages. 22 × 28 cm. Three ring binding. US$20.00.

This study guide is written for the soil and plant analysis courses given at the International Institute of Tropical Agriculture for senior agricultural technologists working in the tropics. It has a large section on basic soil science focusing on tropical soils. Apart from methodology for soil and plant analysis, this book discusses water analysis, equipment required in a modern laboratory, operating and managing an analytical services laboratory as well as laboratory safety and precautions.

Conferences

Hartman, E.H., C.H.H. ter Kuile

Current and Future Trends in Tillage in the Humid and Subhumid Tropics

1983. 8 pages. Saddle stitched.

Presented at Fifth Session of the Food and Agriculture Organization (FAO) Panel of Experts on Agricultural Mechanization. Rome, Italy, 20-22 April 1983.

S.K. Hahn

Tropical Root Crops Their Improvement and Utilization

1984. 29 pages. Saddle stitched.

Based on a paper presented at a conference organized by the Commonwealth Agricultural Bureau on "Advancing Agricultural Production in Africa" held 13-17 February, 1984 at Arusha, Tanzania.

E.H. Hartman

Evolution and Changes at IITA The Way Ahead

1984. 14 pages. Saddle stitched.

Presentation to International Centers Week.

B.T. Kang, G.F. Wilson, and T.L. Lawson

Alley Cropping A Stable Alternative to Shifting Cultivation

1984. 22 pages. US$5.00.

In English.

B.T. Kang, G.F. Wilson, and T.L. Lawson

La Culture en Couloirs un Substitut d'Avenir a la Culture Itinerante

1984. 22 pages. US$5.00.

In French.

Singh, S.R., H.F. van Emden, and T. Ajibola (Editors)

Pests of Grain Legumes — Ecology and Control

1978. 454 pages. 16 × 23 cm. Hard cover. US$25.00.

Based on papers from 46 contributors at a symposium held at the International Institute of Tropical Agriculture, this book is valuable to entomologists, breeders and agronomists. It presents 40 papers covering legume work in Asia, the Far East, Australia, Africa, the United States, Latin America, the Caribbean and dicusses bionomics, assessment management and international cooperation.

Lal, R.

Soil Tillage and Crop Production

1979. 361 pages. 17 × 25 cm. Wrap around. US$15.00.

This book is based on conference papers on the "Role of soil physical properties in maintaining productivity of tropical soils." It reports research on tillage in different tropical regions.

Crop Genetic Resources in Africa

1980. 231 pages. 17 × 25 cm. Wrap around. US$5.00.

Proceedings of a workshop organized jointly by the Association for the Advancement of Agricultural Sciences in Africa (AAASA) and the International Institute of Tropical Agriculture (IITA) held in January

1978 at IITA, Ibadan, Nigeria. Presents 24 papers from 30 authors on cocoa, coffee, fibers, oil seeds, vegetables, food legumes, roots, tubers, plantains, cereals and forests. It also has section on the role of international organizations and recommendations for action to conserve crop genetic resources in Africa.

Nwanze, Kanayo F. and Klaus Leuschner (Editors)

Proceedings on the International Workshop on the Cassava Mealybug *Phenacoccus manihot* Mat-Ferr (Pseudococcidae)

First published in 1978. Reprinted in 1983. 85 pages. 21 × 27 cm. Saddle stitched. Free.

Papers given at a workshop in Zaire covering topics such as the biology, taxonomy and biological control of the cassava mealybug. Also included is a paper on the biology and ecology of the sugar cane mealybug.

Proceedings of IITA Collaborators' Meeting on Grain Legume Improvement

1975. 179 pages. 21 × 28 cm. Wrap around. Free.

Contains 42 papers from 40 authors covering plant improvement, entomology, pathology, seed quality as well as biochemistry, physiology, agronomy, and recommendations from the participants.

Expert Consultation on the Mechanization of Rice Production

1975. 280 pages. 21 × 28 cm. Wrap around. Free.

Presents 35 papers on rice mechanization, soil tillage, rice harvesting and threshing as well as mechanization and cropping systems. It gives recommendations on how rice agriculture should be mechanized.

Proceedings of IITA Collaborators' Meeting on Grain Legume

1975. 179 pages. 21 × 28 cm. Wrap around. Free.

Presents 42 papers on various aspects of research carried out by the International Institute of Tropical Agriculture and its collaborators. It covers plant improvement, entomology, pathology, seed quality, biochemistry, physiology and agronomy. It also gives recommendations for participants.

Proceedings of Physiology Program Formulation Workshop

1975. 114 pages. 21 × 28 cm. Wrap around. Free.

Presents 22 papers from 31 authors whose aim at the meeting was to try and determine the most effective ways in which plant physiology can have the greatest and most rapid impact on crop improvement programs.

Miscellaneous

IITA Journal Reprints

Since 1972.

Over the years, IITA scientists have published over 400 articles in professional and research journals and have contributed to books on topics concerning their work. For details, refer to the publication, "International Institute of Tropical Agriculture — Record of Publications, Volume 1 and 2."

Tasks for the Eighties — A long Range Plan

1981. 60 pages. 18 × 25 cm. Saddle stitched. Free.

The book charts the general direction to follow to fulfill the mandate of the International Institute of Tropical Agriculture. It analyzes policy issues and suggests answers. It also outlines agricultural problems the Institute can reasonably attempt to tackle indicating additional scientific manpower and finances required for the next 5 to 10 years.

Toward Nigeria's Success in Agricultural Development

1981. 13 pages. 14 × 21 cm. Saddle stitched. Free.

Contains the speeches of Nigeria's President, Alhaji Shehu Shagari and Dr. E.H. Hartmans, the Director General of the Institute of Tropical Agriculture on the occasion of President Shagari's visit to IITA on Wednesday, 19 November 1980.

The Challenge of the Tropics

1977. 32 pages. 18 × 23 cm. Saddle stitched. Free.

Highlights the development and objectives of the International Institute of Tropical Agriculture. It also outlines constraints farmers and scientists face, reviews IITA's contribution in the Institute's first decade and presents a brief look at what IITA might achieve in the near future.

In English and French.

Hartmans, E.H.

Land Development and Management in Tropical Africa

1981. 21 pages. 21 × 28 cm. Saddle stitched. Free.

This is a concise description of current land development and management practices in tropical Africa. It covers soil and land characteristics, land clearing and development, and postclearing management and cropping systems.

An Evaluation of Production Training Courses of the International Institute of Tropical Agriculture, Ibadan, Nigeria

1977. 72 pages. 21 × 28 cm. Wrap around. Free.

This survey is an evaluation of how relevant training courses at the International Institute of Tropical Agriculture are to course participants when they return to their home countries. The survey covers West African course participants.

The Food Crisis in Tropical Africa

1983. 24 pages. 10 × 23 cm. Saddle stitched. Free.

Gives an account of IITA's recent achievements in farming systems, agricultural engineering, cassava, sweet potato, maize, rice, cowpea, and soybean work. It points out the major problems in food production in Africa and indicates how the International Institute of Tropical Agriculture is trying to overcome them.

In English and French.

Maize and Cowpea Improvement in Semi-Arid Africa

1983. 44 pages. 18 × 23 cm. Saddle stitched. Free.

Focuses on IITA's contribution in the Semi-Arid Food Grain Research and Development project. Coordinated by the Scientific, Technical and Research Commission of the Organization of African Unity, SAFGRAD is a regional project in which 25 OAU member states participate to improve production of maize, sorghum, millet, cowpea, and groundnuts.

In English and French.

Upland Rice in Africa

1982. 43 pages. 21 × 28 cm. Saddle stitched. Free.

This is a concise description of upland rice production in Africa.

Rice for Self-Sufficiency in Sierra Leone

1980. 30 pages. 18 × 23 cm. Saddle stitched. Free.

A brief report on the cooperative rice research and development project carried out between 1974 and 1980. It involved the United Nations Development Program, the Food and Agriculture Organization of the United Nations, the Sierra Leone Government and the International Institute of Tropical Agriculture. The report describes the project, its achievements and what remains to be done to achieve self-sufficiency.

High-Yielding Varieties Technology — A Multilocational Crop Improvement Program for Rice, Maize, Cowpeas and Soybeans in Tropical Africa

Revised 1981. 65 pages. 21 × 28 cm. Saddle stitched. Free.

Describes a multilocational crop improvement program proposed by IITA for funding by the Commission of European Communities.

IITA Visitors' Guide

1983. Single sheet 23 × 49 cm folded to 10 × 23 cm. Free.

Provides practically all information a visitor to IITA would like to have from the time of planning a trip to IITA to recreation facilities such as tennis courts.

In English and French.

International Institute of Tropical Agriculture

1981. Single sheet 40 × 46 cm folded to 10 × 23 cm. Free.

A general information publication covering IITA's staffing, location, research and training programs.

Training and Postgraduate Research Opportunities at IITA

1983. 24 pages. 10 × 23 cm. Saddle stitched. Free.

This booklet describes training opportunities at International Institute of Tropical Agriculture. The Institute offers several types of degree-related research programs. The booklet describes the objectives for training, the program and facilities available.

In English and French.

Profile of IITA International Programs

1982. 16 pages. 10 × 23 cm. Saddle stitched. Free.

This publication describes IITA's programs carried out in various parts of Africa.

In English and French.

Recommendations for Lima Bean Uniform Cultivar Trial

1976. 24 pages. 21 × 28 cm. Spiral binding. Free.

Researchers at the International Institute of Tropical Agriculture identified high-yielding lines among the climbing lima beans. These were to be tested at different locations. The booklet describes standard crop management, trial layout and data collection procedures all locations were to follow to get comparable results.

Recommendations for Pigeonpea Uniform Cultivar Trial

1976. 15 pages. 21 × 28 cm. Spiral binding. Free.

Pigeonpea cultivars from elite stock identified or developed at the International Institute of Tropical Agriculture were to be tested at different locations. This booklet describes standard crop management, standard layout and data collection procedures all locations were to follow to get comparable results.

Recommendations for Cowpea Uniform Cultivar Trial

1976. 25 pages. 21 × 28 cm. Spiral binding. Free.

Cowpea cultivars from elite stock identified or developed at the International Institute of Tropical Agriculture were to be tested at different locations. The booklet describes standard crop management, trial layout and data collection procedures at the different locations that were to be followed to get comparable results.

Bambara Groundnut *Voandzeia subterranea* Thouars — Abstracts of World Literature, 1900-78

1978. 55 pages. 21 × 28 cm. Saddle stitched. Free.

This book presents 260 entries covering agronomy, physiology, growth, development, chemical composition, botany, taxonomy, geographical distribution, genetics, breeding, pest control, pathology and nutrition studies.

Sidik Sadki

Methods for Seed Germination and Seedling Establishment of Yam *Dioscorea rotundata* (Poir)

1976. 5 pages. 12 × 23 cm. Saddle stitched. Free.

Lack of well developed embryos and seed dormancy limited the capacity to improve the yield and quality of white yam, *Dioscorea rotundata* through sexual propagation. Having overcome these problems, seed of over 40,000 genotypes has been produced. This booklet describes the methods used for seed germination and seedling establishment.

Okonmah, L.U.

Rapid Multiplication of Yams

1980. 21 pages. 21 × 28 cm. Saddle stitched. Free.

This booklet describes two methods of multiplying rapidly planting sets for yam. The technique of rooted vine cuttings is useful for research, extension and other agricultural institutions whereas the Anambra State production system is suitable for both agricultural institutions and farmers.

Recommendations for the Establishment, Cultivation and Evaluation of Root Crop Test Material

1976. 18 pages. 21 × 28 cm. Saddle stitched. Free.

This booklet gives recommendations for testing root crops in different locations. It is a guide for collaborating in establishing, cultivating and evaluating test materials distributed by the International Institute of Tropical Agriculture.

Singh, S.R., B.B. Singh, L.E.N. Jackai, and B.R. Ntare

Cowpea Research at IITA

1983. 20 pages. 14 × 21 cm. Saddle stitched. Free.

Describes briefly the advances made over the past few years by the cowpea program at the International Institute of Tropical Agriculture at Ibadan, Nigeria.

Tissue Culture-Improved Cultivars for Africa's National Programs

1982. Poster. 29 × 43 cm. Free.

The International Institute of Tropical Agriculture produces tissue culture material free from pests and diseases for distribution to various African countries. This publication describes the Institute's culture program.

In English and French.

Leatherdale, Donald

Thesaurus on Tropical Grain and Forage Legume

1977. 352 pages. 21 × 28 cm. Wrap around. Free.

Suitable for information retrieval purposes, this thesaurus covers terms found in tropical grain legume research and covers terms found in tropical legume and development literature.

Bibliographies

Winged beans — Abstracts of World Literature, 1900-1977

1978. 68 pages. 21 × 28 cm. Wrap around. Free.

Presents 293 abstracts of literature on the winged bean and others of the genus Psophocarpus. Most entries are published literature, but some unpublished writeups such as essays and research notes are included.

Lawani, S.M. and M.O. Odubanjo

A Bibliography of Yams and the Genus *Dioscorea*

1976. 192 pages. 21 × 28 cm. Wrap around. Free.

This is a bibliography of yams from the early 1800s to 1975.

Lawani, S.M. and S.B. Akande

International Institute of Tropical Agriculture — Record of Publications

Vol. 1, 1969-1980. 89 pages; Vol. 2, 1981-1982. 54 pages. 1981 for Vol. 1 and 1983 for Vol. 2. Both 21 × 28 cm. Saddle stitched. Free.

These two volumes list all publications of the International Institute of Tropical Agriculture from the time the Institute was established until 1982. Unpublished conference and seminar papers are not included. Includes more than 1100 entries.

Ibekwe, G.O.

Sweet Potato *Ipomoea batatas* (L) — Abstracts of Selected Research and Development Literature, 1949-1979

1979. 260 pages. 21 × 28 cm. Wrap around. Free.

Presents 551 abstracts on different aspects of sweet potato production considered relevant to research and developing the crop. About 75% of the papers listed in the publication are available in the library of the International Institute of Tropical Agriculture. On request, individuals can get a limited number of photocopies of these papers.

Lawani, S.M., F.M. Alluri, and E.N. Adimorah

Farming Systems in Africa — A Working Bibliography, 1930-1978

1979. 251 pages. 16 × 24 cm. Hard cover. US$20.00. G.K. Hall & Co.

This bibliography contains 1,989 entries in 29 subject categories. It covers studies of existing farming systems, the rationale of these systems as well as their environmental, economic, social and technical constraints.

Cowpeas *Vigna unguiculata* L. Walp — Abstracts of World Literature

Vol. 1, 1950-1973; Vol. 2, 1900-1949; Vol. 3, 1974-1980; Vol. 4, 1981-1982. Vol. 1, 1977, 343 pages; Vol. 2, 1979, 194 pages; Vol. 3, 1981, 435 pages; Vol. 4, 1984, 147 pages.

Most of the publications listed are available on request from the International Grain Legume Information Centre of the International Institute of Tropical Agriculture, Ibadan, Nigeria.

Ibekwe, G.O.

Mealybugs — Abstracts of Selected Literature

1977. 148 pages. 21 × 28 cm. Wrap around. Free.

Presents over 500 abstracts covering taxonomy, morphology, physiology, anatomy, histology, reproduction including development genetics, biology, ecology, biological and chemical control.

Lawani, S.M. and T.A.B. Seriki (Compilers)

Serials for Rice Research

1972. 34 pages. 21 × 28 cm. Stapled. Free.

This compilation is primarily for librarians and documentalists. It lists periodicals and journals which regularly carry articles on rice. The serials include journals, reports, proceedings, transactions, bulletins and memoirs published between 1966 and 1970.

Ibekwe, G.O.

Cassava Bacterial Blight — Abstracts of Literature

1978. 71 pages. 21 × 28 cm. Wrap around. Free.

Presents 187 entries grouped in general section, etiology, symptomatology, epidemiology, yield losses, cultural control and breeding for resistance. Most of the papers in this compilation are available on request from the library of the International Institute of Tropical Agriculture, Ibadan, Nigeria.

Ibekwe, G.O.

Cassava Bacterial Blight — Abstracts of Literature (Supplement)

1978. 30 pages. 21 × 28 cm. Wrap around. Free.

Presents 50 entries grouped into general section, etiology, symptomatology, epidemiology, yield losses, cultural control, chemical control, breeding and integrated control. Most of the papers in this compilation are available on request from the Library of the International Institute of Tropical Agriculture, Ibadan, Nigeria.

Periodicals

IITA Research Briefs

Four issues a year since 1980. Pages vary. 21 × 28 cm. Saddle stitched. Free.

In this series, individual scientists of the International Institute of Tropical Agriculture describe various aspects of ongoing research projects. These briefs also carry news about major events in the Institute.

Paradisiaca

Annually since 1976. Pages vary. 21 × 28 cm. Saddle stitched. Free.

This is a newsletter of the International Association for Research on Plantain and Other Cooking Bananas.

In English and French.

Tropical Grain Legume Bulletin

Published quarterly since 1975. Pages vary. 21 × 28 cm. Saddle stitched. Free.

This series is a forum for research and extension workers for exchanging information on tropical legumes particularly on cowpeas and soybeans. Each issue covers a variety of articles by different authors around the world.

Asnani, V.L., J.A. Otoo, and S.K. Hahn

Improved Seed Yam Production Technology

1985. 9 pages. Saddle stitched. Free.

IITA is continuing its research to make further improvements in seed yam production technology and develop new methods for rapid multiplication and improved storage. The research described in the following paragraphs is showing much promise.

Biological Control of Cassava Mealybug and Green Spider Mites

25 pages. Information Series No. 16.

In English.

Lutte Biologique la Cochenille et les acariens verts du Manioc en Afrique

25 pages. Serie Bulletin d'information no. 16.

In French.

ILCA

International Livestock Centre for Africa

Address: P. O. Box 5689 • Addis Ababa
Cable Address: ILCAF ADDIS ABABA
Telex Address: ILCA 21207 Telephone: 18-32-15

ILCA

General Information

Mandate: To assist national efforts which aim to effect a change in production and marketing systems in tropical Africa so as to increase the sustained yield and output of livestock products and improve the quality of the life of the people of this region.

Centre Objectives

ILCA is an international centre of livestock research, training and documentation. Through research, it seeks to increase the production and sale of livestock and livestock products. This research effort is supported by an extensive documentation service and by the provision of training programmes for national research workers. ILCA's goal of increasing livestock production originates in the premise that the output of the arable and pastoral land of Africa is the major determinant of economic growth, and that inadequate increase in livestock and crop output in most African countries in recent years are the key reason for the unsatisfactory rate of economic growth in these countries. The role of livestock in stimulating economic growth derives from their capacity to generate rapid increases in the cash income of subsistence farmers, enabling these farmers to purchase the agricultural inputs needed to generate large increase in food grain production. In the traditional agricultural systems of Africa, there is a strong complementarity between livestock output and crop production, with increase in the former quickly leading to an upturn in the latter. It is this relationship which ILCA specifically seeks to exploit.

Annual Reports

CIPEA

Rapport Annuel 1982 Une Année au Service de l'élevage en Afrique

1984. viii, 70 pages. 28 × 21 cm. Paperback.

This report summarizes the work of ILCA during the 1982 calendar year. The foreword gives the background to ILCA's research approach and the main body of the report highlights the work of ILCA's field research programmes in Ethiopia, Nigeria, Mali, Kenya, and Botswana. Further sections highlight ILCA's work on livestock policy, trypanotolerance, forage legume agronomy, nutrition, aerial surveys, computer services, library and documentation, training and publications. A centre spread focuses on the state of the livestock sector in countries of sub-Saharan Africa. The report includes a full staff list, details of ILCA's Board of Trustees, Donors and publications.

Available in English and French.

ILCA

ILCA Annual Report

Annually since 1981; latest 1984. Varies (approx. 80 pages). 28 × 21 cm. Saddle stitched. ISSN 0255-0040 (E) and 0255-3473 (F).

Highlights the work of ILCA's field research programmes in Ethiopia, Nigeria, Mali, Kenya, and Botswana as well as work on livestock policy, trypanotolerance, forage legume agronomy, small ruminants and camels, nutrition, aerial survey, computer services, library and documentation, training and publications.

Available in English and French.

Monographs

No. 1
ILCA and Institut d'économie rurale du Mali

Evaluation of the Productivities of Maure and Peul Cattle Breeds at the Sahelian Station, Niono, Mali

1978. iv, 109 pages. 29.5 × 21 cm. Paperback. ISBN 92-9053-003-0.

This report provides information on the productivity of two African cattle breeds and some of their crosses used at the Station d'Elevage et de Recherches Zootechniques du Sahel in Mali. Among the traits which are analysed here are reproductivity, milk yield and composition, mortality, body weight and various body measurements. The results are compared to those from production systems found in comparable environments elsewhere in Africa.

Available in English and French.

No. 2

J. C. M. Trail, C. H. Hoste and Y. J. Wissocq;
Y. J. Wissocq, P. H. Hoste, and I. L. Mason

Le Bétail Trypanotolérant d'Afrique Occidentale et Centrale Tome 1: Situation Générale, Tome 2: Situations Nationales

1979. x, 155 et vi, 311 pages. 29.5 × 21 cm. Encollé. Disponible en francais. ISBN 92-9053-009-X and 92-9053-010-3.

Cette monographie présente l'examen fait de la classification, de la distribution, des paramétres de production et de la productivité des bovins, ovins et caprins trypanotolérants étudiés dans 18 pays d'Afrique occidentale et centrale. Des recommendations et quelques situations intéressantes sont présentées pour évaluer le potentiel et l'utilisation future des races trypanotolérantes et pour assurer la conservation de celles en danger d'extinction.

No. 3
J. C. M. Trail and K. E. Gregory

Sahiwal Cattle: An Evaluation of their Potential Contribution to Milk and Beef Production in Africa

1981. vii, 128 pages. 29.5 × 21 cm. Paperback. ISBN 92-9053-011-1.

In this Monograph the potential for both milk and meat production of the Sehiwal cattle breed and its crosses with Bos taurus and indigenous Bos indicus breeds in various African environment has been evaluated. The data was derived from five ranches in Kenya where cattle is reared under different environmental conditions and with different production objectives.

Available in English and French.

No. 4
J. J. R. Grimsdell and S. Westley (Editors)

Low Level Aerial Survey Techniques

1981. vi, 243 pages. 29.5 × 21 cm. Paperback. ISBN 92-9053-012-X.

Based on workshop papers and discussion summaries on the development of low-level aerial survey techniques and their present application to livestock, wildlife and land-use evaluation, this Monograph reviews survey designs and sampling procedures as well as problems of bias, information transfer and coordination with information collected at other survey levels. It includes recommendations for further research and cooperation in this field.

Available in English.

No. 5
E. Mukasa - Mugerwa

The Camel (*Camelus dromedarius*): A Bibliographical Review

1981. vii, 147 pages. 29.5 × 21 cm. Paperback. ISBN 92-9053-013-8.

Written with the aim of assessing the present performance and productive potential of the dromedary, this last Monograph in ILCA's series describes the origins and distribution of camel breeds, their reproductive performance and nutritional requirements as well as common camel diseases. The camel's suitability for milk and meat production and transport, and the use of camel hair and hides are also discussed alongside with management and socioeconomic factors.

Available in English.

No. 6
M. Murray, J. C. M. Trail, D. A. Turnera, and Y. Wissocq

Livestock Productivity and Trypanotolerance, Network Training Manual

1983. 198 pages. 29.5 × 21 cm. Paperback. ISBN 92-9053-024-4.

This training manual describes the parameters and techniques which are used in the collection of data in the area of animal health, tsetse, and animal productivity, and indicates how relevent information is extracted, analysed and interpreted.

Available in English and French.

Conference, Workshop, and Symposia Proceedings

ILCA

Evaluation and Mapping of Tropical African Rangelands

1981. xvi, 399 pages. 30 × 22 cm. Paperback. ISBN 92-9053-000-6.

This report summarizes ideas and suggestions relating to the evaluation of sub-Saharan rangeland resources, which was the main theme of the ILCA sponsored seminar held in Bamako, Mali in 1975. It presents a review of rangeland survey and evaluation techniques, discusses site development methods, sampling and data processing, transcription of data from aerial photographs to base maps, and finally, recommends guidelines for rangeland survey and evaluation in the future.

Available in English and French.

S. B. Westley (Editor)

East African Pastoralism: Anthropological Perspectives and Development Needs

1982. vi, 207 pages. 29 × 21 cm. Paperback. ISBN 92-9053-022-7.

This selection of 11 conference papers addresses topics ranging from ethnographic information on specific forms of pastoralism in eastern Africa, methodological problems in the collection and analysis of anthropological information, theoretical issues in the study of society, to the various ethical and policy problems facing development planners and administrators. The need to maintain an ecological balance in pastoral production systems under development is also discussed.

Available in English.

R. M. Gatenby and J. C. M. Trail (Editors)

Small Ruminant Breed Productivity in Africa

December 1982. 96 pages. 29.5 × 21 cm. Paperback. ISBN 92-9053-023-5.

This is a collection of 11 papers presented at the Seminar on Small Ruminant Breed Productivity which was held at ILCA, Ethiopia in 1982. The topics covered include breeding and production objectives, current research approaches in sub-Saharan Africa and other tropical areas, methods of collection and analysis of small ruminant production data, and a review of current research findings in Africa.

Available in English.

H. N. Le Houerou (Editor)

Browse in Africa: The Current State of Knowledge

1983. 491 pages. 30 × 21 cm. Paperback. ISBN 92-9053-025-1.

These proceedings present a unique review of the current knowledge of browse species and their role in the development of livestock production. The papers included in this volume deal with the ecology of browse trees and shrubs, their geographical distribution, relative palatability for different animals and their composition and theoretical value. The primary and secondary production of browse, its intensification through technically possible, economically viable and socially acceptable operations, and the need for the regeneration of browse populations through appropriate management methods are also discussed, as well as future research priorities to eliminate the various gaps in the knowledge on the subject.

Available in English and French.

Pastoral Systems Research in Sub-Saharan Africa. Proceedings of the Workshop Held at ILCA, Addis Ababa, Ethiopia, 21-24 March 1983

1983. vii, 480 pages. 28.5 × 20 cm. Paperback. ISBN 92-9053-038-3.

This document summarizes the proceedings of the workshop on Pastoral Systems Research in sub-Saharan Africa held at ILCA's headquarters in Addis Ababa from 21 to 24 March 1983. The workshop focused on the techniques and processes in pastoral systems research with particular reference to ILCA's experience in sub-Saharan Africa.

Presentations appear in their source language. Summaries of discussion sessions presented in both English and French.

D. H. Hill

Peste des Petits Ruminants (PPR) in Sheep and Goats Proceedings of the International Workshop held at IITA, Ibadan, Nigeria, 24-26 September 1980

1983. v, 104 pages. 29.5 × 21 cm. Paperback. ISBN 92-9053-040-5.

These proceedings are a summary of the current state of knowledge on the epidemiology, symptomatology, pathology and diagnosis of *peste des petits ruminants* (PPR) in West Africa. The physicochemical and biological characters of the PPR virus and the role of other agents responsible for the pathogenesis of PPR are also discussed and clear recommendations are given as to the further direction of policy and research on the problem.

Available in English.

J. E. Sumberg and K. Cassady (Editors)

Sheep and Goats in Humid West Africa

1985. vi, 74 pages. 29.5 × 21 cm. Paperback. ISBN 92-9053-058-8.

These proceedings summarize the results of research on dwarf sheep and goat production in humid West Africa. Among the topics discussed are disease profiles of village sheep and goats; feed intake, growth and reproduction rates of West African Dwarf goats; integration of the small ruminant enterprise into the wider farming system of the zone; and the economics of the improved production systems of ruminants.

Available in English.

Miscellaneous, Reports, and Brochures

ILCA Systems Studies

No. 1
F. M. Anderson et J. C. M. Trail

Modélisation Mathématique des Systèmes de Production Animale: Application au Botswana du Modèle de Production Bovine de la Texas A & M University

1982. iii, 57 pages. 29.5 × 21 cm. Encolle. ISBN 92-9053-026-X.

Ce rapport présente l'application du modèle de production bovine mis au point à Texas A & M University, aux systèmes de production traditionnel et de ranching au Botswana. On y identifie également les problèmes pour lesquels les techniques de l'application de modèlisation pourraient faciliter les activités de recherche et de développement sur la production animale in Botswana.

Disponible en Francais.

C. de Haan, S. B. Westley et S. Chater

Systems Studies No. 3: L'élevage des Petits Ruminants dans les Régions Tropicales Humides

1983. v, 69 pages. 29.5 × 20.5. Encollé. ISBN 92-9053-035-9.

Ce document rend compte des recherches effectuées sur la production des petits ruminants dans la zone humide de l'Afrique de l'Ouest. Il comprend une analyse portant sur les paramètres de production des races naines africaines, un inventaire des stratégies alimentaires et une évaluation de la fréquence de certaines maladies. Por ailleurs, l'un des volets de l'étude a été consacré à l'examen de la production des petits exploitations et des exploitations à grande échelle.

No. 2
ILCA

La Production Animale dans la Zone Subhumide de l'Afrique de l'Ouest: Une Etude Régionale

1984. iii, 101 pages. 29.5 × 20.5 cm. Encollé. ISBN 92-9053-050-2.

Après une brève description de l'environnement et des éleveurs Peuls auxquels appartient la majeure partie du bétail de la zone, cette étude fait l'inventaire des ressources fourragères et animales, se penche ensuite sur la trypanosomiase et autres zoonoses. Enfin, elle analyse les stratégies d'utilisation et de mise en valeur des ressources territoriales, en puisant largement dans l'experience du Nigéria en matière de production de ruminants.

Disponible en francais.

ILCA Research Reports

No. 2
P. A. Konandreas and F. M. Anderson

Cattle Herd Dynamics: An Integer and Stochastic Model for Evaluating Production Alternatives

June 1983. vii, 95 pages. 25.5 × 18 cm. Paperback. ISBN 92-9053-016-2.

This report presents the rational and formal specifications of a simulation model used for evaluating cattle herd production dynamics. In the model a herd is simultaneously represented as both a biological and an economic unit. The analytical background is presented and the model is explained in detail, with the simulation of feed intake, growth and milk production, reproduction and mortality being discussed. The policy options in herd management are also discussed.

Available in English and French.

No. 3
A. Fall, M. Diop, S. Sandford, Y. J. Wissocq, J. Durkin, and J. C. M. Trail

Evaluation of the Productivities of Djallonke Sheep and N'Dama Cattle at the Centre de Recherches Zootechniques, Kolda, Senegal

September 1982. vi, 70 pages. 25.5 × 17 cm. Paperback. ISBN 92-9053-020-0.

This publication reports on the analysis of reproductive performance, viability and growth performance traits of Djallonke sheep and N'Dama cattle at the Centre de Recherches Zootechniques, Kolda in Senegal. These traits are used to build up overall indices for the two species. The report presents an analysis of carefully recorded productivity data collected up to 1981.

Available in English and French.

No. 4
G. Gryseel and F. M. Anderson

ILCA Research Report 4 — Research on Farm and Livestock Productivity in the Central Ethiopian Highlands: Initial Results 1970-1980

December 1983. vii, 52 pages. 25.5 × 17.5 cm. Paperback. ISBN 92-9053-044-8.

This report of ILCA's studies on the traditional smallholder production systems of the highlands, productivity aspects of the system's crop and livestock components are analyzed, results of research on innovations are given and the approach to Farming Systems Research with special reference to livestock is discussed together with the implications for adoption of the approach and its results by a national agency.

Available in English.

No. 5
R. T. Wilson, P. N. de Leeuw et C. de Haan (Editors)

Recherches sur les Systèmes des Zones Arides du Mali: Résultats Préliminaires

Mai 1983. vii, 189 pages. 25.5 × 17.5 cm. Encollé. ISBN 92-9053-031-6.

Ce rapport décrit les systèmes extensifs traditionnels de production agropastorale et pastorale dans le delta intérieur du Niger, ainsi que les constraintes à l'accroissement de leur productivité. Les ressources fourragerès, la production animale, la production agricole, l'economie villageoise et l'organisation socio-territoriale sont analysées et des conclusions générales sur l'orientation des recherches futures sont esquissées.

Available in French.

No. 6
G. A. Classen, K. A. Edwards, and E. H. J. Schroten

The Water Resource in Tropical Africa and Its Exploitation

1984. vii, 104 pages. 25.5 × 17.5 cm. Paperback. ISBN 92-9053-043-X.

This report deals with the mechanics of the hydrological cycle, the origins of the wide variations in rainfall, the potential for water resources development in pastoral areas and low-cost methods of exploiting these resources. Problems of water quality are also covered in outline. The final chapter makes recommendations.

Available in English.

No. 7
J. M. King

Livestock Water Needs in Pastoral Africa in Relation to Climate and Forage

1982. ix, 95 pages. 25.5 × 17 cm. Paperback. ISBN 92-9053-037-5.

In this report a detailed discussion of the physiology of water and energy use in the herbivore is given and a diagram of the interrelation between the two has been constructed. The information obtained on water and energy use by livestock in the traditional pastoral system has been applied to water and livestock development in the context of range, woodland and livestock management and implications for research are presented notably for hypothesis and modelling, as well as component research for which priority topics and species are given.

Available in English.

No. 8
S. Sandford
Organization and Management of Water Supplies in Tropical Africa
1983. vi, 45 pages. 25.5 × 17.5 cm. Paperback. ISBN 92-9053-042-1.

In this report traditional and modern strategies used to overcome water shortages are discussed. The technical, administrative and environmental problems experienced in the past development of water supplies are outlined as is the relationship between technology, equity, management and control. The implications of past experience for planning water development in the future are considered. Proposals are made for future research.

Available in English.

No. 9
J. C. M. Trail, K. Sones, J. M. C. Jibbo, J. Durkin, D. E. Light and Max Murrary
Productivity of Boran Cattle Maintained by Chemoprophylaxis under Trypanosomiasis Risk
1985. 76 pages. 25.5 × 17.5 cm. Paperback. ISBN 92-9053-056-1.

Detailed analysis of the performance of Boran cattle has shown that prophylactic drugs can be used successfully in an area of high tse-tse challenge. Twenty thousand calving records collected over a 10-year period at Mkwaya Ranch in Tanzania were analyzed. The results of the analysis are published in this report.

Available in English.

No. 10
P. A. Konandreas, F. M. Anderson and J. C. M. Trail
Economic Trade-Offs Between Milk and Meat Production Under Various Supplementation Levels in Botswana.
1983. v, 52 pages. 25.5 × 17 cm. Paperback. ISBN 92-9053-034-0.

A dynamic and stochastic cattle simulation model is briefly described and validated for production conditions in a study area in Botswana using Tswana and Simmental × Tswana cattle. The performances of the two genotypes under various milking and supplementation policies are compared, and the economic trade-offs between milk and meat production presented. An optimum production strategy is defined, and policy options for the development of Botswana's dairying sector are outlined.

Available in English.

No. 11

G. H. Kiwuwa, J. C. M. Trail, M. Y. Kurtu, G. Worku, F. M. Anderson, and J. Durkin

Crossbred Dairy Cattle Productivity in Arsi Region, Ethiopia

1983. vi, 29 pages. 25.5 × 17 cm. Paperback. ISBN 92-9053-036-7.

This publication reports on analyses carried out on a range of performance traits and productivity estimates for indigenous Arsi and Zebu cattle and eight different grades of these crossed with Jersey and Friesian, maintained for milk production. Data covered the period 1968 to 1981, and the animals were kept at Asela station and on surrounding smallholder farms in the Arsi Region of Ethiopia.

Available in English.

ILCA Brochures

ILCA

ILCA

1981. 24 pages. 16 × 24 cm. Brochure.

This brochure gives a broad overview of the work of ILCA. It explains the establishment and evolution of the Centre, the focus of ILCA's research work and where it is conducted, and summarizes the early results.

Available in English and French.

ILCA

ILCA: The First Years

1980. 127 pages. 29.5 × 21 cm. Paperback.

This brochure briefly describes the origins and mandate of ILCA and highlights the Centre's achievements in the first years of its existence.

Available in English and French.

ILCA

Studies of Range Livestock Production Systems Under Induced Changes

1980. 16 pages. 30 × 21 cm. Brochure.

This booklet describes the impact of ILCA's research on national development schemes. It focuses on the design and application of monitoring techniques for the study of the traditional livestock production systems, and on their response to development inputs.

Available in English and French.

Bibliographies, Directories and Abstracts

Microfiche indexes

This series of publications lists non-conventional literature collected from African countries. A bibliographic index and a subject index constitute the main sections of the document. Author and institution or collective author indexes are also given. The information provided is unavailable elsewhere.

Catalogue des Documents Microfilmés par l'équipe CIPEA/CRDI; Mission du Burundi

Février 1982. vii, 20 pages. 28 × 21 cm. Agrafé. ISBN 92-9053-014-6.

Catalogue des Documents Microfilmés par l'equipe CIPEA/CRDI; Mission du Cameroun

Février 1982. vii, 23 pages. 28 × 21 cm. Agrafé. ISBN 92-9053-015-4.

Catalogue of Documents Microfiched by ILCA/IDRC Team: Tanzania Mission

June 1982. vii, 19 pages. 28 × 21 cm. Brochure. ISBN 92-9053-017-0.

Index to Livestock Literature Microfiched in Zambia 1981

September 1982. ix, 22 pages. 24 × 17 cm. Brochure. ISBN 92-9053-024-3.

Index des Documents Microfichés au Zaire

Décembre 1982. ix, 24 pages. 24 × 17 cm. Agraféz. ISBN 92-9053-045-6.

Index to Livestock Literature Microfiched in Zimbabwe (Part 2)

March 1983. ix, 54 pages. 24 × 17 cm. Brochure. ISBN 92-9053-028-6.

Index to Livestock Literature Microfiched in Ghana 1981

1982. ix, 43 pages. 24 × 17 cm. Brochure. ISBN 92-9053-033-2.

Index des Documents Microfichés au Sénégal

1983. ix, 39 pages. 24 × 17 cm. Brochure. ISBN 92-9053-032-4.

Index des Documents Microfichés par l'équipe CIPEA/CRDI en Haute Volta

1983. xi, 26 pages. 24 × 17 cm. Brochure. ISBN 92-9053-051-0.

Index des Documents Microfichés par l'équipe CIPEA/CRDI en Côte d'Ivoire

1984. xi, 32 pages. 24 × 17 cm. Brochure. ISBN 92-9053-049-9.

Index des Documents Microfichés par l'équipe CIPEA/CRDI au Rwanda

1984. xi, 34 pages. 24 × 17 cm. Brochure. ISBN 92-9053-053-7.

Index to Livestock Literature Microfiched by the ILCA/IDRC team in Malawi

1984. xii, 72 pages. 24 × 17 cm. Brochure. ISBN 92-9053-055-3.

Compiled by Michael R. Goe and Michael Hailu

Animal Traction: A Selected Bibliography

February 1983. vi, 42 pages. 24 × 17 cm. Paperback. ISBN 92-9053-027-8.

This bibliography is a comprehensive listing of documents relating to the research or application of animal traction throughout the world. The bibliography is divided into 13 subject sections, and includes both species and author indexes. It constitutes essential reference material for anyone working on animal traction.

ILCA Documentation

Index to Livestock Literature Microfiched in Zimbabwe (Part 2)

March 1983. ix, 54 pages. 24 × 17 cm. Brochure. ISBN 92-9053-028-6.

This index lists the titles and references of the documents microfiched by ILCA's documentation team who visited Zimbabwe in 1981.

ILCA

Index to Livestock Literature Microfiched in Ghana 1981

1982. ix, 43 pages. 24 × 17 cm. Brochure. ISBN 92-9053-033-2.

This index lists the titles and references of the documents microfiched by ILCA's documentation team who visited Ghana in 1981.

ILCA

Index des Documents Microfichés au Sénégal

1983. ix, 39 pages. 24 × 17 cm. Brochure. ISBN 92-9053-032-4.

This index lists the titles and references of the documents microfiched by ILCA's documentation team who visited Senegal.

Periodicals

ILCA

ILCA Bulletin

Quarterly. 30-36 pages. 28 × 21 cm. Brochure. ISBN 0255-0008.

This publication presents articles summarized from research papers written by ILCA staff members and provides an up-to-date account of aspects of the centre's work.

Available in English and French.

ILCA

ILCA Newsletter

Quarterly. 8 pages. 28 × 21 cm. Brochure. ISBN 0255-0024.

The ILCA Newsletter reports on ILCA and national activities in the context of African livestock research and development. It reaches an audience of 4000, 70% of which are in Africa.

Available in English and French.

ILRAD

International Laboratory for Research on Animal Diseases

P.O. Box 30709, Nairobi, Kenya
Cable: ILRAD Nairobi Kenya
Telex: 22040 ILRAD
Telephone: Nairobi 592311

General Information

Objectives:

The International Laboratory for Research on Animal Diseases (ILRAD) was established in 1973. The laboratory complex was constructed on a 70-hectare site donated by the Kenya Government at Kabete, on the outskirts of Nairobi, Kenya.

ILRAD's mandate is to conduct basic research leading towards the development of safe, effective and economically feasible measures to control livestock diseases which seriously limit world food production. Emphasis is on disease control by immunological means. However, this does not preclude research on other promising disease control measures, such as the strategic use of chemotherapy, genetic improvements and vector control.

ILRAD also carries out training designed to support the development of scientific and field personnel who can extend research on pressing animal health problems — primarily in Africa, but also in other parts of the world. Scientists and technicians are trained to carry out the most effective disease control programs possible with the means currently available. At the same time, they are prepared to initiate new field programs when improved control methods are ready for introduction.

Major research programs:

Two important livestock diseases caused by protozoan parasites were chosen as the first targets for research at ILRAD: East Coast fever, which is a virulent form of theileriosis, and African animal trypano-somiasis.

Annual Reports

Published for 1976, 1977, 1979, 1980, 1981, 1982, 1983. Approx. 60 pages. From 1981 24.5 × 17.5 cm. Saddle stitched. From 1981 English and French. Free.

Conferences, Workshop, and Symposia Proceedings

J. B. Henson and Marilyn Campbell (Editors)

Theileriosis: Report of Workshop

1977. 112 pages. 24.5 × 17 cm. Perfect. ISBN 0-88936-124-X. Publisher: International Development Research Centre, Box 8500, Ottawa, Canada K1G 3H9.

ILRAD has a few copies which we distribute free to LDC users. For bulk orders by surface mail, contact publisher.

Report of a workshop on theileriosis, a tick-borne/parasitic disease/ animal disease/occurring in Africa and the Middle East — examines the effects of the disease on cattle production, various means of effective disease control, incidence in other species of bovidae; discusses/research/activities.

Miscellaneous

ILRAD

Trypanosomiasis Research at ILRAD

1982. 10 pages. 21.5 × 10 cm. Concertina. English and French. Free.

This brochure, written in a non-technical style, describes African animal trypanosomiasis, its importance and present control measures. It gives general information about ILRAD and summarizes ILRAD's trypanosomiasis research program.

ILRAD

Theileriosis Research at ILRAD

1982. 10 pages. 21.5 × 10 cm. Concertina. English and French. Free.

This brochure, written in a non-technical style, describes theileriosis, particularly the virulent East African form, East Coast fever, and explains its importance and present control measures. It gives general information about ILRAD and summarizes ILRAD's East Coast fever research program.

ILRAD

Training Opportunities at ILRAD

1982. 10 pages. 21.5 × 10 cm. Concertina. English and French. Free.

This brochure gives a brief description of ILRAD's institutional history and research on trypanosomiasis and theileriosis. The training program is described in detail, with information for potential participants at the technical, postgraduate and post-doctoral level.

ILRAD

Scientific Publications from ILRAD

1983. 23 pages. 21.5 × 10 cm. Saddle stitched. Free.

This brochure gives a brief description of ILRAD's research and training programs. Scientific articles and other publications by ILRAD staff members are listed for the period 1975-1981.

Periodicals

ILRAD Reports

1983-. 6 pages. 29 × 21 cm. Free.

This quarterly newsletter, published in English and French, was launched in July 1983. It features descriptions of ILRAD's research and training activities. Courses, conferences, and other events at ILRAD are also announced and recent articles published by ILRAD scientists are listed.

IRRI

International Rice Research Institute

P.O. Box 933, Manila, Philippines
Cable: RICEFOUND, MANILA
Telex: 45365 RICE PM via ITT; 22456 IRI PH via RCA; 63786 RICE PN
via EASTERN

General Information

Rice is life itself to almost a third of the world's 4.5 billion people. The average annual income of rice consumers in the developing nations is less than US$300. Rice is a secondary staple for another 450 million people.

Rising populations and the demand for more food put increasing burdens on the world's small-scale rice farmers.

The International Rice Research Institute (IRRI) has as its objective the improvement of the quality and quantity of rice. Established in 1960 by the Ford and Rockefeller Foundations, this nonprofit organization is adjacent to the University of the Philippines at Los Baños, 65 km southeast of Manila.

Today, IRRI is funded through the Consultative Group for International Agricultural Research, a group of donor agencies dedicated to the improvement of agriculture in developing nations. IRRI scientists cooperate with scientists across the world to develop improved rice varieties and technology.

Since its inception, IRRI has stressed the communication of rice research and production technolgy through books, conference proceedings, periodicals, and other instructional materials.

Ordering instructions

Because the primary purpose of IRRI publications is to increase the flow of rice information in developing nations, *IRRI provides materials at a 60% discount in developing nations.*

Book dealers in both highly developed countries (HDC) and in less developed countries (LDC) receive a 20% discount.

Prices and airmail and surface mail costs are included for HDC and LDC. Please add appropriate postage fee to the price of each book.

CUSTOMERS IN THE PHILIPPINES MAY PAY WITH PHILIPPINE CURRENCY AT THE OFFICIAL U.S. DOLLAR-PESO EXCHANGE RATE AT THE TIME OF PURCHASE. PESO POSTAGE INFORMATION CAN BE FOUND ON APPENDIX A.

Surface mail orders can only be sent in 5-kg packages. We can only process surface mail orders that total 20 kg or less.

We encourage customers to order publications via *air mail* because:

1. About 30% or more of our publications dispatched via surface mail never reach their destinations, and
2. Surface mail delivery to most destinations takes 5-6 months or longer.

For bulky orders, IRRI strongly advises customers to order via *air freight.* This refers to any order of 20 kg or more (the equivalent of 8 copies of the *IRRI Annual Report,* about 475 pages each; or 32 of *Rice Improvement,* 186 pages). For air freight shipments we *only charge the customer actual air freight charges.* We are not able to process *sea freight* orders.

Prepayment is required

Payment can be made by check in Philippine pesos or US dollars *drawn on any US banks.* **IRRI cannot process checks for US dollars drawn on bank accounts outside of the USA, unless the funds are drawn through U.S. banks and this is indicated on the check.** Checks of selected currencies drawn on their respective country banks such as French francs, pound sterling, yen, HK dollars, Swiss francs, Deutchmarks, Australian dollars, Singapore dollars, are also accepted. *Checks on other currencies cannot be negotiated in the Philippines.* Checks drawn on dollar accounts in US banks by a correspondent foreign bank in special check arrangement are also acceptable. Payment may also be made in US$ cash, UNESCO coupons or traveler's checks. A list of distributing bodies of UNESCO coupons in Asia, Africa, and Latin America is found on pages 265-267.

Please fill the Order Form (Appendix B) and send with prepayment to:

> Communication and Publications Department
> IRRI, P.O. Box 933
> Manila, Philippines

Make checks payable to IRRI-CPD.

Basic Sets

All major IRRI publications are available as *Basic Sets.*

Libraries or individuals in developing nations can purchase a *Basic Set* of all IRRI books and technical publications in stock (except bibliographies) for US$400.00, excluding air freight. In highly developed countries the price is US$900.00, plus air freight. A *Basic Set* is about 55 kg of publications.

For a proforma invoice that includes exact air freight charges write to: Division F, Communication and Publications Department, IRRI, P.O. Box 933, Manila, Philippines.

Distributors

Overseas:
Customers in **North America** can also order IRRI books from: Agribookstore, IADS, Inc., 1611 North Kent St., Arlington, Virginia 22209, U.S.A.

Customers in **Europe** may order from Verlag Josef Margraf, Oberwiesenstrase 32, 7000 Stuttgart 75, F. R. Germany.

Customers in **Japan** may order from Publishers International Corporation, 2nd Newfield Building, 42-3 Ohtsuka 3-Chome Brunk-ku, Tokyo 112, Japan.

Customers in **Taiwan, China** may order from Harvest Farm Magazine, 14 Wenchau Street, Taipei, Taiwan, Republic of China.

Prices are similar but postal rates are lower and delivery is faster.

Philippines:
Customers in Metro Manila can also purchase IRRI publications at:

- Philippine Education Company (PECO)
  768 Aurora Boulevard
  Cubao, Quezon City

- IRRI Makati Office
  605 Doña Narcisa Building
  Ayala Avenue, Makati
  Tel: 88-48-69 or 88-15-14

- National Media Production Center
  Bohol Avenue, Quezon City
  Tel: 98-66-49

- National Bookstore
  In front of Araneta Coliseum
  Cubao, Quezon City

- National Bookstore
  701 Rizal Avenue
  Manila

- Erehwon Bookstore
  Beside Casaje Optical
  MCC, Makati
  Tel: 87-12-19 or 87-97-90

- Casa Linda Bookshop
  2nd Floor, San Antonio Plaza
  Forbes Park, Makati
  Tel: 817-61-56; 85-11-26; 69-95-35

- Melga Enterprises
  6th Floor, Doña Josefa Building
  Remedios St. cor. Taft Avenue, Manila
  Tel: 50-12-61 to 69

IRRI maps are also distributed by

- Gloria Cunanan Map House
  92 Gen. Gil J. Puyat Avenue
  Buendia, Makati, Metro Manila
  Tel. 86-19-54

Customers in the Los Baños area can also order IRRI books from:

- Bookstore Department
  New Agrix Supermarket, Inc.
  Lopez Avenue, Grove
  College, Laguna
  Tel: 50111

- Alvin's Bookstore
  Lopez Avenue, Grove
  College, Laguna

Distributing bodies for UNESCO coupons in Asia, Africa, and Latin America

ASIA

Burma
Burmese National Commission for UNESCO
Dept. of Higher Education, Ministry of Education
Office of the Ministers
Theinbyu Street, Rangoon

India
Indian National Commission for Cooperation
  with UNESCO
Ministry of Education and Social Welfare
"C" Wing, Shastri Bhawan, New Delhi 1

Indonesia
Lembaga National Indonesia Untuk UNESCO
c/o Dept. pf Pendidikan Dan Kebudajaan
Djalan Tjilatjap 4, Djakarta

Japan
Japan Society for the Promotion of Science
Yamato Bldg., 5-3-1 Kojimachi
Chiyoda-ku, Tokyo 102

Korea
Korean Exchange Bank
I.P.O. 2924, Seoul

Nepal
Nepalese National Commission for UNESCO
Min. of Education, Kathmandu

Pakistan
Pakistan National Commission for UNESCO
Min. of Education, Islamabad

Sri Lanka
Sri Lanka National Commission for UNESCO
Min. of Education, Malay Street, Colombo 2

AFRICA

Cameroon
Tresorerie Generale, Yaounde

Central African Republic
Commission Nationale Centrafricaine pour l'UNESCO
Min. de l'Education Nationale de la Jeunesse
des Sports et des Arts, B.P. 791, Bangui

Gabon
Commission Nationale pour l'UNESCO
Ministére de l'Education Nationale et d ela Recherche
 Scientifique
B.P. 264, Libreville

Ghana
Ghana National Commission for UNESCO
Min. of Education, Ringroad East
P.O. Box 2739, Accra

Guinea
Commission Nationale Guineene pour l'UNESCO
Ministére de la Recherche Scientifique
B.P. 561, Conakry

Madagascar
Commission de la République Malagache pour l'UNESCO
11, rue Dussol, Behoririka, Tananarive

Mali
Commission Nationale Maliénne pour l'UNESCO
Min. de l'Enseignement Superieur
Secondaire et de la Recherche Scientifique
B.P. 119, Bamako

Nigeria
Nigerian National Commission for UNESCO
Fed. Min. of Education, Victoria Is., Lagos

Sudan
UNDP in the Sudan, P.O.B. 913, Khartoum

Upper Volta
Commission Nationale de la République de Haute-Volta
 pour l'UNESCO, B.P. 7046, Ouagadougou

Zaire
Commission Nationale Zairoise pour l'UNESCO
Commissariat d'Etat Chargé de l'Education Nationale
Kinshasa

Zambia
UNDP in South-East Africa, P.O. Box 1966, Lusaka

LATIN AMERICA

Argentina
Comision Nacional Argentina para la UNESCO
Avenida Eduardo Madero 235-6 piso, Buenos Aires

Bolivia
Comision Nacional para la UNESCO
Ministerio de Educacion y Cultura
C.P. 4107, La Paz

Brazil
Inst. Brasileiro de Educacâo
Ciencia e Cultura (I.B.E.C.C.), Praia de Botafogo
186 Terreo Salas 101/102, Rio de Janeiro

Chile
(C.O.N.I.C.Y.T.) Direccion de Operaciones y Desarrollo
   de Programas, Canada 308 — Casilla 297-V
Santiago de Chile, Univ. de Concepcion, Casilla 2187
Concepcion

Colombia
I.C.E.T.E.X. Oficina de Relaciones Nacionales
   o Internacionales, Cra. 3A. N° 18-24
Apdo. Aereo 5735, Bogota, D.E.

Uruguay
Conicyt, Sarandi, 444 P. 4
C.C. 1869, Montevideo

Research Highlights

International Rice Research Institute

IRRI Highlights 1984

1985. 102 pages. 17.7 × 22.8 cm. Paperback. ISSN 0155-1142. HDC $7.50, LDC $3.00 plus airmail ($5.40) or surface mail ($2.25) postage.

A yearly report on IRRI progress in areas such as rice breeding for resistance to diseases, insects, drought, adverse soils, and cold and hot temperature; pest control; irrigation water management; soil and crop management; cropping systems; machinery development; finances; and other items. Illustrations in color.

The theme of IRRI Highlights 1984 is "New Directions." The report emphasizes the importance of integrating advanced technology such as biological engineering with adaptive research to meet immediate and long-term needs of the world's rice farmers. National program building through training, cooperative research, and scientific communication is essential to the goal.

Also available, Research Highlights for: 1978, 1979, 1980, 1981, 1982, and 1983.

Annual Reports

International Rice Research Institute

IRRI Annual Report for 1983

1984. 548 pages. 17.7 × 25.4 cm. Paperback. HDC $23.00, LDC $9.20 plus airmail ($25.50) or surface mail ($2.25) postage. (IRRI Annual Report for 1984, in press.)

This is an in-depth annual report of IRRI's overall research progress presented on a problem area basis rather than by scientific discipline. Major sections include Genetic Evaluation and Utilization (problem-oriented plant breeding); Control of Diseases, Insects, and Weeds; Irrigation Water Management; Soil and Crop Management; Environment and Its Influence; Constraints on Rice Yields; Consequences of New Technology; Rice-Based Cropping Systems; and Machinery Development.

Also available: IRRI Annual Report for 1977, 1978, 1979, 1980, 1981, and 1982.

Monographs

T. T. Chang and E. Bardenas

The Morphology and Varietal Characteristics of the Rice Plant

1965. 40 pages. 21.5 × 27.9 cm. Paperback. ISBN 971-104-008-5. HDC $3.00, LDC $1.20 plus airmail ($5.70) or surface mail ($1.00) postage.

Rice workers have long recognized the need for uniformity in genetic nomenclature. This publication: 1) proposes a set of reasonably definitive terms that adequately describe the various parts of the rice plant and its processed products, 2) defines varietal characteristics that are useful in identification and classification, and 3) describes a number of commonly observed mutant traits in both morphological and genetical terms.

K. A. Gomez

Techniques for Field Experiments with Rice

1972. 46 pages. 15 × 22.8 cm. Paperback. ISBN 971-104-049-2. HDC $3.00, LDC $1.20 plus airmail ($5.70) or surface mail ($1.00) postage.

This manual is primarily intended for field researchers conducting experiments on rice. It defines and provides tested solutions to major problems of experimentation technique. The first part of the manual covers field plot techniques, experimental design, and sources of experimental error. The second part deals with sampling and measurement techniques for determining plot yield and other characters.

International Rice Research Institute

Major Research in Upland Rice

1975. 255 pages. 15 × 22.8 cm. Paperback. ISBN 971-104-013-1. HDC $4.50, LDC $1.80 plus airmail ($12.00) or surface mail ($1.25) postage.

Scientific literature on upland, or dryland, rice is scarce and scattered. But upland rice comprises about 10% of the world's rice land. Furthermore, upland rice varieties are often used as parents in breeding programs to develop drought-resistant varieties. IRRI scientists with experience in upland rice wrote chapters related to their particular fields for the book.

T. T. Chang

Manual on Genetic Conservation of Rice Germplasm for Evaluation and Utilization

1976. 77 pages. 15 × 22.8 cm. Paperback. ISBN 971-104-007-7. HDC $3.00, LDC $1.20 plus airmail ($5.40) or surface mail ($1.00) postage.

Designed to help rice researchers conserve, evaluate, and use existing gene pools in rice, this manual outlines operations related to the scientific management of genetic resources for varietal improvement.

S. Yoshida, D. A. Forno, J. H. Cock, and K. A. Gomez

Laboratory Manual for Physiological Studies of Rice

1976. 83 pages. 21.5 × 27.9 cm. Paperback. ISBN 971-104-035-2. HDC $3.00, LDC $1.20 plus airmail ($7.30) or surface mail ($1.25) postage.

This manual is primarily intended for studies of crop physiology and agronomy. It includes procedures particularly suited for routine chemical analysis and physiological studies of the rice plant.

Vo Tong Xuan and Vernon E. Ross

Training Manual for Rice Production

1976. 140 pages. 21 × 26.5 cm. Paperback. ISBN 971-104-054-9. HDC $5.25, LDC $2.10 plus airmail ($10.50) or surface mail ($1.25) postage.

Rice research has added greatly to the storehouse of knowledge on how to increase rice yields. Unfortunately, the flow of this knowledge to the rice farmer has been slow. The need exists to train rice extension workers in this new technology so they in turn can teach farmers. This manual was developed to serve those involved in teaching the new rice technology, and can serve as a guide to tasks involved in producing irrigated rice.

Frans R. Moormann and Nico van Breemen

Rice: Soil, Water, Land

1978. 185 pages. 15 × 22.8 cm. Paperback. ISBN 971-104-031-X. HDC $7.50, LDC $3.00 plus airmail ($12.00) or surface mail ($1.25) postage.

This volume helps find a void in the available knowledge on soils and rice. The authors strongly emphasize — and reply to — the need for study of the interrelationships between performance of the rice and its natural or man-modified environment.

Yujiro Hayami

Anatomy of a Peasant Economy

1978. 149 pages. 15 × 22.8 cm. Paperback. ISBN 971-104-039-5. HDC $5.25, LDC $2.10 plus airmail ($8.10) or surface mail ($1.25) postage.

Dr. Yujiro Hayami analyzed the impact of new rice technology in a typical rice village in the Philippines. The intense case study involved the use of both household record keeping and interviews surveys. These data are among the most comprehensive on the village economy that have yet been collected, assembled, and analyzed.

S. K. De Datta, K. A. Gomez, R. W. Herdt, and R. Barker

A Handbook on the Methodology for an Integrated Experiment Survey on Rice Yield Constraints

1978. 60 pages. 17.7 × 22.8 cm. Paperback. ISBN 971-104-036-0. HDC $3.00, LDC $1.20 plus airmail ($4.80) or surface mail ($1.00) postage.

A multitude of factors have been identified as possible causes of low levels of rice production in Asian countries, but few have been empirically verified. This handbook was prepared for anyone interested in the methodology that IRRI and associated national research programs use in the yield constraints project. The purpose of that project is to determine the importance of factors contributing to a "yield gap" in areas where improved technology is available but farmers have not adopted it.

P. R. Jennings, W. R. Coffman, and H. E. Kauffman

Rice Improvement

1979. 186 pages. 17.7 × 22.8 cm. Paperback. ISBN 971-104-003-4. HDC $8.25, LDC $3.30 plus airmail ($12.60) or surface mail ($1.75) postage.

A practical manual, illustrated with color photos, for scientists to consult on each step in the development of improved rice varieties. As the title indicates, rice improvement is broader than plant breeding. The collaboration of plant pathologists, entomologists, agronomists, and scientists from other disciplines in the varietal development effort is reflected throughout the book.

R. F. Chandler, Jr.

Rice in the Tropics

1979. 256 pages. 15 × 22.8 cm. Paperback. ISBN 971-104-058-1. HDC $8.25, LDC $3.30 plus airmail ($12.60) or surface mail ($1.75) postage.

Rice in the tropics discusses the importance of rice as a world food crop; the current stage of rice research and technology; promising research projects that merit continuing attention; postharvest problems, including ways to handle surplus rice production; and how to put the important elements of a successful rice production program into a workable scheme. The appendix describes international agencies that offer technical and financial assistance for agricultural development programs.

Dr. Chandler, the first director of IRRI, wrote the book under the auspices of the International Agricultural Development Service and IRRI. Westview Press made paperback editions available to IRRI for distribution.

B. S. Vergara

A Farmer's Primer on Growing Rice

1979. 221 pages. 15 × 22.8 cm. Paperback. ISBN 971-104-051-4. HDC $6.00, LDC $2.40 plus airmail ($11.10) or surface mail ($1.25) postage.

A farmer's primer was written to help progressive rice farmers and technicians understand why and how the improved rice varieties and farm technology increase production. Dr. B. S. Vergara, IRRI plant physiologist, explains agricultural practices such as why a farmer incubates seed, why he applies fertilizer, and how and when that fertilizer should be incorporated. Farmer's primer illustrations are available to encourage translations.

A farmer's primer on growing rice is now available from IRRI in:

English	— ISBN 971-104-051-4
French	— ISBN 971-104-090-5
Hiligaynon	— ISBN 971-104-091-3
Ilokano	— ISBN 971-104-092-1
Pilipino	— ISBN 971-104-093-X
Cebuano	— ISBN 971-104-118-9
Pampango	— ISBN 971-104-094-8
Warai	— ISBN 971-104-
Spanish	— ISBN 971-104-125-1
Bikol	— ISBN 971-104-132-4
Kiswahili	— ISBN 971-104-135-9
Creole	— ISBN 971-104-142-1

P. A. Roger and S. A. Kulasooriya

Blue-Green Algae and Rice

1980. 112 pages. 15 × 22.8 cm. Paperback. ISBN 971-104-028-X. HDC $6.00, LDC $2.40 plus airmail ($7.50) or surface mail ($1.25) postage.

Blue-green algae and rice is a comprehensive compilation and synthesis of 369 available articles on the specialized relationship between BGA and rice, and on the potential of BGA as an effective and cheap source of nitrogen.

A section on BGA ecology examines the effect of physical, biotic, soil, and agronomic factors on the growth of BGA. The physiology section provides information on the ecological implications — beneficial and, sometimes, detrimental — of the photosynthetic and nitrogen-fixing processes of BGA. The algalization chapter discusses inoculation methods that farmers and researchers use; the effect of algalization on rice, soil, and soil microflora; algalization technology; and factors that inhibit algalization.

Adelita C. Palacpac

World Rice Statistics

1980. 130 pages. 22.5 × 28 cm. Paperback. Free to rice improvement programs.

This report is a compilation of basic data on rice production, consumption, trade, prices, and other related statistics from a number of sources. It is the only publication of its kind that draws together such a wide breadth of information from secondary sources and presents that information in a handy reference form.

S. Yoshida

Fundamentals of Rice Crop Science

1981. 269 pages. 15 × 22.8 cm. Paperback. ISBN 971-104-052-2. HDC $12.50, LDC $5.00 plus airmail ($12.60) or surface mail ($1.25) postage.

Fundamentals of rice crop science, authored by Dr. Shouichi Yoshida, IRRI plant physiologist, provides the reader with a well-integrated view of rice science. Although plant physiology provides the book's skeleton, Dr. Yoshida has added current information from agronomy and soil science. Dr. Yoshida drew not only on his own experience in the writing of Fundamentals, but also that of rice scientists in Japan and other countries. Fundamentals include extensive research findings generated in Japan that are not widely available in English.

S. K. De Datta

Principles and Practices of Rice Production

1981. 618 pages. 15 × 22.8 cm. Paperback. ISBN 971-104-053-0. HDC: Send orders to J. Wiley, 605 Third Ave., New York, USA. LDC $7.20 plus airmail ($15.00) or surface mail ($1.50) postage.

Principles and practices of rice production contains comprehensive coverage of significant research results on various aspects of rice production. Chapters include the climate and its effect on rice production; landscape and soils on which rice is grown; chemical changes in submerged soils; morphology; growth and development of the rice plant; rice varietal development and seed production; types of rice culture; land preparation; water use and water management; mineral nutrition and fertilizer management; weeds, insects, diseases, and their control; postproduction technology; and the effect of modern rice technology on the world's food supply.

Principles and Practices of Rice Production was published by John Wiley & Sons of New York. IRRI distributes low cost paperback edition in developing nations.

R. B. Alocilja, E. P. Cervantes, and L. D. Haws

A Continuous Rice Production System . . . the Rice Garden

1981. 16 pages. 17.7 × 22.8 cm. Paperback. ISBN 971-104-055-7. HDC $3.00, LDC $1.20 plus airmail ($3.00) or surface mail ($0.75) postage.

This brochure describes a method of growing rice in garden-like plots within a rice farm. The farmer staggers the planting of the plots. Thus, subsequent operations such as fertilization, weeding, and harvesting are also staggered. The farmer's labor inputs and income are spread across the year, rather than being seasonal. Risks of crop failure because of typhoons or other hazards may be lessened because portions of the crop are always at varying growth stages.

E. A. Heinrichs, S. Chelliah, S. L. Valencia, M. B. Arceo, L. T. Fabellar, G. B. Aquino, and S. Pickin

Manual for Testing Insecticides on Rice

1981. 131 pages. 17.7 × 22.8 cm. Paperback. ISBN 971-104-024-7. HDC $13.75, LDC $5.50 plus airmail ($10.20) or surface mail ($1.25) postage.

Insecticides play a major role in integrated systems of rice pest management. This manual describes methodologies that IRRI regularly uses to evaluate pesticides. Most of the methodologies have not been published previously. The manual should promote the evaluation of insecticides in rice, resulting in safer and more effective farmer use of chemicals in an integrated approach to insect management. Color photos.

Keith Moody

Major Weeds of Rice in South and Southeast Asia

1981. 86 pages. 10.2 × 17.8 cm. Paperback. ISBN 971-104-025-5. HDC $5.50, LDC $2.20 plus airmail ($3.90) or surface mail ($1.00) postage. English and Urdu.

This book includes colorplates and descriptions of the most widespread weeds of rice in South and Southeast Asia, based on information available in the literature and a survey of weed scientists. The weed species are listed alphabetically by family and by scientific name within each family. A minimum of botanical terms is used. A comprehensive glossary is included for those who are unfamiliar with the botanical language.

H. G. Zandstra, E. C. Price, J. A. Litsinger, and R. A. Morris

A Methodology for On-Farm Cropping Systems Research

1981. 155 pages. 17.7 × 22.8 cm. Paperback. ISBN 971-104-045-X. HDC $10.00, LDC $4.00 plus airmail ($12.00) or surface mail ($1.75) postage.

Cropping systems research seeks the technology that will intensify food production by improving crop yields or growing an extra crop during the year.

This handbook describes a methodology for analyzing the crop production environment, selecting sites with potential for introducing an extra crop or increasing crop yield, and establishing techniques for increasing crop production or cropping intensity. The research methods were developed for rice-based cropping systems, but are already used in nonrice-growing areas and can readily be adapted to include perennial crops.

International Rice Research Institute

Evolution of the Gene Rotation Concept for Rice Blast Control

1982. 136 pages. 15 × 22.8 cm. Paperback. ISBN 971-104-012-3. HDC $6.00, LDC $2.40 plus airmail ($6.30) or surface mail ($1.00) postage.

This collection of 10 research papers documents the evolution of the gene rotation concept. The rotation of specific genes for resistance is a new, exciting, and effective method of blast control. The method is based on local studies of the evolution of pathogenicity within the blast fungus population.

Antonio J. Ledesma

Landless Workers and Rice Farmers: Peasant Subclasses Under Agrarian Reform in Two Philippine Villages

1982. 237 pages. 15 × 22.8 cm. Paperback. ISBN 971-104-043-3. HDC $8.75, LDC $3.50 plus airmail ($11.10) or surface mail ($1.25) postage.

Modern technology has increased rice production, and agrarian reform aimed for a more equitable distribution of income and land resources for the tillers of the soil in the Philippines in the early 1970s.

Antonio J. Ledesma, a Jesuit priest and agricultural scholar, assessed the impact of both institutional and technological changes on all

peasant groups including landless workers, within the same rice-growing villages. The study examined:
1. The basic differences between landless workers and tenant farmers in terms of labor allocation, household economy, and security of tenure.
2. The formation of three major peasant groups at the bottom strata of rural society under agrarian reform.
3. Problem areas in agrarian reform, particularly as they pertain to interactions between landless workers and tenant farmers.

C. Barlow, S. Jayasuriya, and E. C. Price

Evaluating Technology for New Farming Systems: Case Studies from Philippine Rice Farms

1982. 119 pages. 15 × 22.8 cm. Paperback. ISBN 971-104-070-4. HDC $6.00, LDC $2.40 plus airmail ($7.50) or surface mail ($1.25) postage.

This publication provides a detailed view of factors that affect the adoption of multiple cropping techniques recently developed by the IRRI Cropping Systems program. The authors provide a helpful insight into cropping systems research and the economics of small rice farms in Asia.

R. F. Chandler, Jr.

An Adventure in Applied Science: a History of the International Rice Research Institute

1982. 236 pages. 15 × 22.8 cm. Paperback. ISBN 971-104-063-8. HDC $9.50, LDC $3.80 plus airmail ($11.10) or surface mail ($1.25) postage.

Written by IRRI's first director, Dr. Robert F. Chandler, Jr., Adventure details the development of IRRI from 1955, when a team of scientists and research administrators began to think about the establishment of "an international rice research institute."

Chandler's narration of the adventure of building and staffing IRRI — completed with interesting and often humorous anecdotes about the people and problems he dealt with — takes the reader through his years (1960-1972) as director.

R. Herdt and C. Capule

Adoption, Spread, and Production Impact of Modern Rice Varieties in Asia

1983. 54 pages. 15 × 22.8 cm. Paperback. ISBN 971-104-083-2. HDC $3.25, LDC $1.30 plus airmail ($5.40) or surface mail ($1.00) postage.

The development and spread of modern rice varieties have contributed substantially to increased rice production achieved by Asian countries since 1965. This book provides a comprehensive view of the contribution of modern rices in Bangladesh, Burma, India, Indonesia, the Republic of Korea, Malaysia, Nepal, Pakistan, Philippines, Sri Lanka, and Thailand.

James E. Wimberly

Technical Handbook for the Paddy-Rice Postharvest Industry in Developing Countries

1983. 200 pages. 17.7 × 22.8 cm. Paperback. ISBN 971-104-075-1. HDC $8.75, LDC $3.50 plus airmail ($9.30) or surface mail ($1.25) postage.

Postharvest losses during drying, processing, and storage of rice may amount to from 7 to 26% of total rice production. James Wimberly has compiled a clear, comprehensive, and cohesive work that explains rice handling, transport, drying, cleaning, storage, parboiling, and milling. Appropriate equipment, performance, and design criteria are discussed for each postharvest step.

International Rice Research Institute

Field Problems of Tropical Rice: Revised Edition

1983. 172 pages. 10.2 × 17.8 cm. Paperback. ISBN-971-104-080-8. HDC $5.00, LDC $2.00 plus airmail ($3.90) or surface mail (US$1.00) postage.

Field Problems of Tropical Rice has been one of IRRI's most popular publications. The 1983 revised edition has 153 color plates to help rice workers identify common production problems such as insects, diseases, weeds, and problem soils.

Although the original edition focused on problems in Asia, the revision also identifies common problems of Africa and Latin America.

Like the original edition, the revision is designed to facilitate its low-cost translation and copublication by agricultural programs, international agencies, and publishers in developing nations. IRRI also arranges the printing of bulk orders of nonEnglish Field Problems (2,000 copies or more), using IRRI's color plates, at actual printing and paper cost, plus freight.

Field problems of tropical rice: revised edition is now available from IRRI in

English	— ISBN 971-104-080-8
French	— ISBN 971-104-087-5
Spanish	— ISBN 971-104-086-7
Vietnamese	— ISBN 971-104-085-9
Cebuano	— ISBN 971-104-088-3
Pilipino	— ISBN 971-104-084-0
Warai	— ISBN 971-104-089-1
Pampango	— ISBN 971-104-127-8
Ilokano	— ISBN 971-104-129-4
Bengali	—
Punjabi	— ISBN 971-104-136-7
Hiligaynon	— ISBN 971-104-134-0
Bikol	— ISBN 971-104-128-6

IRRI is holding a limited number of nonEnglish copies for international distribution. Editions in other languages are in preparation.

T. R. Hargrove, R. C. Cabrera, and F. E. Manto

Copublication: IRRI Design, Procedures, and Policies for Multilanguage Publication in Agriculture

1983. 16 pages. 17.7 × 22.8 cm. Paperback. ISBN 971-104-117-09. HDC $2.50, LDC $1.00 plus airmail ($1.80) or surface mail ($0.25) postage.

IRRI publishes almost exclusively in English, our working language. But IRRI tries to alleviate the language problem through copublication — cooperative endeavors whereby the original publisher of a book grants a second agency permission to translate, publish, and disseminate that publication in another language. More than 600,000 copublished IRRI books had been printed in 31 languages by 1984.

IRRI's most successful copublication efforts have presented simple, basic information in books designed for easy, inexpensive copublication. Two such books, *Field Problems of Tropical Rice* and *A Farmer's Primer on Growing Rice,* account for 86% of the total IRRI books printed in 29 nonEnglish languages.

This highly illustrated booklet describes design techniques to encourage inexpensive copublication of materials published in B&W and in color.

K. A. Gomez and A. A. Gomez

Statistical Procedures for Agricultural Research (Second Edition)

1984. 680 pages. 15 × 22.8 cm. Paperback. ISBN 971-104-048-4. HDC: send orders to J. Wiley, 605 Third Ave., New York, USA. LDC $8.00 plus airmail ($24.50) or surface mail ($1.95) postage.

The availability of trained statisticians for consultation on the design and data analysis of crop experiments is a luxury in most developing countries.

The second edition of *Statistical Procedures for Agricultural Research* is designed to help agricultural researchers use the most appropriate statistical techniques for their experiments.

The authors strived to make the content usable by any subject material specialist. They chose the simpler and more commonly used statistical procedures, with special emphasis on field experiments with crops. Most examples are concerned with rice, but have applicability to a wide range of annual crops.

International Rice Research Institute

Basic Procedures for Agroeconomic Research

1984. 236 pages. 15 × 22.8 cm. Paperback. ISBN 971-104-081-6. HDC $8.75, LDC $3.50 plus airmail ($10.80) or surface mail ($1.75) postage.

This textbook-like publication includes chapters on statistical concepts, crop budgets, agricultural wage rates, factor shares, cropping systems research methodology, farm household systems, optimum fertilizer recommendations, economic analysis of new technologies, markets, data collection, and farm level surveys.

O. Mochida and associates

Insecticide Evaluation 1983

1984. 150 pages. 21.5 × 28 cm. Paperback. Free to rice improvement programs.

Includes the performance of coded and commercial insecticides tested under greenhouse and field conditions against insect pests of rice, mungbean, cowpea, maize, and sorghum. A separate section discusses the use of insecticides in integrated pest management studies involving yield loss assessments, economics of insect control, and selective toxicity. Contains 114 tables, 7 figures, and an appendix listing the 104 insecticides tested and the companies providing them.

M. Z. Hoque

Cropping Systems in Asia On-Farm Research and Management

1984. 196 pages. 15 × 22.8 cm. Paperback. ISBN 971-104-106-5. HDC $7.50, LDC $3.00 plus airmail ($4.00) or surface mail ($0.50) postage.

Dr. M. Zahidul Hoque presents the highlights of on-farm cropping systems research and development by the member country programs in the Asian Cropping Systems Network (ACSN) coordinated by IRRI. Included are detailed data-based presentations and discussion of farmers' existing cropping systems and testing of improved cropping patterns and component technologies in eight well-defined agro-ecological environments. Also briefly described are the concepts, approaches, and methodologies developed and used by the ACSN scientists. The use of site-specific on-farm research findings in target area development is discussed with specific ACSN examples.

W. H. Reissig, E. A. Heinrichs, J. A. Litsinger, K. Moody, L. Fiedler, T. W. Mew, and A. T. Barrion

Illustrated Guide to Integrated Pest Management in Rice in Tropical Asia

1985. 390 pages. 22 × 28 cm. Paperback. ISBN 971-104-120-0. HDC $16.25, LDC $6.50 plus airmail ($20.00) or surface mail ($2.25) postage.

The widespread introduction of high-yielding rice varieties in Asia during the last 20 years and associated changes in production practices have improved conditions for insects, diseases, weeds, and rodents. Pesticide use is expensive and may destroy beneficial organisms. This guide to integrated pest management (IPM) provides practical, comprehensive information for IPM workers throughout Asia. It includes sections on insect pests, biology and control of weeds, biology and management of riceland rats, and disease management. The publication is highly illustrated and designed for multilanguage copublication.

E. A. Heinrichs, F. G. Medrano, and H. R. Rapusas

Genetic Evaluation for Insect Resistance in Rice

1985. 356 pages. 17.7 × 22.8 cm. Paperback. ISBN 971-104-110-3. HDC $15.00, LDC $6.00 plus airmail ($9.00) or surface mail ($1.00) postage.

Rice varieties with multiple insect and disease resistance are grown on more than 20 million hectares. Resistant varieties provide pest control at almost no cost to the farmer. This book describes procedures developed at the International Rice Research Institute (IRRI) and elsewhere for rearing insects, screening rice varieties for insect resistance, and studying the nature of resistance. A list of varieties with insect resistance is included. Much of this heretofore unpublished information is based upon the experience of IRRI's Genetic Evaluation and Utilization Program.

S. H. Ou

Rice Diseases

1985. 450 pages. Paperback. 24.4 × 18.7 cm. HDC: send orders to Commonwealth Mycological Institute, Ferry Lane, Kew, Richmond Surrey TW9 3AF, England. LDC, $6.00 plus airmail ($24.50) or surface mail ($1.95) postage.

Rice Diseases was written by Dr. S. H. Ou, former head of IRRI's Plant Pathology Department. This book, a revision of the original Rice Diseases published in 1972, embodies the experience of Dr. Ou's 35 years in rice pathology.

The book provides an up-to-date review of knowledge on the diseases of rice caused by fungi, bacteria, viruses, mycoplasmas, and physiological factors. The world literature on each disease is reviewed, and illustrations of symptoms and causal organisms are provided.

Rice Diseases was published by Commonwealth Mycological Institute, U.K. The IRRI Communication and Publications Department purchased a stock of paperback copies for distribution in the Third World only.

Conferences

International Rice Research Institute

Changes in Rice Farming in Selected Areas of Asia

1975. 377 pages. 15 × 22.8 cm. ISBN 971-104-040-9. HDC $6.75, LDC $2.70 plus airmail ($16.50) or surface mail ($1.75) postage.

In early 1971, a group of development specialists in agricultural economics, rural sociology and related fields at IRRI and the University of the Philippines at Los Baños (UPLB) considered research to improve our understanding of changes occurring and the problems associated with the adoption of the new rice technology at the farm level. They saw a need to collect and analyze data from a number of locations in different rice farming areas in South and Southeast Asia that would be useful to research workers, program planners, policy makers in these countries.

This volume includes reports of changes associated with new rice technology in villages in India, Indonesia, Malaysia, Pakistan, Philippines, and Thailand.

International Rice Research Institute

Climate and Rice

1976. 565 pages. 15 × 22.8 cm. Paperback. ISBN 971-104-034-4. HDC $16.50, LDC $6.60 plus airmail ($27.00) or surface mail ($2.25) postage.

This publication includes papers by internationally known biological and physical scientists at a 1974 IRRI symposium on the interactions between climate and rice. Included are papers on the climatic environment in which rice is grown commercially or experimentally; the influence of each climatic variable on different phases of rice growth and on insect and disease incidence; and ways to further increase and stabilize rice yields under variable climatic conditions.

International Rice Research Institute

Deep-Water Rice

1977. 239 pages. 15 × 22.8 cm. Paperback. ISBN 971-104-015-8. HDC $6.00, LDC $2.40 plus airmail ($11.10) or surface mail ($1.25) postage

This volume contains 25 papers presented at the Deep-Water Rice Workshop in Bangkok in 1976. Sections include: basic studies on deep-water rice; screening methods; reports on progress in deep-water rice improvement in countries of South and Southeast Asia and West Africa; and recommendations for future progress.

International Rice Research Institute

Constraints to High Yields on Asian Rice Farms: an Interim Report

1977. 235 pages. 15 × 22.8 cm. Paperback. ISBN 971-104-037-9. HDC $4.50, LDC $1.80 plus airmail ($11.10) or surface mail ($1.25) postage.

The average rice yields on farmers' fields, even in areas where the adoption of modern rice varieties is high, are lower than those commonly harvested from scientists' experimental plots. The International Rice Agro-Economic Network (IRAEN) was organized in 1974 to enable cooperating scientists in Asia to identify and study the factors that constrain rice yields in farmers' fields. This volume contains reports on factors that influence the use of modern rice technology in study areas in India, Indonesia, Malaysia, Thailand, and the Philippines.

International Rice Research Institute

Cropping Systems Research and Development for the Asian Rice Farmer

1977. 454 pages. 15 × 22.8 cm. Paperback. ISBN-971-104-044-1. HDC $7.50, LDC $3.00 plus airmail ($22.20) or surface mail ($1.75) postage.

Farmers in Asia generally have two alternatives for increasing food production. They can increase crop or increase the number of crops grown each year. This volume of papers from an IRRI sysmposium on cropping systems research deals with alternatives, but emphasizes efforts to increase cropping intensities.

International Rice Research Institute

Interpretive Analysis of Selected Papers from Changes in Rice Farming in Selected Areas of Asia

1978. 166 pages. 15 × 22.8 cm. Paperback. ISBN 971-104-041-7. HDC $4.50, LDC $1.80 plus airmail ($9.30) or surface mail ($1.25) postage.

Nine previously unpublished papers on the effects of the adoption of semidwarf rices and related technology on income, employment, and other factors in six South and Southeast Asian coutries.

International Rice Research Institute

Economic Consequences of the New Rice Technology

1978. 402 pages. 15 × 22.8 cm. Paperback. ISBN 971-104-042-5. HDC $8.25, LDC $3.30 plus airmail ($16.80) or surface mail ($1.75) postage.

New technology is essential to agricultural development, but its effects can be harmful as well as beneficial. Research to identify socio-economic consequences of the introduction of new technology provides important information for the strategy and design of biological and engineering research. This volume contains selected papers presented by eminent social scientists at the IRRI conference on the "Economic Consequences of the New Rice Technology." The book reflects the high degree of complementarity essential between research efforts in the biological and the social sciences.

International Rice Research Institute

International Agricultural Machinery Workshop

1978. 203 pages. 15 × 22.8 cm. Paperback. ISBN 971-104-047-6. HDC $6.75, LDC $2.70 plus airmail ($10.20) or surface mail ($1.25) postage.

IRRI's Agricultural Engineering program helps small farmers increase food production by developing machines to eliminate production bottlenecks such as harvesting, threshing, and land preparation. This volume contains papers on the transfer of the technology to local manufacturers, presented at an IRRI Industrial Extension Workshop of agricultural engineers, manufacturers, and scientists from 16 countries.

International Rice Research Institute

Irrigation Policy and Management in Southeast Asia

1978. 198 pages. 17.7 × 25.4 cm. Paperback. ISBN 971-104-026-3. HDC $6.75, LDC $2.70 plus airmail ($12.00) or surface mail ($1.25) postage.

The supply of water — more than any other single factor — controls the production of field crops in the tropics. Year-round cropping systems are generally possible in areas where irrigation systems are well-conceived and efficiently managed. This volume includes selected papers authored by scientists and engineers in national irrigation planning and management programs and presented at an IRRI seminar to identify shortcomings in existing irrigation distribution systems and ways to improve them.

International Rice Research Institute

Soils and Rice

1978. 825 pages. 15 × 22.8 cm. Paperback. ISBN 971-104-030-1. HDC $16.50, LDC $6.60 plus airmail ($36.60) or surface mail ($2.25) postage.

This compilation of papers presented at the 1977 Soils and Rice Symposium at the International Rice Research Institute (IRRI) contains 8 major sections on rice soils: rice soils of the world; composition, genesis, morphology, and classification; physical properties; chemical and electrochemical changes; microbiology; fertility; management; and problem soils.

International Rice Research Institute

Chemical Aspects of Rice Grain Quality

1979. 390 pages. 15 × 22.8 cm. Paperback. ISBN 971-104-010-7. HDC $9.00, LDC $3.60 plus airmail ($16.70) or surface mail ($1.75) postage.

Grain quality largely determines the market price for rice and its acceptance by consumers. The Workshop on the Chemical Aspects of Grain Quality, held at IRRI in 1978, was attended by 28 breeders and chemists from 11 countries and IRRI. These workshop proceedings are likely to be the basic reference on rice grain quality for many years.

International Rice Research Institute

Rice Blast Workshop

1979. 222 pages. 15 × 22.8 cm. Paperback. ISBN 971-104-011-5. HDC $7.50, LDC $3.00 plus airmail ($10.20) or surface mail ($1.25) postage.

Blast, caused by the fungus *Pyricularia oryzae,* is one of the most widespread diseases of rice. Its control and management is difficult because of the instability of the fungus and the marked variability in pathogenicity, which results in different races.

The 1977 Rice Blast Workshop was held at IRRI to summarize knowledge about blast and its control, and to develop research strategies to improve that control. This volume contains workshop papers.

International Rice Research Institute

Rainfed Lowland Rice

1979. 341 pages. 15 × 22.8 cm. Paperback. ISBN 971-104-014-X. HDC $8.25, LDC $3.30 plus airmail ($14.40) or surface mail ($1.75) postage.

A third of the world's rice area is planted to rainfed lowland, or rainfed wetland rice. But the improved rice technology has largely bypassed the rainfed lowland rice farmer and comparatively little research attention has been directed to his needs. The 1978 International Rice Research Conference, held at IRRI, focused on this important subject. This publication contains selected conference papers focusing on research to develop improved rainfed lowland technology.

International Rice Research Institute

1978 International Deep-Water Rice Workshop

1979. 300 pages. 15 × 22.8 cm. Paperback. ISBN 971-104-016-6. HDC $6.75, LDC $2.70 plus airmail ($13.50) or surface mail ($1.75) postage.

Thirty to forty percent of Asia's rice lands are subject to annual monsoon floods. Water depths of 0.5 to 1.0 meter are common over wide areas of Asia. Even deeper water levels occur in the "floating" rice areas of South and Southeast Asia. Modern rice technology has bypassed farmers in those regions — the improved semidwarf rice varieties cannot tolerate deep-water conditions. This volume includes selected papers presented at a 1978 Deep-water Rice Workshop held in Calcutta, India, where 60 scientists from 6 nations planned strategies for developing improved technology for the millions of farmers who grow deep-water rice.

International Rice Research Institute and the Office of Rural Development, Suweon, Korea

Rice Cold Tolerance Workshop

1979. 139 pages. 15 × 22.8 cm. Paperback. ISBN 971-104-018-2. HDC $6.75, LDC $2.70 plus airmail ($7.20) or surface mail ($1.25) postage.

Rice cold tolerance is an important reference for scientists involved in the improvement of rice for the mountainous regions of the tropics, and for temperate regions where cold temperatures often prohibit farmers from growing improved rice varieties. It contains the proceedings of the first International Rice Cold Tolerance Workshop, held at the Office of Rural Development (ORD) in Suweon, Korea, and jointly sponsored by ORD and IRRI.

International Rice Research Institute

Brown Planthopper: Threat to Rice Production in Asia

1979. 369 pages. 15 × 22.8 cm. Paperback. ISBN 971-104-022-0. HDC $9.00, LDC $3.60 plus airmail ($15.90) or surface mail ($1.75) postage.

The brown planthopper has been an important rice pest in China, Japan, and Korea for centuries. In recent years, epidemic numbers have attacked wetland rice crops across the tropics.

Leading rice scientists met in a 1977 IRRI symposium to bring together all the known information on the brown planthopper, to identify priority research areas, and to strengthen scientific collaboration on its control. These symposium proceedings will hopefully serve as a stimulus to accelerate the development of suitable methods for hopper control.

International Rice Research Institute

Nitrogen and Rice

1979. 499 pages. 15 × 22.8 cm. Paperback. ISBN 971-104-029-8. HDC $13.50, LDC $5.40 plus airmail ($19.20) or surface mail ($1.75) postage.

Most of the research on nitrogen transformation and biological nitrogen fixation processes in soils has been oriented toward well-aerated dryland soils. Little has been reported on nitrogen transformation and fixation in wetland rice soils. Nitrogen and rice is a compilation of papers from the Nitrogen and Rice Symposium at IRRI in 1978, at which 98 scientists from 28 countries working on nitrogen fixation in paddy soils exchanged research results.

International Rice Research Institute

Farm-Level Constraints to High Rice Yields in Asia: 1974-77

1979. 411 pages. 15 × 22.8 cm. Paperback. ISBN 971-104-038-7. HDC $10.50, LDC $4.20 plus airmail ($17.40) or surface mail ($1.75) postage.

Asian rice yields would be substantially higher if farmers took full advantage of the new rice technology. But even in areas where modern varieties are used, farmers' yields are often lower than they could be. The constraints studies in this volume, authored by researchers from across Asia and from IRRI, discuss why farmers continue to produce less than the known potential.

The United Nations University and the International Rice Research Institute

Interfaces Between Agriculture, Nutrition, and Food Science

1979. 143 pages. 15 × 22.8 cm. Paperback. ISBN 971-104-050-6. HDC $3.00, LDC $1.20 plus airmail ($7.80) or surface mail ($1.25) postage.

A priority in eliminating world hunger is greater understanding of the interrelations and interactions among agricultural production, food science, and nutrition. This workshop volume represents an interchange among scientists and program planners and implementers concerned with those vital issues and this can help lead to improvements in world food supplies, health, and nutrition.

International Rice Research Institute

Irrigation Water Management

1980. 170 pages. 15 × 22.8 cm. Paperback. ISBN 971-104-027-1. HDC $7.50, LDC $3.00 plus airmail ($8.00) or surface mail ($1.25) postage.

Irrigation expansion has aided agricultural development in the rice-growing countries in the past decade, but technical, economic, and socio-institutional problems impede the full realization of its potential benefits. A 1979 planning workshop was held at IRRI to identify management factors that constrain irrigation water efficiency. Participants included 46 research scientists, administrators, and irrigation policy makers and project implementers from 9 countries. Irrigation water management is a compilation of the issue papers.

International Rice Research Institute and the New York State College of Agricultural and Life Sciences, Cornell University

Soil-Related Constraints to Food Production in the Tropics

1980. 468 pages. 15 × 22.8 cm. Paperback. ISBN 971-104-032-8. HDC $13.50, LDC $5.40 plus airmail ($17.40) or surface mail ($1.75) postage.

A priority in eliminating world hunger is greater understanding of the interrelations and interactions among agricultural production, food science, and nutrition. This workshop volume represents an interchange among scientists and program planners and implementers concerned with those vital issues and thus can help lead to improvement in world food supplies, health, and nutrition.

World Meteorological Organization and International Rice Research Institute

Agrometeorology of the Rice Crop

1980. 254 pages. 17.7 × 25.4 cm. Paperback. ISBN 971-104-033-6. HDC $9.00, LDC $3.60 plus airmail ($12.00) or surface mail ($1.15) postage.

Rice grows in more diverse agrometeorologic conditions than any other food crops. This volume is from a December 1979 IRRI symposium to determine the status of weather records in rice-growing regions and plan their collection, analysis, and dissemination to users; to determine essential meteorological variables to be monitored in rice-weather experiments and to plan such future experiments; and to develop plans and priorities for rice-weather data analysis such as mapping and crop modeling.

Agricultural Development Council, Inc., and the International Rice Research Institute

Communication Responsibilities of the International Agricultural Research Centers

1980. 41 pages. 17.7 × 25.4 cm. Paperback. ISBN 971-104-056-5. HDC $3.00, LDC $1.20 plus airmail ($5.70) or surface mail ($1.00) postage.

A summary of recommendations made at a conference on communication responsibilities on international agricultural centers, held at IRRI. The publication focuses on facilitating the flow of scientist-to-scientist information; keeping donors, policy makers and the public informed; training of national communication staff; communication research and evaluation; and communication relationships with national agricultural programs.

International Rice Research Institute

Innovative Approaches to Rice Breeding

1980. 182 pages. 15 × 22.8 cm. Paperback. ISBN 971-104-001-8. HDC $7.50, LDC $3.00 plus airmail ($8.10) or surface mail ($1.25) postage.

Efforts to develop improved rice varieties for diverse agroecologic conditions require new as well as traditional plant breeding techniques. This volume contains technical papers on breeding innovations for resistance to insects, and to adverse conditions such as low temperature, saline soil, and drought. Other papers focus on hybrid rice breeding, distant hybridization, mutation breeding, and tissue culture.

International Rice Research Institute and Chinese Academy of Agricultural Sciences

Rice Improvement in China and Other Asian Nations

1980. 307 pages. 15 × 22.8 cm. Paperback. ISBN 971-104-004-2. HDC $8.25, LDC $3.30 plus airmail ($13.50) or surface mail ($1.75) postage.

The papers in this volume analyze the current status of rice research in China and other Asian nations. The emphasis throughout the volume is on the breeding of varieties with genetic resistance to insects and diseases, and on cultural practices to enable farmers to control pests without costly chemicals.

International Rice Research Institute

Rice Research Strategies for the Future

1982. 559 pages. 15 × 22.8 cm. Paperback. ISBN 971-104-061-1. HDC $21.25, LDC $8.50 plus airmail ($15.60) or surface mail ($2.50) postage.

The first 10 years of the International Rice Research Institute was clearly the decade of IR8, the first of the modern semidwarf varieties that helped initiate the green revolution. The second decade was one of broadened research and training activities as IRRI's program focused on the less-advantaged farmers — such as those with rainfed crops, deep water, and adverse soils — and on assisting the development of national rice improvement programs.

IRRI started its third decade with a research pipeline of technology that will soon serve rice farmers, and held a symposium on Rice Research Strategies for the Future. This volume contains the symposium papers and discussions, focusing on strategies and plans to adapt and move this technology to millions of small-scale farmers.

Institute of Tropical Agriculture and International Rice Research Institute

Japan's Role in Tropical Rice Research

1982. 46 pages. 15 × 22.8 cm. Paperback. ISBN 971-104-065-4. HDC $3.75, LDC $1.50 plus airmail ($5.40) or surface mail ($1.00) postage.

The Institute of Tropical Agriculture, Kyushu University, Japan, and the International Rice Research Institute cosponsored a seminar on Japan's role in tropical rice research in September 1980. The main purpose of the seminar was to plan strategies for the chanelling of Japan's accumulated rice research knowledge and pool of rice scientists toward the improvement of rice production in the tropics. This volume contains summaries of major papers given and recommendations made at the seminar.

International Rice Research Institute and the Division for Global and Interregional Projects, United Nations Development Programme

Report of an Exploratory Workshop on the Role of Anthropologists and Other Social Scientists in Interdisciplinary Teams Developing Improved Food Production Technology

1982. 108 pages. 15 × 22.8 cm. Paperback. ISBN 971-104-064-6. HDC $5.00, LDC $2.00 plus airmail ($4.00) or surface mail ($0.50) postage.

This volume contains six discussion papers presented at the Exploratory Workshop on the Role of Anthropologists and other Social Scientists in Interdisciplinary Teams Developing Improved Food Production Technology, held in Los Baños 23-26 March 1981.

Included are policy recommendations on future interdisciplinary approaches and explorations of the strengths and limitations of anthropology and related social sciences in developing improved agricultural technology.

International Rice Research Institute

Drought Resistance in Crops with Emphasis on Rice

1982. 416 pages 15 × 22.8 cm. Paperback. ISBN 971-104-078-6. HDC $16.25, LDC $6.50 plus airmail ($17.40) or surface mail ($1.75) postage.

This compilation of papers presented at the May 1981 IRRI symposium on drought resistance attempts to explain drought stress across a range of environmental, agroclimatic, and genetic conditions in cereal crops, with emphasis on rice.

International Rice Research Institute

1981 International Deep-Water Rice Workshop

1982. 520 pages. 15 × 22.8 cm. Paperback. ISBN 971-104-017-4. HDC $19.50, LDC $8.00 plus airmail ($15.60) or surface mail ($2.50).

These proceedings of the 1981 International Deep-water Rice Workshop, Bangkhen, Thailand, are concrete evidence of the significant progress made in deep-water rice research. IRRI and the Department of Agriculture, Ministry of Agriculture and Cooperatives, Thailand, cosponsored the workshop.

International Rice Research Institute

Rice Tissue Culture Planning Conference

1982. 120 pages. 15 × 22.8 cm. Paperback. ISBN 971-104-077-8. HDC $5.50, LDC $2.20 plus airmail ($6.60) or surface mail ($1.75) postage.

This volume contains papers presented at the Rice Tissue Culture Planning Conference held at IRRI in 1982. The papers explore the potential benefits of applying tissue culture techniques to rice breeding and emphasize the problems that first must be solved. Special emphasis is placed upon the need for collaborative work to incorporate this new plant breeding technique into rice improvement programs.

International Rice Research Institute

Report of a Workshop on Cropping Systems Research in Asia

1982. 762 pages. 15 × 22.8 cm. Paperback. ISBN 971-104-076-X. HDC $27.50, LDC $11.00 plus airmail ($36.60) or surface mail ($2.25) postage.

The 1980 cropping systems conference at IRRI served as a forum for researchers from the Asian Cropping Systems Network. This volume includes 60 papers presented at the conference, which represent the wide range of research that has resulted in workable cropping patterns that are acceptable to Asian farmers and can increase food production.

International Union of Pure and Applied Chemistry (IUPAC) and International Rice Research Institute

Chemistry and World Food Supplies: the New Frontiers

1982. 664 pages. 17.2 × 24.8 cm. Paperback. HDC: send orders to Pergamon Press, Suite 104, 150 Consumers Road, Willowdale, Ontario, M2J1P9, Canada. LDC $5.00 plus airmail ($13.80) or surface mail ($1.25) postage.

In December 1982, IRRI and the International Union of Pure and Applied Chemistry (IUPAC) jointly sponsored the second international conference on Chemical Research Applied to World Food Needs (CHEMRAWN II). CHEMRAWN II recognized increasing worldwide nutritional needs and sought to define areas of chemical research that may increase agricultural production potential and improve food processing systems in developing nations.

Chemistry and World Food Supplies: the New Frontiers contains 56 invited CHEMRAWN lectures by leading agricultural scientists, chemists, biochemists, and biologists representing the developing and developed nations.

International Rice Research Institute

Weed Control in Rice

1983. 432 pages. 15 × 22.8 cm. Paperback. ISBN 971-104-014-3. HDC $15.00, LDC $6.00 plus airmail ($17.40) or surface mail ($1.75) postage.

Weed control in rice is a compilation of papers presented at an IRRI-International Weed Science Society conference held in August 1981. Papers underscore the importance of weeds to crop production, identify gaps in knowledge, and emphasize the need for maintaining an ecological balance between weeds and crops.

International Rice Research Institute

Consequences of Small-Farm Mechanization

1983. 192 pages. 15 × 22.8 cm. Paperback. ISBN 971-104-082-4. HDC $7.50, LDC $3.00 plus airmail ($10.20) or surface mail ($1.25) postage.

This book contains selected papers presented at a 1981 IRRI workshop that reviewed the progress of a USAID-funded project, The Consequences of Small-Farm Mechanization on Production, Income, Rural Employment in Selected Countries of Asia. Countries include Pakistan, Thailand, Bangladesh, Indonesia, Nepal, and the Philippines.

The Institute of Genetics, Academia Sinica, and International Rice Research Institute

Cell and Tissue Culture Techniques for Cereal Crop Improvement

1983. 455 pages. 15 × 22.8 cm. Paperback. ISBN 971-104-008-1 HDC $32.50, LDC $13.00 plus airmail ($15.00) or surface mail ($1.80) postage.

China has used tissue culture techniques to develop a number of new rice varieties in the past decade. IRRI began to apply tissue culture to rice breeding in 1979. Significant progress has also been made in the laboratories of many nations.

The workshop on Potentials of Cell and Tissue Culture Techniques in the Improvement of Cereals was held in 19-23 October 1981 in Beijing, China. Cosponsors were the Institute of Genetics, Academia Sinica, and IRRI. Scientists from basic and applied research areas met to identify potential areas in which cell and tissue culture could particularly aid in varietal development of cereals, particularly of rice.

This proceedings volume includes reports of progress made by scientists from a dozen countries and the recommendations of 50 participants.

International Rice Research Institute

1983 Rice Germplasm Conservation Workshop

1983. 110 pages. 15 × 22.8 cm. Paperback. ISBN 971-104-115-4. HDC $6.25, LDC $2.50 plus airmail ($2.00) or surface mail (0.75) postage.

Rice workers of many nations have played an active role in collecting and conserving their indigenous rice germplasm since the 1930s or even earlier. The exchange and cooperative use of rice germplasm are unique examples of the power and value of cooperative endeavor.

The International Rice Research Institute (IRRI) and the International Board of Plant Genetic Resources (IBPGR) are concerned with revitalizing the collection activities toward completion of the conservation of diverse rice germplasm. The second Rice Germplasm Conservation Workshop, held at IRRI 25-26 April 1983, represents another joint venture of IRRI and IBPGR. It is hoped that the discussions and planning sessions of the Workshop, summarized in this publication, will lead to achievement of the goal of preserving for posterity the fruits of evolution in rice.

International Rice Research Institute

Potential Productivity of Field Crops Under Different Environments

1983. 530 pages. 15 × 22.8 cm. Paperback. ISBN 971-104-114-6. HDC $17.50, LDC $7.00 plus airmail ($25.50) or surface mail ($2.25) postage.

In the past two decades, much progress has been made in crop physiology dealing with the growth and potential yield of field crops under different environments. The Symposium on the Potential Productivity of Field Crops Under Different Environments, which was held in Los Baños, Philippines, 22-26 September 1980, offered a unique opportunity for physiologists working on different crops to meet, share their research progress, and discuss their problems. This volume contains key papers presented at that symposium.

International Union of Pure and Applied Chemistry and
International Rice Research Institute

CHEMRAWN II, Chemistry and World Food Supplies: Perspectives and Recommendations

1984. 169 pages. 15 × 22.8 cm. Paperback. ISBN 971-104-105-7. HDC $7.50, LDC $3.00 plus airmail ($9.60) or surface mail ($1.40) postage.

The CHEMRAWN II (Chemical Research Applied to World Needs) conference explored worldwide nutritional needs and defined areas of chemical research that may increase agricultural production and improve food processing systems in developing nations.

Two publications were produced from CHEMRAWN II. *Chemistry and World Food Supplies: The New Frontiers* was published by Pergamon Press of Canada; Third World Customers may order paperback editions from IRRI ($16.50 by airmail; $6.25 surface).

CHEMRAWN II, Chemistry and World Food Supplies: Perspectives and Recommendations volume was published by IRRI and contains eight plenary lectures by experts from both developed and developing countries who examined from a world perspective the scientific, economic, political, and cultural issues that affect food production, processing, storage, and distribution. Three of the plenary lectures were by Nobel Laureates.

Orders from North America and Europe should be sent to:
>Office of International Activities
>American Chemical Society
>1155 Sixteenth Street, N.W.
>Washington, D.C. 20036, USA

International Rice Research Institute

Workshop on Research Priorities in Tidal Swamp Rice

1984. 220 pages. 15 × 22.8 cm. Paperback. ISBN 971-104-102-2. HDC $7.50, LDC $3.00 plus airmail ($4.25) or surface mail ($0.75) postage.

Rice farmers in the tidal swamps of Bangladesh, India, Indonesia, Sri Lanka, Thailand, and Vietnam face problems different from those of farmers in more favored environments.

The Workshop on Research Priorities in Tidal Swamp Rice, held 22-25 June 1981 in Banjarmasin, South Kalimantan, Indonesia, marked a departure from the international rice research community's usual approach to tidal swamp rice culture. But scientists have increasingly begun to recognize the tidal swamps as a unique environment, and to consider the rice culture practiced in those areas important enough to merit a workshop exclusively on tidal swamp areas and attempted to identify new technologies that, by a concerted effort, could be generated for the fragile environment of tidal swamps.

International Rice Research Institute

Organic Matter and Rice

1984. 631 pages. 15 × 22.8 cm. Paperback. ISBN 971-104-104-9. HDC $17.50, LDC $7.00 plus airmail ($11.50) or surface mail (1.00) postage.

Organic matter can play an increasingly important role in rice production, particularly considering the rising prices of chemical fertilizers.

Organic Matter and Rice contains the papers presented at the International Conference on Organic Matter and Rice, which was held at IRRI 27 September to 1 October 1982. Topics covered included the potential of organic manures in rice production, organic sources of plant nutrients, decomposition of organic matter in wetland rice soils, organic matter and soil physical properties, organic matter and plant growth, and management and evaluation of organic manures.

International Rice Research Institute

Judicious and Efficient Use of Insecticides on Rice

1984. 182 pages. 15 × 22.8 cm. Paperback. ISBN 971-104-100-6. HDC $7.50, LDC $3.00 plus airmail ($3.25) or surface mail ($0.50) postage.

Insecticides are not always essential for stabilized rice production, and sometimes may be dangerous to human health and the environment. The 1983 Workshop on Judicious and Efficient Use of Insecticides on Rice focused on integrated pest control, including biological, cultural, and varietal control measures. Topics range from efficacy and economic evaluation to resistance and resurgence. Workshop recommendations for new directions that insecticide technology in rice should follow are included.

Rockefeller Foundation and International Rice Research Institute

Biotechnology in International Agricultural Research

1985. 435 pages. 15 × 22.8 cm. Paperback. ISBN 971-104-124-3. HDC $8.75, LDC $3.50 plus airmail ($6.00) or surface mail ($0.50) postage.

The International Agricultural Research Centers (IARCs) are concerned with the development of economical, ecologically sound, and socially acceptable technologies. Another important aspect of their work is the diffusion of benefits from new technologies to all farmers, irrespective of farm size or farmer input purchasing and risk-taking capacity.

This publication includes papers presented and recommendations drafted by a distinguished group of international experts. Three broad areas discussed are: in vitro propagation, disease elimination, germ-plasm conservation and exchange; wide crosses, somatic hybridization and embryo culture in plants, and molecular biology and genetic engineering in plants and animals; animal diseases and livestock productivity; and general recommendations for future IARC research.

International Rice Research Institute

An Overview of Upland Rice Research Proceedings of the 1982 Bouaké, Ivory Coast Upland Rice Workshop

1984. 566 pages. 15 × 22.8 cm. Paperback. ISBN 971-104-121-9. US$17.50 plus airmail ($11.50) or surface mail ($1.00) postage.

This is perhaps the first major compilation of papers on upland rice that includes comprehensive information on varietal improvement, agronomic management, environment, and cultural practices for Africa, Asia, and Latin America. Included are 6 base papers, 2 from each continent, and 32 scientific papers.

International Rice Research Institute

Multilanguage Publication in Agriculture: Workshop Report and Description of Participating Agencies

1985. 59 pages. 15 × 22.8 cm. Paperback. ISBN 971-104-123-5. US$2.75 plus airmail ($2.00) or surface mail ($0.50) postage.

Language differences that block effective communication in agriculture are a world tragedy.

About 60 key communicators of agricultural information in developing nations met at the International Rice Research Institute in December 1983 at an international workshop on Copublication: Strategies for Multilanguage Publication in Agriculture. Participants discussed strategies whereby educational materials, particularly those for extension agents and farmers, could be systematically translated, copublished, and distributed by national agricultural agencies and private publishers.

This volume includes recommendations of the workshop delegates for an international network of international and national agencies and publishers to systematically inform members of key agricultural materials that are available for translation and copublication.

The volume includes descriptions and addresses of 49 Third World agencies and publishers that have expressed specific interest in the translation and copublication of agricultural materials in nonEnglish languages.

International Rice Research Institute

Education for Agriculture

1985. 204 pages. 15 × 22.8 cm. Paperback. ISBN 971-104-126-X. HDC $8.75, LDC $3.75 plus airmail ($3.50) or surface mail ($0.50) postage.

In November 1984 the Symposium on Education for Agriculture was held at IRRI, jointly sponsored by the Committee on Science and Technology in Developing Countries (COSTED) and the Asian Association of Agricultural Colleges and Universities (AAACU). Its purpose was to review the present state of agricultural education in the context of the opportunities now existing for a learning revolution. Participants were from leading agricultural universities in Asia, Africa, North America, and Europe as well as from ministries of agriculture and education, international and national research centers, funding agencies, agribusiness entities, and farmers' organizations.

The objectives were to assess present needs of agricultural education and training and to suggest methods for bringing together the resources and capabilities of all sectors involved in this great endeavor.

This volume includes 19 papers presented at the symposium.

Miscellaneous

International Rice Research Institute

The IRRI Cropping Systems Training Program with Analysis of the 1980-81 CSTP

21.8 × 28 cm. Free to rice improvement and cropping systems programs.

This booklet describes the history of the IRRI Cropping Systems Training Program (CSTP) from its inception in 1969. Included are discussions of the course objectives, course content, training methodology, and training activities. Included is an analysis of the 1980-81 CSTP based on responses of the trainees to a questionnaire covering various aspects of the program.

International Rice Research Institute

IRTP Monitoring Tour Reports

21.7 × 27.9 cm. Free to rice improvement programs by subscription.

Rice monitoring tours are organized yearly by the International Rice Testing Program (IRTP). They provide the opportunity for scientists from national programs and IRRI to jointly visit and review local and IRTP rice evaluation trials. Tour participants share their research experiences, discuss and plan future research strategies, and make specific recommendations on aspects relating to the focal subject of the tours. IRRI publishes the monitoring tour reports, containing the observations and recommendations.

International Rice Research Institute

Annual Report of IRTP Nurseries

21 × 27.8 cm. Free to rice improvement programs by subscription.

The International Rice Testing Program (IRTP), coordinated by IRRI, represents a community effort of worldwide scientists in different disciplines to speed the development of improved varieties for the diverse environments in which rice is grown. The most promising breeding materials from IRRI and the national programs are exchanged annually for uniform, worldwide evaluation. The annual report of IRTP nurseries contains results of such international testing.

International Board for Plant Genetic Resources and International Rice Research Institute

Genetic Conservation of Rice

1970. 54 pages. 15 × 22.8 cm. Paperback. ISBN 971-104-005-0. HDC $3.00, LDC $1.20 plus airmail ($5.40) or surface mail ($1.00) postage.

A report on the varietal diversity of rice, including a review of the status of rice germplasm collections in Asia, Africa, and Latin America; the extent of genetic erosion in rice; and recommendations for long-range plans for a global program of rice genetic resources collection and conservation.

International Rice Research Institute and International Board for Plant Genetic Resources

Descriptors for Rice *Oryza Sativa* L

1980. 21 pages. 15 × 22.8 cm. Paperback. ISBN 971-104-000-X. HDC $3.00, LDC $1.20 plus airmail ($3.90) or surface mail ($1.00).

This bulletin stems from discussions of interested rice researchers during the 1977 Workshop on the Genetic Conservation of Rice, sponsored by IRRI and the International Board for Plant Genetic Resources.

International Rice Research Institute

Parentage of IRRI Crosses IR1-IR50,000

1980. 306 pages. 10.2 × 24 cm. Paperback. ISBN 971-104-001-6. HDC $10.00, LDC $4.00 plus airmail ($8.70) or surface mail ($1.25) postage.

Rice researchers often need to know the ancestry of IRRI varieties and breeding lines. Computerization of IRRI breeding records has made this compact summary of the first 50,000 IRRI crosses possible.

International Rice Research Institute

Beyond IR8/IRRI's Second Decade

1980. 27 pages. 21.5 × 27.9 cm. Paperback. ISBN 971-104-057-3. HDC $8.25, LDC $3.30 plus airmail ($4.80) or surface mail ($1.00) postage.

IR8 was to tropical rice as what the Model T Ford was to automobiles . . . a rugged variety that could go almost anywhere. IRRI released IR8 in the late 1966; the variety revolutionized production in tropics and inspired such terms as miracle rice and green revolution. But as IRRI entered its second decade in the 1970s it was becoming obvious that the worldwide impact of the IR8-type varieties was less than expected; only 25-30% of the world's rice farmers have adopted the new rices.

This nontechnical bulletin describes progress made at IRRI during the 1970s and IRRI's programs to develop new ranges of improved rice technology for the regions bypassed by the Green Revolution. It includes 66 color photos.

N. C. Brady

A Global Experiment in Agricultural Development

1982. 24 pages. 15 × 22.8 cm. Paperback. ISBN 971-104-062-X. HDC $2.50, LDC $1.00 plus airmail ($2.75) or surface mail ($0.40) postage.

This publication describes the establishment, accomplishments, and future challenges of a bold new concept in agricultural development — the worldwide network of International Agricultural Research Centers. The author, Dr. Nyle C. Brady, served as IRRI director general from 1973 to 1981 and is now Senior Assistant Administrator, Bureau of Science and Technology, US Agency for International Development.

International Rice Research Institute

A Plan for IRRI's Third Decade

1982. 79 pages. 17.7 × 22.8 cm. Paperback. ISBN 971-104-066-2. HDC $5.00, LDC $2.00 plus airmail ($5.40) or surface mail ($1.00) postage.

IRRI's contributions to the alleviation of hunger have been significant, but the Institute's greatest contributions may be expected in the coming decade. IRRI's strategy for the coming decade is outlined in this plan. It should provide the basis for a continuing program of rice research to meet the world's rice requirements for the coming decade and lay foundations for future increases in rice production that will be needed in the year 2000.

Food and Agriculture Organization (FAO) and International Rice Research Institute (IRRI)

Natural Enemies of Insect Pests of Rice

1983. 79 × 22.8 cm. Enamel stock paper. HDC $2.00, LDC $0.50 plus airmail ($3.90) or surface mail ($0.50) postage. (Add $1.00 or ₱10.00 for cost of mailing tube.) Available in Bahasa Indonesian, Bahasa Malaysian, Bengali, Bikolano, Cebuano, English, Hindi, Ilongo, Ilokano, Maguindanao, Pilipino, Tamil, Thai, Warai. 49 color photos, 5 diagrams.

Natural enemies control populations of all rice insect pests. Insecticides kill natural enemies and can cause outbreaks of major pests such as brown planthoppers and whitebacked planthoppers.

The poster, available from IRRI in 14 languages, was designed to help farmers and rice workers recognize these natural enemies and preserve them by using insecticides only when rice pest populations are so high that they will damage the rice crop. Smaller numbers of pests will not damage rice and provide food for their natural enemies.

Library and Documentation Center, International Rice Research
Institute

International Directory of Rice Workers

1983. 221 pages. 21.5 × 27.9 cm. Paperback. ISBN 971-104-098-0.
HDC $8.00, LDC $3.20 plus airmail ($8.20) or surface mail ($0.90)
postage.

The *International Directory of Rice Workers* is the first comprehensive
listing of scientists, administrators and extension specialists of the rice
world. The directory includes the following information: name, date of
birth, marital status, children, nationality, field of specialization,
education, membership in professional organizations, awards, professional experience, and address.

The first listing gives complete biographic information on rice workers.
The second listing is by country of residence and the third, by field of
specialization.

International Rice Research Institute

Terminology for Rice Growing Environments

1984. 40 pages. 15 × 22.8 cm. Paperback. ISBN 971-104-119-7. HDC
$2.75, LDC $1.10 plus airmail ($5.30) or surface mail ($1.00) postage.

This publication briefly describes the IRRI terminology system for rice
growing environments. It includes two papers. The first paper names
and describes environments ranging from uplands to tidal wetlands.
The second integrates the IRRI system with existing and proposed
environmental classification systems.

International Rice Research Institute

International Rice Research: 25 Years of Partnership

1985. 188 pages. 17.7 × 22.8 cm. Paperback. ISBN 971-104-130-8.
HDC $7.50, LDC $3.00 plus airmail ($3.50) or surface mail ($0.50)
postage.

On the occasion of its 25th anniversary, the International Rice
Research Institute takes pride in releasing *International Rice Research:
25 Years of Partnership.*

In this book IRRI presents the fruits of 25 years of cooperation with
national rice research programs and with advanced scientific institutions in both developed and developing countries. The record of
progress in rice science is an outstanding example of the power of

purposeful collaboration. The past 25 years have seen marked increases in the quantity and quality of rice production.

Chapter titles are:
1. Rice and the role of IRRI
2. IRRI and national programs
3. Genetic resources
4. Varietal improvement
5. Advances in rice production technology
6. Integrating rice technology into farmers' production systems
7. Impact of modern rice technology
8. Training and communication
9. Looking ahead

Maps

R. E. Huke/Maps 1

Agroclimatic and Dry-Season Maps of South, Southeast, and East Asia

1982. 21.7 × 27.9 cm. ISBN 971-104-069-7. HDC $25.00, LDC $10.00 plus airmail ($10.80) or surface mail ($1.25) postage.

Agroclimatic and Dry Season Maps of South, Southeast, and East Asia is a series of five maps, with a booklet, portraying the length, time, and intensity of the water-deficit period for the rice-producing area from Pakistan through Korea.

Each map set presents a climate-based regional division of the area producing the vast majority of the world's rice. The regions are generic to enable ready identification of areas of similar climatic regime. Data from about 3,000 research stations were used to develop the dry-season maps.

The maps were developed by Dr. Robert E. Huke, an IRRI visiting scientist from the Geography Department, Dartmouth College, New Hampshire, USA.

R. E. Huke/Maps 2

Rice Area by Type of Culture: South, Southeast, and East Asia

1982. 21.7 × 27.9 cm. ISBN 971-104-068-9. HDC $25.00, LDC $10.00 plus airmail ($10.20) or surface mail ($0.80) postage.

The three maps and accompanying tables in *Rice Area by Type of Culture: South, Southeast, and East Asia* are the first attempt to unify terms and to standardize the data base for Asia's rice-producing areas. The maps present the data in a spatial perspective; the tables provide a base for statistical analysis.

Dr. Robert E. Huke, IRRI visiting scientist from the Department of Geography, Dartmouth College, Hanover, New Hampshire, USA, collected and standardized the data and plotted the maps.

R. E. Huke/Maps 3

Complete Series of the Above 8 Maps Plus 2 Accompanying Booklets and Folders

ISBN 971-104-070-0. HDC $42.50, LDC $17.00 plus airmail ($20.40) or surface mail ($1.50) postage.

The map series were developed by Dr. Robert E. Huke, an IRRI visiting scientist from the Geography Department, Dartmouth College, Hanover, New Hampshire, USA.

International Rice Research Institute/Maps 4

Agroclimatic Map of the Philippines

ISBN 971-104-067-0. HDC $3.00, LDC $1.20 plus airmail ($3.00) or surface mail ($.075) postage.

Bibliographies

Requests and purchases for bibliographies may be addressed to:

> Library and Documentation Center
> IRRI, P.O. Box 933
> Manila, Philippines.

Milagros C. Zamora, The Library and Documentation Center

International Bibliography of Rice Research (Annual Supplements)

1983. 690 pages. 21.7 × 27.9 cm. Paperback. ISSN 0074-2021. HDC $60.00, LDC $24.00 plus airmail ($51.00) or surface mail ($2.25) postage.

This annual bibliography contains worldwide references to rice research published each year. The arrangement is classified. An author index and a manually produced keyword index are provided.

Also available for: 1977, 1978, 1979, 1980, 1981, and 1982.

Dionisia T. Capaya, The Library and Documentation Center

International Bibliography on Azolla

1980. 66 pages. 21.5 × 27.9 cm. Paperback. HDC $12.75, LDC $5.10 plus airmail ($4.50) or surface mail ($1.50) postage.

The water fern Azolla is economically important because, through its symbiotic relationship with the blue-green algae Anabaena, it can "fix" nitrogen in rice paddies. Research reports on Azolla are widely scattered in the scientific literature, in diverse languages. This bibliography attempts to gather all Azolla literature; citations date from 1783 through 1979.

Library and Documentation Center

International Bibliography on Azolla (1983 Supplement)

1984. 106 pages. 22 × 28 cm. Paperback. US$13.75 plus airmail ($5.40) or surface mail ($1.60) postage.

This bibliography supplements and updates the International Bibliography on Azolla published in 1979. It includes literature published from 1980 to 1983 and earlier literature that was not included in the 1979 volume.

The entries follow a classified arrangement. Author and keyword indexes are provided at the end.

Most of the items included are available at the IRRI Library and Documentation Center. These items are made available on request to working scientists of all countries at the following nominal charges:

> Photoprints: US$0.20 for each page or a fraction
> copied from an article or book, plus postage.

Mila R. Ramos, The Library and Documentation Center

International Bibliography on Cropping Systems (Annual Supplements)

1979. 336 pages. 21.5 × 27.9 cm. Paperback. HDC $20.00, LDC $8.00 plus airmail ($22.20) or surface mail ($1.75) postage.

This bibliography supplements and updates the *International Bibliography on Cropping Systems, 1973-1977.* The entries are classified. Author and keyword indexes are provided at the end.

Also available for: 1973-1974, 1975, 1976, 1977, 1978.

Fe M. Alluri, The Library and Documentation Center

A Bibliography of Rice Literature (Translations Available in the International Rice Research Institute Library and Documentation Center)

1976. 191 pages. 21.5 × 27.9 cm. Paperback. HDC $15.00, LDC $6.00 plus airmail ($14.40) or surface mail ($1.25) postage.

A compilation of all translations of rice literature available in the IRRI Library and Documentation Center as of October 1975. Each translation entry includes: author and title of article, title of publication, volume number, issue number, pagination, publication date, translators' names, IRRI Library numbers, and a keyword index.

International Rice Testing Program, International Rice Research Institute

Upland Rice Research: an International Bibliography 1965-1982

1983. 129 pages. 21.7 × 27.9 cm. Paperback. HDC $7.50, LDC $3.00 plus airmail ($8.60) or surface mail ($1.00) postage.

Most upland rice farmers continue to grow locally selected varieties under low input levels, and have not benefited from the fruits of modern rice technology.

Production problems of upland rice demand specially tailored varieties and management techniques. Satisfactory progress will be achieved only with active international cooperation.

Upland rice researchers must know about each other's results and experiences. Unfortunately, their work is widely scattered in the literature.

This bibliography was compiled to make world literature specifically focused on upland rice conveniently available. The *Bibliography* contains 1,222 citations representing research in 44 countries.

Periodicals

International Rice Research Institute

IRRI Research Paper Series (IRPS)

Number of pages varies. 21.5 × 27.9 cm. Saddle stitched. ISSN 0115-3862.

The *IRRI Research Paper Series* is a vehicle for timely publication of research findings by IRRI and associated scientists. The papers go regularly to libraries and certain institutions. Individuals may request *for only one or two selected papers* on a complimentary basis. Requests for complimentary copies must be made on organizational letterheads. Additional copies may be purchased at prices indicated below.

Overseas				Philippines	
Per copy		Air-mail	Surface mail	Air-mail	Surface mail
HDC	LDC				
$1.25	$0.50	$4.30	$0.85	₱15.80	₱0.80

Bound volumes of IRPS, 20 issues/volume (IRPS 1-20, IRPS 21-40, IRPS 41-60, IRPS 61-80, IRPS 81-100) are available at the following prices:

Overseas				Philippines	
Per copy		Air-mail	Surface mail	Air-mail	Surface mail
HDC	LDC				
$46.00	$18.50	$36.00	$2.10	₱108.00	₱3.20

No. 1 Recent studies on rice tungro disease at IRRI
No. 2 Specific soil chemical characteristics for rice production in Asia
No. 3 Biological nitrogen fixation in paddy field studied by in situ acetylene-reduction assays
No. 4 Transmission of rice tungro virus at various temperatures: a transitory virus-vector interaction
No. 5 Physiochemical properties of submerged soils in relation to fertility
No. 6 Screening rice for tolerance to mineral stresses
No. 7 Multi-site tests environments and breeding strategies for new rice technology
No. 8 Behavior of minor elements in paddy soils
No. 9 Zinc deficiency in rice: a review of research at the International Rice Research Institute
No. 10 Genetic and sociologic aspects of rice breeding in India
No. 11 Utilization of the azolla-anabaena complex as a nitrogen fertilizer for rice
No. 12 Scientific communication among rice breeders in 10 Asian nations
No. 13 Rice breeders in Asia: a 10-country survey of their backgrounds, attitudes, and use of genetic materials
No. 14 Drought and rice improvement in perspective
No. 15 Risk and uncertainty as factors in crop improvement research
No. 16 Rice ragged stunt disease in the Philippines
No. 17 Residues of carbofuran applied as a systemic insecticide in irrigated wetland rice: implications for insect control

International Rice Research Institute

International Rice Research Newsletter (IRRN)

Number of pages varies. 21.5 × 27.9 cm. Saddle stitched. Free. ISSN
0115-0944.

The International Rice Research newsletter (IRRN) is published by
IRRI in the interest of improving the communication of relevant rice
research among scientists in many nations. The IRRN is published
every 2 months and has an annual subject index.

The IRRN invites all scientists to contribute concise summaries of
significant research on rice or rice-based cropping systems for review.
Contributions should be limited to no more than two double-spaced
pages and no more than two graphics (tables, figures, or photographs).
Authors are identified by name, title, and research organization.

International Rice Research Institute

IRRI Reporter

Usually 4 pages. 21.5 × 27.9 cm. Single fold. ISSN 0115-2467. Free.

The IRRI Reporter is a quarterly publication describing IRRI's research
program and directions.

Complimentary subscriptions to the *IRRI Reporter* and the IRRN are
available. Please use the tear-up Mailing List Access Form (Appendix C).

Audiovisual Materials

International Rice Research Institute

IRRI Audiovisual Production Training Modules

A 63-module series of audiovisual lessons in rice production training is available from IRRI. The modules are organized into seven topical units: Production Management; Growth and Morphology of Rice; Production Problems and Techniques: Weeds, Diseases, and Their Control; Pests and Their Control; Research Design and Analysis; and Soil Relationships. Developed to supplement IRRI's 6-month rice production training course, the audiovisual series is an aid to rice production in the tropics.

LDC countries: $35.00/module (includes airmail postage). For orders of 5 or more modules, $25.00/module plus actual airfreight. Request a pro forma invoice for orders of 5 or more modules. For complete 63-module series, $1500.00 (includes airfreight).

Other countries: $45.00/module (includes air mail postage). For orders of 5 or more modules, $35.00/module plus actual airfreight. Request a pro forma invoice for order of 5 or more modules. For complete 63-module series, $2000.00 (includes airfreight).

International Rice Research Institute

PM-1 Fertilizer and Fertilizer Management

71 colorslides, script, cassette, 47 minutes.

From this module, the student learns to:
1. Describe fertilizer requirements of rice varieties in a given soil condition.
2. Explain why basal fertilizer is incorporated into the soil.
3. Explain the effects of trace elements on the utilization of N, P, and K.
4. List and explain significant research findings on fertilizer rate, type, and method of application.
5. Explain the reasons for basal application and topdressing of fertilizer and their effects on yield.
6. Describe the seasonal effect on fertilizer response.

International Rice Research Institute

PM-2 The Two-Crop System of Growing Rice in Rainfed-Lowland Areas

61 colored slides, script, cassette. 27 minutes.

From this module, the student learns to:
1. Recognize the possibilities of growing two-crop of rice.
2. Identify each step necessary for establishing the first and second rice crops under this system.
3. Describe the cultural and management practices to successfully raise 2 crops under rainfed-lowland conditions.
4. Perform the skills in each practice involved in the two-crop system of growing rice.

International Rice Research Institute

PM-3 Fertilizer Materials and Calculations

76 colored slides, script, cassette. 36 minutes.

From this module, the student learns to:
1. Define a fertilizer and recognize the forms of fertilizers.
2. List the common fertilizer materials and describe their characteristics.
3. Calculate the amount of fertilizer materials needed given the area, recommended rate, and kind of fertilizer materials.
4. Determine least expensive fertilizer combinations given the recommended rate and prices of various materials.

International Rice Research Institute

PM-4 Irrigation Use and Management

73 colored slides, script, cassette. 39 minutes.

From this module, the student learns to:
1. Compute the volume of water required to till and puddle a given land area and to grow a rice crop.
2. Identify water sources that can be used for irrigation.
3. Discuss the effects of flooding on rice.
4. Describe irrigation practices and their advantages and disadvantages.

International Rice Research Institute

PM-5 Solar Radiation in Relation to Rice Production

43 colored slides, script, cassette. 30 minutes.

From this module, the student learns to:
1. Explain solar radiation and how it is measured.
2. Explain the effect of solar radiation on photosynthesis.
3. Describe the effect of solar radiation on spikelet number, dry matter accumulation, and grain yield.
4. Explain the influence of solar radiation on the reproductive and ripening stages.
5. Explain the effect of solar radiation on the nitrogen response in rice.
6. Explain the effect of solar radiaticn on rice grown in monsoon areas.
7. Describe differences in the effects of temperature and solar radiation on grain yield.

International Rice Research Institute

PM-6 Yield Components

53 colored slides, script, cassette. 38 minutes.

From this module, the student learns to:
1. List the components that influence the total yield of rice.
2. Suggest ways to maximize the contribution of each component to total yield.
3. Describe the effects of management practices such as fertilizer application, water control, and weeding on yield components.
4. Describe the procedure for determining the yield components in rice.

International Rice Research Institute

PM-7 Lodging in Rice

79 colored slides, script, cassette. 39 minutes.

From this module, the student learns to:
1. Recognize the different types of lodging.
2. Identify the potential economic losses due to lodging.
3. Identify the factors that lead to lodging; explain the mechanics involved.
4. Identify the plant characteristics associated with resistance to lodging; explain how they are affected by environmental factors.
5. List the ways to minimize lodging.

International Rice Research Institute

PM-8 Enterprise Budgeting

59 colored slides, script, cassette. 40 minutes.

From this module, the student learns to:
1. Explain enterprise budgeting and how it differs from whole farm budgeting.
2. Evaluate gross benefits and estimate effective crop prices and variable costs.
3. Estimate the cost of labor, capital, and land.
4. Estimate returns to land, labor, and capital.
5. Evaluate data sources.

International Rice Research Institute

PM-9 Irrigation Management and Techniques

54 colored slides, script, cassette. 28 minutes.

From this module, the student learns to:
1. Discuss the conveyance of irrigation waters.
2. Describe how water is lost through conveyance and how the loss can be minimized.
3. List how water in a field is lost and describe ways to minimize the water losses.
4. Describe techniques to measure the flow rate of irrigation water and use common measuring procedures.
5. Identify types of water pumps; describe their operation and maintenance.

International Rice Research Institute

PM-10 Introduction to Azolla

41 colored slides, script, cassette. 14 minutes.

From this module, the student learns to:
1. Identify Azolla.
2. Identify the structures and important morphological characteristics of Anabaena.
3. Describe the symbiotic relation between Azolla and Anabaena.
4. Identify the reproductive structures of Azolla.
5. Describe sexual reproduction in Azolla.
6. Describe the physiological response of Azolla to temperature, light, and mineral nutrients.

International Rice Research Institute

GM-1 Morphology of the Rice Plant

75 colored slides, script, cassette. 25 minutes.

From this module, the student learns to:
1. Name the three eco-geographic races of rice and recognize their traits.
2. Relate the structures and morphological characteristics of the rice plant to varietal differences.
3. Describe the milling of rice and the physicochemical characteristics of the milled kernels.

International Rice Research Institute

GM-2 Dormancy and Germination of the Rice Seed

39 colored slides, script, cassette. 27 minutes.

From this module, the student learns to:
1. Define seed dormancy and explain its significance.
2. Explain the possible causes of dormancy in rice seed and other plant seeds.
3. Describe methods of breaking dormancy in rice seed.
4. List and explain the factors that enhance or retard seed germination.
5. Identify the steps that occur in seed germination.

International Rice Research Institute

GM-3 Growth Stages of the Rice Plant

80 colored slides, script, cassette. 33 minutes.

From this module, the student learns to:
1. Explain how a rice plant grows and develops.
2. Recognize growth stages in rice.
3. Identify the growth stage of a given rice plant according to the 0 to 9 numerical scale.
4. Relate major nutrient requirements and timing of fertilization to specific stages of growth.

International Rice Research Institute

GM-4 Photosynthesis and Respiration of the Rice Plant

76 colored slides, script, cassette. 53 minutes.

From this module, the student learns to:
1. Define photosynthesis and respiration.

2. Explain the photosynthesis/respiration balance and how it affects yield.
3. Explain the significance of competition for light in a rice population.
4. Define mutual shading.
5. Explain the relations between plant type and mutual shading and between mutual shading and grain yield.
6. Recognize the environmental factors that affect photosynthesis/ respiration.

International Rice Research Institute

GM-5 Effects of Temperature and Photoperiodism on Rice Growth

68 colored slides, script, cassette. 41 minutes.

From this module, the student learns to:
1. Describe the effect to temperature on plant growth.
2. Describe the effect to temperature on the morphology and reproduction of the rice plant.
3. Define photoperiodism; explain how a photoperiod sensitive variety is affected by daylength.
4. List the prerequisites for photoperiod induction.
5. Describe the effect of daylength on the growth stages of different rice varieties.

International Rice Research Institute

GM-6 Physicochemical Properties of the Rice Grain

55 colored slides, script, cassette. 26 minutes.

From this module, the student learns to:
1. List the composition and distribution of constituents of brown rice.
2. List factors affecting milling, cooking, and eating quality of rice.
3. Describe the properties of rice starch and how they affect cooking.
4. Describe the effect of temperature during ripening on the amylose content and gelatinization temperature of starch.
5. List screening procedures for cooking and eating quality.
6. List changes during grain storage that affect cooking and eating quality.

International Rice Research Institute

PPT-1 Why Rice Yields are Low in the Tropics

73 colored slides, script, cassette. 53 minutes.

From this module, the student learns to:
1. Identify the causes of low rice yields in the tropics.
2. Describe approaches for increasing rice production in the tropics.

International Rice Research Institute

PPT-2 Weather Factors and Their Relations to One Another

55 colored slides, script, cassette. 48 minutes.

From this module, the student learns to:
1. Define agricultural meteorology.
2. Differentiate weather from climate.
3. List and describe the weather elements in the tropics.
4. List the common meteorological instruments used to measure weather elements.
5. Explain the relations among weather elements.
6. Determine practical applications of climatic information in crop production.

International Rice Research Institute

PPT-3 Methods of Growing Rice

74 colored slides, script, cassette. 38 minutes.

From this module, the student learns to:
1. Describe the two general methods of growing rice and the advantages and disadvantages of each method.
2. Discuss the techniques and requirements for each method.
3. Describe the three methods of raising seedlings for transplanting; list their advantages and disadvantages.
4. Explain the factors affecting proper spacing in transplanted rice and their importance.

International Rice Research Institute

PPT-4 Land Preparation

71 colored slides, script, cassette. 47 minutes.

From this module, the student learns to:
1. Explain the reasons for land preparation; list and describe the operations involved in primary and secondary tillage.
2. List and describe the key sequential steps in land preparation.
3. Identify the types of tillage equipment and their use.
4. Define terms used in plowing.
5. Describe various tilling patterns for a rectangular field; identify some common causes of plowing difficulties.
6. Compare animal power and tractor power, their cost, advantages and disadvantages, power, speed, and ease of operation.
7. Calculate the time needed to plow and/or harrow a field with a specific piece of tillage equipment.

International Rice Research Institute

PPT-5 Economic Decisions in Rice Farm Production

78 colored slides, script, cassette. 41 minutes.

From this module, the student learns to:
1. List the primary management decisions a farmer must make in a rice production enterprise.
2. Apply the marginal principle of economics to rice production; use the concepts of opportunity cost to allocate limited cash.
3. Calculate the costs of and returns on rice production.

International Rice Research Institute

PPT-6 Harvesting and Threshing

63 colored slides, script, cassette. 39 minutes.

From this module, the student learns to:
1. Explain how to determine when a rice crop should be harvested.
2. Give several reasons why a paddy must be harvested at optimum maturity.
3. List the methods of harvesting rice.
4. Explain the factors that affect the method of harvesting rice.
5. Describe the types of paddy threshers.
6. List factors that affect the efficiency of paddy threshers.

International Rice Research Institute

PPT-7 Drying, Storing, and Milling of Rice

78 colored slides, script, cassette. 50 minutes.

From this module, the student learns to:
1. List the reasons why paddy rice must be dried.
2. Explain the essential steps in drying paddy.
3. Identify the various grain drying systems and their operation.
4. Identify the steps necessary for the proper operation of a dryer.
5. Analyze the economics of the purchase and operation of a batch dryer; determine the break-even-point.
6. Determine the advantages and disadvantages of the various storing and milling procedures.

International Rice Research Institute

PPT-8 Stored Grain Pests of Rice

46 colored slides, script, cassette. 24 minutes.

From this module, the student learns to:
1. Identify the grain pests that attack stored rice.
2. Describe the mode of attack of these pests on the rice grain.
3. Discuss sources of infestation and methods of detecting stored grain insects.
4. Identify methods of preventing and controlling insect pests of stored rice grain.

International Rice Research Institute

PPT-9 Sprayer Calibration

40 colored slides, script, cassette. 20 minutes.

From this module, the student learns to:
1. Calibrate a sprayer and determine the application rate as volume of water per unit area.
2. Measure nozzle delivery, walking speed, and swath.
3. Compute the total amount of chemical and water needed to treat a given area.
4. List modifications that can be made to the knapsack sprayer to make it more efficient.
5. Maintain sprayer equipment.

International Rice Research Institute

WDC-1 Weed Identification — Lowland and Upland Rice

66 colored slides, script, cassette. 34 minutes.

From this module, the student learns to:
1. Define a weed and describe how weeds affect crop growth.
2. Recognize when weeds in rice should be controlled.
3. Recognize botanical taxa and their taxonomic relations.
4. Interpret the scientific nomenclature of the weeds described in this module.
5. Classify weeds into broadleaf weeds, grasses, and sedges.
6. Identify the most common weeds in lowland and upland rice.

International Rice Research Institute

WDC-2 Methods of Weed Control

62 colored slides, script, cassette. 34 minutes.

From this module, the student learns to:
1. List and explain the methods (indirect and direct) of controlling weeds, with special emphasis on lowland rice.
2. Explain the principles of chemical weed control: why and under what conditions herbicides kill weeds but not the rice plants.
3. Differentiate systemic from contact herbicides and selective from nonselective herbicides.
4. List the requirements for effective use of herbicides to control rice weeds.
5. List according to their efficiency, effectiveness, and physical forms the herbicides recommended for controlling weeds in lowland and upland soils; give the rate and time of application.
6. List criteria for selecting herbicides when more than one is available.

International Rice Research Institute

WDC-3 Bacterial Blight and Bacterial Leaf Streak of Rice in Tropical Asia

65 colored slides, script, cassette. 18 minutes.

From this module, the student learns to:
1. Describe the symptoms of bacterial blight and bacterial leaf streak.
2. Identify bacterial blight and bacterial leaf streak both in the field and from specimens or slides.
3. Describe the development and transmission of bacterial blight and bacterial leaf streak.
4. Give most effective and practical control measures for both diseases.

International Rice Research Institute

WDC-4 Fungal Diseases of Rice

80 colored slides, script, cassette. 55 minutes.

From this module, the student learns to:
1. Describe and identify the symptoms of blast disease in the leaf, stem, and panicle.
2. Name the causal organism of blast; describe its life cycle and the factors which favor its development.
3. List control measures against blast disease.
4. Describe and identify the symptoms of all other fungal diseases discussed in this module; describe their life cycles.
5. List the factors favorable for the spread and development of these diseases and control measures.

International Rice Research Institute

WDC-5 Virus and Virus-Like Diseases of Rice

71 colored slides, script, cassette. 29 minutes.

From this module, the student learns to:
1. Describe the symptoms of 14 virus and virus-like diseases of rice.
2. Use a simple key for classifying virus diseases.
3. Identify these diseases from slides, specimens, or in the field.
4. Give the geographical distribution of each disease.

International Rice Research Institute

WDC-6 Nutritional and Physiological Disorders of the Rice Plant

79 colored slides, script, cassette. 43 minutes.

From this module, the student learns to:
1. Identify factors affecting nutrient uptake.
2. Describe and identify the symptoms caused by deficiency of each essential element.
3. Describe procedures to solve nutritional problems.
4. Describe the toxicity symptoms of those elements that cause toxicity when they are present in excess.

International Rice Research Institute

WDC-7 Transmission and Control of Virus and Virus-Like Diseases of Rice

65 colored slides, script, cassette. 44 minutes.

From this module, the student learns to:
1. Identify the methods of rice virus transmission.
2. Categorize each of the virus and virus-like diseases by their transmission method.
3. Identify the known insect vectors for each of the virus and virus-like diseases of rice.
4. List the six major differences between a transitory and a persistent virus.
5. Draw a transmission cycle for each virus and virus-like diseases.
6. Describe the most effective control measures for each of the virus and virus-like diseases of rice.

International Rice Research Institute

WDC-8 Herbicide Calculations

37 colored slides, script, cassette. 22 minutes.

From this module, the student learns to:
1. Give the recommended rate and label information from a herbicide product.
2. Compute the amount of herbicide needed to treat a given area.
3. Convert content designations of any herbicide product to a percent of active ingredient or its equivalent.

International Rice Research Institute

PC-1 Chemical Control of Rice Insect Pests

80 colored slides, script, cassette. 54 minutes.

From this module, the student learns to:
1. Describe how insecticides kill insect pests of rice.
2. Explain the modes of action of contact and systemic insecticides.
3. Explain why foliar spray chemical control is often ineffective in the tropics.
4. List the common contact and systemic insecticides and give their recommended rates for important insect pests.
5. List precautions to be observed when using insecticides.
6. Explain the criteria used to determine whether to follow preventive or protective measures; if a protective measure is taken, explain how to decide the appropriate application time.

International Rice Research Institute

PC-2 Rice Gall Midge and Whorl Maggot

60 colored slides, script, cassette. 52 minutes.

From this module, the student learns to:
1. Identify the eggs, larvae, and adults of the gall midge and the whorl maggot.
2. Describe the life history of the gall midge and whorl maggot.
3. Identify and describe the damage caused by the gall midge and the whorl maggot.
4. Describe the advances in breeding for resistance to these pests.

International Rice Research Institute

PC-3 Stem Borers of Rice

79 colored slides, script, cassette. 28 minutes.

From this module, the student learns to:
1. Describe the life cycle of the major stem borers.
2. Identify the egg masses, larvae, and adults of the most important rice stem borers.
3. Identify damage to rice caused by rice stem borers.
4. Describe the most effective methods of controlling the major stem borers of rice.

International Rice Research Institute

PC-4 Leafhoppers and Planthoppers of Rice

71 colored slides, script, cassette. 38 minutes.

From this module, the student learns to:
1. Identify the egg masses, nymphs, and adults of the major leafhoppers and planthoppers of rice.
2. Describe the life cycle of the major leafhoppers and planthoppers of rice.
3. Identify the damage caused by the leafhoppers and planthoppers of rice.
4. List the rice virus diseases transmitted by each of the leafhoppers and planthoppers.
5. Describe the most effective methods of controlling leafhoppers and planthoppers.

International Rice Research Institute

PC-5 Minor Pests of Rice

79 colored slides, script, cassette. 43 minutes.

From this module, the student learns to:
1. Identify the egg masses, larvae, nymphs, and adults of the rice bug, stink bug, leaffolder, caseworm, greenhorned caterpillar, skipper, rice hispa, and thrips.
2. Identify the damage caused by these insects.

International Rice Research Institute

PC-6 Insecticide Calculations

52 colored slides, script, cassette. 42 minutes.

From this module, the student learns to:
1. Calculate the amount of insecticide required to make a given parts-per-million (ppm) suspension of an insecticide for root soaking.
2. Calculate the weight of a dust formulation to be applied to a given area.
3. Calculate the weight of a granular formulation to be broadcast over a given area.
4. Make a spray solution of a given percentage concentration.
5. Calculate the weight or volume of an insecticide formulation to mix with water to spray a given area at a given rate.

International Rice Research Institute

PC-7 Identification and Biology of Rats

72 colored slides, script, cassette. 57 minutes.

From this module, the student learns to:
1. Identify the four species of rats which cause damage to Philippine agricultural crops.
2. List the physical characteristics of the rat.
3. Describe the life history, reproductive pattern, and population increase potential of the rat.
4. Describe the habits and behavior of the common rice field rat.
5. Recognize the signs of rat infestations.
6. Estimate the number of rats in a given area by interpreting the "signs".

International Rice Research Institute

PC-8 Methods of Rat Control

61 colored slides, script, cassette. 42 minutes.

From this module, the student learns to:
1. Describe rat control methods.
2. Cite the advantages and disadvantages of each method.
3. Differentiate between acute and chronic poisons.
4. Describe the factors that affect the choice of a rodenticide.
5. Explain "bait shyness" of rats; list techniques to overcome this problem.
6. Describe baiting techniques; explain why prebaiting is done when using acute poison.
7. Explain the construction and use of electric rat control fences.

International Rice Research Institute

PC-9 Insect Pest Management in Rice

64 colored slides, script, cassette. 41 minutes.

From this module, the student learns to:
1. Identify the factors that contribute to the need for an insect pest management program.
2. Recognize the four information elements basic to initiating an insect pest management program.
3. Describe sampling techniques and determine the proper technique to use in a given situation.
4. Determine the economic threshold levels for the insects discussed in the module.
5. Name the four components of a complete pest management program; discuss how they are used in an effective control program.

International Rice Research Institute

PC-10 Varietal Resistance of Rice to Insect Pests

63 colored slides, script, cassette. 36 minutes.

From this module, the student learns to:
1. List the advantages to a pest management program of planting a resistant rice variety.
2. Identify the major resistant varieties and the insects to which they are resistant.
3. Recognize the mechanisms operating in the stem borer-resistant varieties described.
4. Discuss gall midge resistance and the expression of the resistance components found in the varieties discussed.

5. Determine the expression of the resistance components operating in brown planthopper-resistant varieties; explain the value and the limitations of seedling test for resistance.
6. Determine the expressions of the resistance components operating in whitebacked planthopper- and green leafhopper-resistant varieties.

International Rice Research Institute

PC-11 Cultural Control of Rice Insect Pests

78 colored slides, script, cassette. 55 minutes.

From this module, the student learns to:
1. Define and characterize cultural control of insect pests.
2. List the cultural practices that contribute to control of insect pests.
3. Evaluate the effectiveness of cultural practices in insect pest control.
4. Designate the control practices that can be implemented on the farm level; compare them with those that must be accomplished on a community level.

International Rice Research Institute

PC-12 Biological Control of Rice Insect Pests

80 colored slides, script, cassette. 30 minutes.

From this module, the student learns to:
1. Identify the major parasites, predators, and pathogens that naturally attack rice insect pest in tropical Asia.
2. Describe the life cycle of the parasites.
3. Identify pests that are parasitized.
4. List the natural enemies of stem borers and hoppers.
5. Describe the methods of conserving and augmenting the effectiveness of indigenous natural enemies.
6. Describe the steps in the importation, quarantine, and release of natural enemies.

International Rice Research Institute

RD-1 Introduction to Field Experimentation on Rice

44 colored slides, script, cassette. 19 minutes.

From this module, the student learns to:
1. Describe the steps in the "Experimental Approach."
2. Describe the pitfalls that may cause an experiment to fail.
3. Define a "good researcher" and list the qualities he must possess.

International Rice Research Institute

RD-2 Developing Field Experiments in Rice

66 colored slides, script, cassette. 29 minutes.

From this module, the student learns to:
1. Plan a field research problem by describing how to prepare an acceptable statement of the problem, conduct a literature survey, and define the research objectives.
2. Outline a basic research proposal.
3. Explain the importance of a control treatment.
4. Identify and avoid common pitfalls in field experimentation.

International Rice Research Institute

RD-3 Principles of Experimental Design

37 colored slides, script, cassette. 29 minutes.

From this module, the student learns to:
1. Define experimental error and its effects on treatment comparison.
2. Discuss why it is necessary to control (or reduce) experimental error.
3. Discuss why it is necessary to estimate (or measure) experimental error.
4. Describe the three ingredients required of a good experimental design.

International Rice Research Institute

RD-4 Experimental Designs for Rice Field Experiments

78 colored slides, script, cassette. 42 minutes.

From this module, the student learns to:
1. Describe the experimental designs commonly used in rice field experiments.
2. Differentiate between single-factor and factorial experiments.
3. Choose the most suitable design for an intended experiment.

International Rice Research Institute

RD-5 Establishing Field Experiments

76 colored slides, script, cassette. 53 minutes.

From this module, the student learns to:
1. Select an appropriate experimental site.
2. Choose the proper plot size, shape, and orientation.

3. Demonstrate proper use of randomization and blocking.
4. Lay out treatments in experimental blocks and plots according to the design chosen.
5. Prepare field record books.
6. Apply techniques to minimize the effects of soil heterogeneity, border plants, and missing hills on experimental results.

International Rice Research Institute

RD-6 Collection of Experimental Data

64 colored slides, script, cassette. 28 minutes.

From this module, the student learns to:
1. Avoid sampling biases by following the procedures essential to proper data collection.
2. Select proper sampling procedures in accordance with the experimental objectives.
3. Measure and record the rice characteristics in a given experiment to achieve the desired degree of precision with minimum cost.

International Rice Research Institute

RD-7 Statistical Analyses of Experimental Data

59 colored slides, script, cassette. 42 minutes.

From this module, the student learns to:
1. Compute analysis of variance for designs commonly used in rice field experiments.
2. Compare treatments statistically.
3. Measure degree of precision in the form of coefficient of variation.

International Rice Research Institute

RD-8 Rice Breeding Methods

78 colored slides, script, cassette. 47 minutes.

From this module, the student learns to:
1. Differentiate between a genetic and environmental variation.
2. Contrast a qualitative variation with a quantitative variation.
3. Discuss the importance of breeding objectives; describe how they are determined.
4. Describe how a new variety is developed by mass or pureline selection from a land variety, introduction, and hybridization and selection.

International Rice Research Institute

RD-9 Economic Analyses of Data from Rice Experiments

75 colored slides, script, cassette. 28 minutes.

From this module, the student learns to:
1. List experiments that can be analyzed economically and those that cannot.
2. Distinguish between experiments that can be used to determine which of several input alternatives are most profitable and experiments that determine what level of a single output is most profitable.
3. Demonstrate by partial budgeting how an experiment can be used to decide which of several treatments is the most economical.
4. Apply additional costs and additional returns to determine what level of inputs gives the highest net return.

International Rice Research Institute

SR-1 The Soil Body

77 colored slides, script, cassette. 49 minutes.

From this module, the student learns to:
1. Define soil and explain its role as a medium for plant growth.
2. Discuss briefly the contributions of the four major soil components to the soil-plant relation.
3. Make a schematic drawing of a soil profile; describe the characteristics of the various horizons.
4. Discuss the process of soil formation.
5. Describe the role that environmental factors play in the soil formation process.
6. Discuss the importance of soils classification and soil survey reports.

International Rice Research Institute

SR-2 Physical Properties of Soils

78 colored slides, script, cassette. 45 minutes.

From this module, the student learns to:
1. Describe the physical components of soil as a porous mixture.
2. Define soil texture, list the common textural classes, and describe the physical characteristics.
3. Define soil structure; describe types of soil structure.
4. Identify common indices of soil structural condition; discuss their effects on soil aeration and on water infiltration, percolation, and retention.
5. Define bulk density, particle density, and soil porosity.

6. Describe a simple procedure for determining bulk density.
7. Describe soil physical conditions associated with low and high bulk density.
8. Describe soil consistencies at various moisture levels; identify the soil consistency suitable for land preparation.
9. Define soil tilth and physical conditions associated with good tilth.

International Rice Research Institute

SR-3 Chemical Properties of Soils

79 colored slides, sript, cassette. 45 minutes.

From this module, the student learns to:
1. Identify the kinds of soil colloids and predict their effects on chemical properties of the soil.
2. Describe the structure and the composition of 1:1 type and 2:1 type clays.
3. Explain why soils are chemically reactive.
4. Define soil pH; identify soil pH levels and state their effects on nutrient availability and the occurrence of toxic elements.
5. Define cation exchange capacity (CEC); identify CEC levels in soils; and relate nutrient-holding capacity to CEC levels.
6. Define percentage base saturation (% BS); identify percentage BS levels in soils; state the relation of soil pH to % BS.
7. Specify the soil conditions that lead to phosphate fixation and adsorption of mobile anions.

International Rice Research Institute

SR-4 Soil Fertility

50 colored slides, script, cassette. 29 minutes.

From this module, the student learns to:
1. Define and differentiate soil fertility and soil productivity.
2. Recognize the essential nutrient elements derived from the soil; classify them as macro or micronutrient elements.
3. Discuss the fate of essential elements in the soil with emphasis on N, P, and K.
4. Discuss the methods of assessing soil fertility.

International Rice Research Institute

SR-5 Soil Sampling

29 colored slides, script, cassette. 16 minutes.

From this module, the student learns to:
1. Identify and describe sampling areas by their differences in land features such as slope, depth of surface soil, drainage, soil texture, and other visible soil characteristics.
2. Collect a representative soil sample.
3. Provide pertinent information about the area sampled by filling in the prescribed information sheet.
4. Interpret a soil analysis report.

International Rice Research Institute

SR-6 Introduction to Microbiology of Flooded Soils

60 colored slides, script, cassette. 32 minutes.

From this module, the student learns to:
1. Identify the 5 major groups of microorganisms found in the soil; give their relative sizes, numbers, and biomass.
2. Describe the characteristics of each group of soil microorganisms including their nutritional patterns when given.
3. Relate soil microorganisms to food chains and to the decomposition of organic matter.
4. Describe environmental factors that influence decomposition.
5. Explain the nitrogen cycle.

International Rice Research Institute

SR-7 Role of Microorganisms in Flooded Soils

66 colored slides, script, cassette. 35 minutes.

From this module, the student learns to:
1. Describe oxidation-reduction reactions, the accumulation or depletion of specific substances, and changes in microbial populations when soils are newly flooded.
2. Describe the unique features of mature flooded soils by their oxidized and reduced conditions and by the aerobic and anaerobic organisms found in them.
3. Identify the oxidation-reduction pathways for aerobic respiration, anaerobic respiration, and fermentation.
4. Describe the process and organisms responsible for nitrogen fixation.
5. Explain the Azolla-Anabaena complex and its role in nitrogen fixation.

International Rice Research Institute

SR-8 Electrochemical Changes in Flooded Soils

38 colored slides, script, cassette.

From this module, the student learns to:
1. Identify and describe the two distinct layers of a flooded paddy.
2. Describe soil reduction and list 6 processes that cause it.
3. Describe the electrochemical properties of soils and how they are measured.
4. Describe the electrochemical changes that occur in soils during flooding and give their reasons.
5. List advantages and disadvantages of the decrease in redox potential in terms of the availability of nutrients for rice.
6. List 3 advantages and 2 disadvantages of pH changes in flooded soils.
7. Indicate the electrical conductivity that will cause salt injury to rice.

International Rice Research Institute

SR-9 Chemical Changes in Paddy Soils

38 colored slides, script, cassette.

From this module, the student learns to:
1. Describe the chemical changes during flooding and give the reasons for them.
2. List 2 factors affecting carbon dioxide partial pressure change.
3. List 5 factors affecting ammonium accumulation.
4. List 3 factors affecting denitrification.
5. List 4 factors affecting iron change.
6. List 3 factors affecting manganese change.
7. List 3 factors affecting sulfate change.
8. Define ammonification, nitrification, denitrification and give their implications on rice growth.
9. Relate the reduction of iron, manganese, and sulfate to rice growth.
10. Cite the role of flooding on the availability of phosphorus, silicon, and zinc.
11. List 4 methods of preventing iron toxicity.
12. List 2 methods of correcting zinc deficiency in rice.

International Rice Research Institute

IRRI Audiovisual Cropping Systems Research Training Modules

A planned 45-module series of audiovisual lessons in cropping systems research; Site Selection; Site Description; Improved Cropping

Systems Design; Cropping Systems Testing; Preproduction Evaluation; Production Programs; Component Technology Development and Evaluation; Environmental Complexes, Resource Base and Present Cropping Systems; and Agroeconomic Monitoring. Two modules are now available.

International Rice Research Institute

CT-01 The Morphology of Maize

55 colored slides, instruction booklet, cassette. 22 minutes.

From this module, the student learns to:
1. Draw and label the parts of the maize kernel.
2. List the four plant organs.
3. Compare seminal and coronal roots.
4. Name the conspicuous structures of the leaf.
5. Describe the six types of maize, the functions of roots, stem macrostructure, the tassel, and the ear.
6. Explain stem function.

International Rice Research Institute

CS-01 An Introduction to Cropping Systems

62 colored slides, instruction booklet, cassette. 26 minutes.

From this module, the student learns to:
1. Define the objectives of cropping systems research, a subsistence farmer, and bottom up research.
2. Define and give an example of a superimposed trial and a researcher-managed trial.
3. List three ways of increasing world food production.
4. List the components of the farmer's environment.
5. Construct a flow diagram for cropping systems methodology.

International Rice Research Institute

Other IRRI Slide Sets

IRRI has two other sets for sale.

International Rice Research Institute

Field Problems of Tropical Rice

Price: HDC, $60.00; LDC, $50.00 (airmail postage included).

All 158 color plates that appear in the Field Problems for Tropical Rice handbook have been made into 5 cm × 5 cm (2 inch × 2 inch) color transparencies. Each slide is numbered to correspond with the number that appears with the plate in each book. The slide set is supplied with a copy of the handbook.

International Rice Research Institute

The Rices of IRRI

Price is same as rice production training modules.

This set of 80 colorslides provides a visual summary of the International Rice Research Institute's programs, accomplishments, goals, and facilities. Included with the slides is a cassette tape narration and a printed script.

Catalog

Publications on International Agricultural Research and Development)

1984. 539 pages. 13 × 21 cm. Paperback. All nations, US$10.20 (includes airmail postage and handling); Philippines, $2.50 (includes surface mail and handling)

The International Agricultural Research Centers (IARCs) focus modern agricultural research on the crops and livestock that provide 75% of the food for developing nations. The Centers are major publishers of the books, periodicals, slide sets, films, and other educational materials on agricultural science and technology for developing nations.

The second exhibition of *Publications on International Agricultural Research and Development* was held at the 1984 Frankfurt Book Fair, featuring about 1,000 titles published by 12 Centers supported by the Consultative Group on International Agricultural Research (CGIAR), 7 other IARCs, the Board on Science and Technology for International Development (BOSTID) of the U.S. National Academy of Science, and the German Agency for Technical Cooperation (GTZ). The Exhibition was organized by GTZ and the International Rice Research Institute.

IRRI published this 539-page catalog for distribution at the Frankfurt Book Fair, and is handling its worldwide distribution. The catalog is the only compilation of the major publications of all centers, GTZ, and BOSTID. Included is a description of each publication, prices, and ordering instructions.

An in-depth index helps the reader locate all publications in certain fields (i.e. cytogenetics, insect resistance, maize).

Appendix A

Peso Postage Information

ISBN No. 971-104:	Title	Price (₱)[*] Add postage Air-mail	Surface mail
GENETIC EVALUATION & UTILIZATION OF RICE			
000-X	Descriptors for rice *Oryza sativa* L. 1980. 21 p.	10.10	0.80
001-8	Innovative approaches to rice breeding. 1980. 182 p.	36.00	0.80
002-6	Parentage of IRRI crosses IR1-IR50,000. 1981. ____ p.	24.50	0.80
003-4	Rice improvement. 1979. 186 p.	38.90	1.60
004-2	Rice improvement in China and other Asian countries. 1980. 307 p.	38.90	1.60
007-8	Rice tissue culture planning conference. 1982. 120 p.	18.70	0.80
008-1	Cell and tissue culture techniques for cereal crop improvement. 1983. 455 p.	46.10	1.60
002-2	Workshop on research priorities on tidal swamp. 1984. 220 p.	13.20	0.80
GERMPLASM COLLECTION AND MAINTENANCE			
005-0	Genetic conservation of rice. 1978. 54 p.	14.40	0.80
007-7	Manual on genetic conservation of rice germplasm for evaluation and utilization. 1976. 77 p.	14.40	0.80
115-4	1983 Rice germplasm conservation workshop. 1983. 110 p.	6.00	1.00
006-9	Manual for field collectors of rice. 1972. 32 p.	5.80	0.40
AGRONOMIC CHARACTERISTICS			
008-5	Morphology and varietal characteristics of the rice plant. 1965. 40 p.	15.80	0.80
GRAIN QUALITY			
010-7	Chemical aspects of rice grain quality. 1979. 390 p.	47.50	1.60
DISEASE RESISTANCE			
011-5	Rice blast workshop. 1979. 222 p.	28.80	0.80
012-3	Evolution of the gene rotation concept for rice blast control. 1982. 136 p.	20.20	0.80

ISBN No. 971-104:	Title	Price (₱)* Add postage Air-mail	Surface mail
	DROUGHT RESISTANCE		
018-1	Major research in upland rice. 1975. 255 p.	34.60	0.80
014-X	Rainfed lowland rice. 1979. 341 p.	41.80	1.60
078-6	Drought resistance in crops with emphasis on rice. 1982. 416 p.	47.50	1.60
121-9	Upland rice: proceedings of the International Upland Rice Conference.	In press	
	DEEPWATER AND FLOOD TOLERANCE		
015-8	Deepwater rice. 1977. 239 p.	31.70	0.80
016-6	1978 International deepwater rice workshop. 1979. 300 p.	38.90	1.60
017-4	1981 International deepwater rice workshop. 1982. 520 p.	55.20	1.60
	TEMPERATURE TOLERANCE		
018-2	Rice cold tolerance workshop. 1979. 139 p.	20.20	0.80
	CONTROL AND MANAGEMENT OF RICE PESTS		
080-8	Field problems of tropical rice: rev. ed. 1983. 172 p.**	10.10	0.80
	INSECTS		
022-0	Brown planthopper: threat to rice production in Asia. 1979. 369 p.	46.10	1.60
024-7	Manual for testing insecticides on rice. 1981. 131 p.	17.30	0.80
100-6	Workshop on judicious and efficient use of insecticides on rice. 1984. 182 p.	20.00	1.00
	Natural enemies of insect pests of rice. 1983. 49 color photos; 5 diagrams.	22.50	1.60
120-0	Integrated pest management.	50.00	8.00
110-3	Genetic evaluation for insect resistance in rice.	55.00	5.00
	WEEDS		
025-5	Major weeds of rice in South and Southeast Asia. 1981. 86 p.	10.00	0.80
074-3	Weed control in rice. 1983. 422 p.	50.40	1.60
	IRRIGATION WATER MANAGEMENT		
026-3	Irrigation policy and management in Southeast Asia. 1978. 198 p.	34.60	0.80
027-1	Irrigation water management. 1980. 170 p.	27.40	0.80
	SOIL AND CROP MANAGEMENT FOR RICE		
028-X	Blue-green algae and rice. 1980. 112 p.	18.75	0.80
029-8	Nitrogen and rice. 1979. 499 p.	56.20	1.60
071-9	Planting rice. 1974. 9 p.	5.80	0.40
072-7	Production of seedlings. 1972. 24 p.	7.20	0.40
031-X	Rice: soil, water, land. 1978. 185 p.	34.60	0.80
030-1	Soils and rice. 1978. 825 p.	108.00	2.40
032-8	Soil-related constraints to food production in the tropics. 1980. 468 p.	50.40	1.60
104-9	Organic matter and rice. 1984. 631 p.	30.00	2.00

ISBN No. 971-104:	Title	Price (₱)*	
		Add postage	
		Air-mail	Surface mail
ENVIRONMENT AND ITS INFLUENCE			
033-6	Agrometeorology of the rice crop. 1980. 254 p.	34.60	0.80
034-4	Climate and rice. 1976. 565 p.	64.80	1.60
035-2	Laboratory manual for physiological studies of rice. 1976. 83 p.	17.30	0.80
MAPS			
068-9	Rice area by type of culture: South, Southeast, and East Asia (3 maps, booklet)	21.60	0.80
069-7	Agroclimatic and dry-season maps of South, Southeast, and East Asia. (5 maps, booklet)	23.00	0.80
070-0	Complete set of the above 2 agroclimatic and rice area maps (8 maps, 2 booklets)	44.40	1.60
067-0	Map: agroclimatic map of the Philippines	7.20	0.40
CONSTRAINTS ON RICE YIELDS			
036-0	A handbook on the methodology for an integrated experiment-survey on yield constraints. 1978. 60 p.	13.00	0.80
037-9	Constraints to high yields on Asian rice farms: an interim report. 1977. 235 p.	31.70	0.80
038-7	Farm-level constraints to high rice yields in Asia. 1974-77. 1979. 411 p.	47.50	1.60
CONSEQUENCES OF RICE TECHNOLOGY			
039-5	Anatomy of a peasant economy. 1978. 149 p.	23.00	0.80
040-9	Changes in rice farming in selected areas of Asia. 1975. 377 p.	47.50	1.60
041-7	Interpretive analysis of selected papers from Changes in rice farming in selected areas of Asia. 1978. 166 p.	47.50	1.60
042-5	Economic consequences of the new rice technology. 1978. 402 p.	49.00	1.60
043-3	Landless workers and rice farmers: peasant subclasses under agrarian reform in two Philippine villages. 1982. 237 p.	31.70	0.80
083-2	Adoption, spread, and production impact of modern rice varieties in Asia. 1983. 54 p.	14.40	0.80
082-4	Consequences of small-farm mechanization. 1983. 192 p.	28.80	0.80
CROPPING SYSTEMS			
044-1	Cropping systems research and development for the Asian rice farmer. 1977. 454 p.	64.80	1.60
045-X	A methodology for on-farm cropping systems research. 1981. 156 p.	24.50	0.80
081-6	Basic procedures for agroeconomic research. 1984. 236 p.	31.70	0.80
114-6	Symposium on potential productivity of field crops under different environments. 1983. 530 p.	102.20	2.40
106-5	Cropping systems in Asia: on-farm research and management. 1984. 196 p.	10.00	1.00
076-X	Report of a workshop on cropping systems research in Asia. 1981. 762 p.	108.00	2.40

ISBN No. 971-104:	Title	Price (₱)[•] Add postage Air- mail	Surface mail
079-4	Evaluating technology for new farming systems: case studies from Philippine rice farms. 1982. 119 p.	18.70	0.80
MACHINERY TESTING AND DEVELOPMENT			
047-6	International agricultural machinery workshop. 1978. 203 p.	28.00	0.80
074-1	Technical handbook for the paddy rice postharvest industry in developing country. 1983. 200 p.	17.00	0.80
STATISTICAL TECHNIQUES			
101-4	Statistical procedures for agricultural research (second edition). 1984. 680 p.	38.20	2.40
049-2	Techniques for field experiments with rice. 1972. 46 p.	8.60	0.80
NUTRITION			
050-6	Interfaces between agriculture, nutrition, and food science. 1979. 143 p.	21.60	0.80
RICE SCIENCE AND PRODUCTION			
051-4	A farmer's primer on growing rice. 1979. 221 p.***	30.20	0.80
052-2	Fundamentals of rice crop science. 1981. 269 p.	37.40	1.60
053-0	Principles and practices of rice production. 1981. 618 p.	66.00	2.40
054-9	Training manual for rice production. 1976. 140 p.	30.20	0.80
055-7	A continuous rice production system . . . the rice garden. 1981. 16 p.	7.20	0.40
COMMUNICATION OF INFORMATION			
056-5	Communication responsibilities of the international agricultural research centers. 1980. 41 p.	15.85	0.80
117-0	Copublication: IRRI design, procedures, and policies for multilanguage publication in agriculture. 1983. 16 p.	7.20	0.40
BOOK CATALOG			
122-7	Publications on international agricultural research and development (catalog of publications of International Agricultural Research Centers) (IARCs). 1984. 560 p.	20.00	3.50
OTHERS			
057-3	Beyond IR8/IRRI's second decade (a non-technical brochure). 1980. 27 p.	13.00	0.80
058-1	Rice in the tropics: a guide to the development of national programs. 1979. 256 p.	36.00	0.80
060-3	Rice research in the 1980s: summary report. 1982. 33 p.	6.00	0.80
061-1	Rice research strategies for the future. 1982. 559 p.	55.20	1.60
062-X	A global experiment in agricultural development. 1982. 24 p.	7.20	0.40
063-8	An adventure in applied science: a history of the International Rice Research Institute. 1982. 236 p.	31.70	0.80

ISBN No. 971-104:	Title	Price (₱)*	
		Add postage	
		Air-mail	Surface mail
064-6	Report of an exploratory workshop on the role of anthropologists and other social scientists in interdisciplinary teams developing improved food production technology. 1982. 108 p.	7.20	0.80
065-4	Japan's role in tropical rice research. 1982. 46 p.	7.20	0.80
066-2	A plan for IRRI's third decade. 1982. 79 p.	14.40	0.80
109-X	Chemistry and world food supplies: the new frontiers (CHEMRAWN). 1982. 664 p.	83.50	2.40
105-7	CHEMRAWN II, Chemistry and world food supplies: perspectives and recommendations. 1984. 169 p.	31.80	1.00
119-7	Terminology for rice growing environments	15.80	0.80
111-1	Inter-Center seminar on IARCs and biotechnology	35.00	3.00
	International Rice Research: 25 years of partnership	10.20	2.00
	Education for agriculture	10.20	2.00

ISBN No. 971-104:	Year covered	Pages	Price (₱)*	
			Add postage	
			Airmail	Surface mail
ANNUAL REPORTS				
	1977	548	100.80	2.40
	1978	478	102.20	2.40
	1979	538	102.20	2.40
	1980	467	102.20	2.40
	1981	585	102.20	2.40
	1982	548	102.20	2.40
	1983	510	102.20	2.40
RESEARCH HIGHLIGHTS				
	1977	122	24.00	0.80
	1978	118	25.90	0.80
	1979	133	28.80	0.80
	1980	122	28.80	0.80
	1981	138	28.80	0.80
	1982	157	28.80	0.80
	1983	124	13.20	0.80
	1984	101	13.20	0.80
INTERNATIONAL BIBLIOGRAPHY OF RICE RESEARCH				
	1976	470	80.40	2.40
	1977	556	105.10	2.40
	1978	609	119.50	3.20
	1979	735	119.50	3.20
	1980	666	119.50	3.20
	1981	634	119.50	3.20
	1982	563	119.50	3.20
	1983	690	119.50	3.20

ISBN No. 971-104:	Year covered	Pages	Price (₱)[*]	
			Add postage	
			Airmail	Surface mail
INTERNATIONAL BIBLIOGRAPHY OF RICE RESEARCH CUMULATIVE INDEXES				
	1961-1965	143	33.10	0.80
	1966-1970	153	37.40	1.60
	1971-1975	232	51.80	1.60
	1976-1980	894	100.80	2.40
INTERNATIONAL BIBLIOGRAPHY ON CROPPING SYSTEMS				
	1973-1974	300	63.40	1.60
	1975	195	47.50	1.60
	1976	181	51.80	1.60
	1977	239	53.30	1.60
	1978	331	64.80	1.60
	1979	336	64.80	1.60

ISBN No. 971-104:	Title	Price (₱)[*]	
		Add postage	
		Airmail	Surface mail
OTHER BIBLIOGRAPHIES			
	International Bibliography on Azolla. 1980. 66 p.	13.20	0.80
	International Bibliography on Azolla. 1983 Supplement. 106 p.	15.80	1.60
	A Bibliography of Rice Literature (translation available in the IRRI Library and Documentation Center). 1976. 191 p.	41.80	1.60
	Upland rice research: an international bibliography 1965-1982. 1983. 129 p.	25.90	0.80
	International directory of rice workers. 1983. 221 p.	53.50	0.80

ISNAR

International Service for National Agricultural Research

P. O. Box 93375, 2509 AJ — The Hague, Netherlands
Cable: ISNAR
Telex: 33746
Telephone: 070/47.29.91

Mandate and Objectives of ISNAR

According to the constitution of the new Service, ISNAR's primary mandate is "to help strengthen national agricultural research capabilities in developing countries." Only if the scientific capacity and research management capability of the countries concerned are adequately developed so as to enable them to test, adapt, and complement the newly developed technologies, can the ultimate aim of the CGIAR system — of helping the developing countries achieve self-sustaining technical and economic growth — be reached.

To help accomplish this, ISNAR will concentrate its assistance to developing countries mainly on program, priorities, policy, organizational and management programs needed to improve the performance of national agricultural research systems. It will be concerned with commodities and other renewable natural resources important to national development objectives including, but not limited to, the food commodities covered by other components of the CGIAR system.

Such assistance will be available to any developing country upon request, ISNAR's available staffing and financial limitations permitting. The goal is to enable developing countries to plan, organize, manage, and execute agricultural research more effectively from their own human, natural, and financial resources as soon as possible. The new Service will seek to complement, and not compete with, other sources of technical assistance and will frame its program to encourage full use of technical expertise from other sources acceptable to the government concerned.

Annual Reports

Annual Reports 1981, 1982, 1983, 1984

18.4 × 24 cm. Sewn. Single copy free; multiple copies US$5.00.

Detailed full reports on ISNAR activities during one year. Available in English, French, and Spanish.

Monographs

Mosher, A. T.

Critical Requirements for Productive Agricultural Research

May 1982. 22 pages. 14.7 × 21 cm. Wire stitched. Single copy free; multiple copies US$2.00. Available in English, French, and Spanish.

Mosher, A. T.

Three Ways to Spur Agricultural Growth

September 1983. Booklet by A. T. Mosher on Agricultural Growth. Available in English and French.

Manpower Planning in a National Agricultural Research System

1984. Study by P. Bennell on agricultural research (English).

Conference, Workshop, and Symposia Proceedings

Strengthening National Agricultural Research Systems in Asia.

Los Banos, Philippines. June 1981.

Strengthening National Agricultural Research Systems in Africa

June 1981.

Proceedings of a conference for African agricultural research managers, sponsored by ISNAR and the International Federation of Agricultural Research Systems for Development (IFARD), held in Nairobi, Kenya, March 1981. Available in English and French.

The Role of International Associations in Strengthening National Agricultural Research

May 1982. 68 pages. 14.6 × 21 cm. Wire stitched. Single copy free; multiple copies US$2.00.

A conference in cooperation with the International Agricultural Development Service (IADS) and the International Federation of Agricultural Research Systems for Development (IFARD), with the report published jointly by ISNAR and IADS. Held in Bellagio, Italy, December 1981 (English).

Agricultural Research for Development: Potentials and Challenges for Asia

March 1983. 60 pages. 21 × 30 cm. Perfect. Single copy free; multiple copies US$2.00.

Proceedings for a conference for Asian agricultural research managers, sponsored by ISNAR and IFARD, in cooperation with the German Foundation for International Development. Held in Jakarta, Indonesia, October 1982 (English).

Selected Issues in Agricultural Research in Latin America

August 1983.

Proceedings of a conference for Latin American Research Directors, sponsored by IFARD, IICA, and ISNAR in cooperation with the Spanish Government. Held in Madrid, Spain, September 26-October 1, 1982 (English).

First Congress of the African Chapter. The International Federation of Agricultural Research Systems for Development (IFARD)

January 1984.

Proceedings of IFARD Congress. Available in English and French.

Strategies to Meet Demands for Rural Social Scientists in Africa

May 1982. 148 pages. 14.7 × 20.5 cm. Glue and tape. Single copy free; multiple copies US$2.00.

A conference held with support from the Ford Foundation and the International Development Research Centre. Focus of the conference was on the study of rural social science resources in nine African countries by an ISNAR consultant, Dr. Gaston Rimlinger. Dr. Rimlinger's report is given in the pamphlet. The conference was held in Bellagio, Italy, November 1981 (English).

International Federation of African Research Directors

1984.

Proceedings of a conference held in Nigeria (English).

Organization and Management of Research with a Farming System Perspective

1984.

Proceedings of a workshop in Holland (English).

Training Needs in National Agricultural Research Planning and Management

1984.

Proceedings of a colloquium held in Holland (English).

Agricultural Research in Rwanda

Proceedings of a seminar held in Rwanda, 1984 (English).

Agricultural Research in Rwanda: Assessment and Perspective. ISNAR and the Government of Rwanda

Kigali, Rwanda. September 1983.

Available in English and French.

Issues in Organization and Management of Research with a Farming Systems Perspective Aimed at Technology Generation

ISNAR and CIMMYT. The Hague, Netherlands. September 1983.

Potentials for Strengthening Management in National Agricultural Research

ISNAR and Ministry of Agriculture of Jordan. Amman, Jordan. April 1985.

Agricultural Research Policy and Organization in Small Countries

ISNAR and IAC. Wageningen, Netherlands. May 1985.

Miscellaneous

Working to Strengthen the National Agricultural Research Systems of Developing Nations

Brochure describing ISNAR aims and activities. Available in English, French, Spanish, and Arabic.

Report of an ISNAR/IITA Mission to the Institut de Recherche Agronomique et Zootechnique de la Communauté Economique des Pays des Grands Lacs (Burundi, Rwanda, Zaire)

July 1981.

Full report of the ISNAR review mission in Burundi, Rwanda, Zaire. Available in English and French.

A Review of the Agricultural Research System of Malawi

August 1982.

Full report of the ISNAR review mission in Malawi (English).

Kenya's National Agricultural Research System

September 1981.

Full report of the ISNAR review mission in Kenya (English).

A Manpower and Training Plan for the Agricultural Research System in Kenya 1983-1987

November 1982.

Report of a study of resources and projected needs for manpower in agricultural research in Kenya, conducted at the request of the National Council on Science and Technology.

Agro-Technological Research in Ivory Coast

Spring 1984.

Report of the ISNAR review mission in Ivory Coast. Available in English and French.

Agricultural Research in Ivory Coast: Presentation, Evaluation, Proposals for Improvement

Spring 1984.

Full report of the ISNAR review mission in Ivory Coast. Available in English and French.

The National Agricultural Research Systems of Rwanda

August 1983.

Full report of the ISNAR review mission in Rwanda. Available in English and French.

Agricultural and Livestock Research in Upper Volta

Spring 1984.

Report of a joint study World Bank/FAO/ISNAR in Upper Volta. Available in English and French.

La Recherche Agricole à Madagascar. Bilan et Perspectives du FOFIFA

Aout 1983.

Full report of the ISNAR review mission to Madagascar (French).

Development of an Agricultural Research System

1983.

Report of the ISNAR review mission in Somalia (English).

South Pacific Agricultural Research Study: Consultants Report to the Asian Development Bank

June 1981.

Full report of a mission in the South Pacific for the Asian Development Bank (English).

In addition to the full report, ISNAR published individual country reports; copies (English) are available for:
> Cook Islands
> Fiji
> Kiribati
> Papua New Guinea
> Solomon Islands
> Tonga
> Western Samoa
> Other countries and organizations

The Agency for Agricultural Research and Development of Indonesia

October 1981.

Report of the ISNAR review mission to Indonesia (English).

The Agency for Agricultural Research and Development of Indonesia. Executive Summary

October 1981.

An executive summary of the ISNAR review mission to Indonesia (English).

Review of the Program and Organization for Crops Research in Papua New Guinea

June 1982.

Report of the ISNAR mission to Papua New Guinea (English).

A Review of the Agricultural Research Division of Fiji

September 1982.

Report of the ISNAR review mission to Fiji (English).

Solomon Islands Agricultural Research, Extension, and Support Facilities

December 1982.

Report of the project preparation mission for the Asian Development Bank (English).

Strengthening Agricultural Research for National Development in the South Pacific

January 1983.

Project proposal prepared in association with the Asian Development Bank (English).

Report to the Pakistan Agricultural Research Council on Selected Aspects of the Pakistan Agricultural Research System

August 1983.

Report of the ISNAR review mission to Pakistan (English).

El Sistema de Investigación Agropecuaria y Transferencia de Technología en Costa Rica

Junio 1981.

Report of the ISNAR review mission in Costa Rica (Spanish).

The Agricultural Research Systems of Guyana

March 1982.

Report of ISNAR review team mission to Guyana (English).

El Sistema de Investigación Agropecuaria en la Dominicana

Julio 1983.

Report of the ISNAR review mission in Dominican Republic (English and Spanish).

Agricultural Research System in Western Samoa

1984.

Full report of the ISNAR review mission in Western Samoa (English).

Agricultural Research System in Western Samoa (Summary)

1984.

Summary of preceding report (English).

Review of Agricultural Research System of Sri Lanka

1984.

Report of the ISNAR review mission in Sri Lanka (English).

Review of Agricultural Research System of Morocco

1984.

Report of the ISNAR review mission in Morocco (French).

Review of Agricultural Research System of Zaire

1984.

Report of the ISNAR review mission in Zaire (French).

Review of Caribbean Agricultural Research and Development Institute (CARDI)

1984.

Report of ISNAR review mission CARDI (English).

Improvement of Agricultural Research Administration in Cameroon

1984.

Report of ISNAR mission to Cameroon. Available in English and French.

Improving the Management of Agricultural Research in Sudan

1984.

Report of ISNAR mission to Sudan (English).

Management Training and the Strengthening of Agricultural Research Management in Africa

1984.

Study on management training in Africa (English).

Training Requirements for Agricultural Research Management in Africa: a Case Study in Zimbabwe

1984.

Case study on training in Zimbabwe (English).

Research Management

Guidelines for ISNAR Reviews and Evaluations

July 1984.

Considerations for the Development of National Agricultural Research Capacities in Support of Agricultural Development

October 1984.

Available in English and Spanish.

Consideraciones para el Desarrollo de la Capacidad Nacional de Investigacion Agricola de Apoyo al Desarrollo Agricola

June 1985.

Traducción del original en inglés.

Etude de la Réorganisation du Systéme National de Recherche Agronomique du Zaire

Groupe National d'Etude et ISNAR. Févier 1985.

Country Reports

Les Recherches l'ISNAR (Rwanda) sur les Productions Végétales

Janvier 1984.

L'Institut National de la Récherche Agronomique du Maroc

Janvier 1984.

Volume 1. — Bilan et Perspectives.
Volume 2. — Annexes.

The Moroccan National Institute of Agricultural Research

May 1985.

Available in English and French.

Volume 1. — Present Situation and Prospects.
Volume 2. — Annexes.

The National Institute of Agricultural Research in Morocco

Résumé in four languages: English, Arabic, Spanish, French. March 1985.

A Training Plan for the Department of Research and Specialist Services, Zimbabwe, 1985 to 1988

ISNAR and Ministry of Agriculture. Harare, Zimbabwe. February 1985.

Proagro

Proagro Paper No. 1. Agricultural Research in the Private Sector: Issues on Analytical Perspectives

March 1985.

Proagro Paper No. 2. Los Consorcios Rurales de Experimentacion Agricola: Evolucion e Impacto

Abril 1985.

Working Papers

Working Paper No. 1. Agricultural Research in the Public Sector of Latin America: Problems and Perspectives

March 1985.

Working Paper No. 2. Using Evaluations for Planning and Management: an Introduction

April 1985.

Reprints

ISNAR Reprint Series No. 1. Improving the Global System of Support for National Agricultural Research in Developing Countries

September 1984.

ISNAR Reprint Series No. 2. India's Coordinated Crop Improvement Projects — Organisation and Impact

May 1985.

Program and Budget for 1985

1984.

Planning and budgeting of ISNAR for 1985 (English).

ISNAR in the Eighties

1984.

Medium term planning of ISNAR activities. Available in English, French, and Spanish.

ISNAR, the Early Years

1984.

The first years of activities of ISNAR (English).

ISNAR Professional Staff

1984.

Brochure giving biographical resumés of ISNAR professional staff (English).

ISNAR Staff and Board Directory

1984.

Brochure giving biographical resumés of ISNAR staff and board members (English).

Periodical

Quarterly Newsletter

Periodical. Started with Volume 1, Number 1, in October 1984. (English)

WARDA

West Africa Rice Development Association

P.O. Box 1010, Monrovia, Liberia
Cable: WARDA; Telex: WARDA 4333 LI
Telephone: 22.14.66 & 22.19.63

General information

The West Africa Rice Development Association which was formed in September 1970 under the auspices of UNDP, FAO and ECA is an intergovernmental organization consisting of 16 members countries namely: Benin, Burkina-Faso (Upper-Volta), Chad, Ivory Coast, The Gambia, Ghana, Guinea, Guinea-Bissau, Liberia, Mali, Mauritania, Niger, Nigeria, Senegal, Sierra Leone and Togo.

The Headquarters of WARDA are located in Monrovia (Liberia) while its four Regional Research Stations are located in Bouake, Ivory Coast (Upland rice); Rokupr, Sierra Leone (mangrove and swamp rice); Richard-Toll, Senegal (irrigated rice); and Mopti, Mali (deep water/ floating rice). In addition, WARDA has a Regional Training Center at Fendall (Liberia).

WARDA is one of the 13 International Agricultural Research Centers (IARCs) of the Consultative Group on International Agricultural Research (CGIAR). The Association has been mandated to assist its members countries to achieve self-sufficiency in rice, a staple food of West Africans.

Research Highlights

WARDA Activities Highlights: 1983-1984

36 pages. 27 × 21 cm. Wrap around.

This first issue gives WARDA's Research and Development Department recent progress and findings in areas such as supervisor varieties introduction through Coordinated Variety Trials, On-Farm Trials and Adaptive farmer Trials; experiments conducted at the Research Stations; technology assessment and Transfer Programme; and Technical Assistance Programme under which Member States are assisted with rice projects preparation and implementation.

Annual Reports

WARDA 1983 Annual Report

1984. 249 pages. 26.5 × 21.5 cm. Wrap around.

This publication gives an in-depth annual report of WARDA's Departments and Divisions activities progress.

Also available: WARDA Annual Reports for 1971-1974, 1979, 1981, 1982.

Monographs

IRAT Crop Protection Division

Weeding of Paddies in West Africa and Catalog of the Main Weeds

1978. 96 pages. 21 × 14.5 cm. Paperback.

The purpose of the present handbook is to provide the most important data on proper herbicide treatment and to describe the weeds noxious to rice in West Africa, in concise form, for ready reference. It is not a treatise, on weeds, but rather a synopsis of notes intended for rice growers.

Also available in French.

IRAT Division de défense des cultures

Le désherbage des rizières en Afrique de l'Ouest et leurs principales adventices

1978. 96 pages. 21 × 14.5 cm. Broché.

Ce manuel se propose de réunir, sous une forme condensée et facile à consulter, les données essentielles à la réalisation correcte des traitements herbicides et de décrire les principales adventices nuisibles à la riziculture. Il ne s'agit donc pas d'un traité de malherbologie, mais plutôt d'un memento destiné aux riziculteurs.

Disponible également en englais.

H. V. Ruiten and H. P. Rozeboom

Post-Production Rice Technology in the WARDA Region

1980. 79 pages. 27.5 × 21.5 cm. Paperback.

A region-wide survey of post-production rice technology was launched in September 1975 to obtain a clear picture of the technology used in the member countries of WARDA. This report is intended to provide the overall view indeed to determine where intervention is most needed to reduce losses in quantity and quality and to increase efficiency. The subjects encompassed by the survey are: harvesting, threshing, winnowing, drying, storage, parboiling and milling.

Also available in French.

H. V. Ruiten and H. P. Rozeboom

Technologie du riz pendant et aprés la récolte dans la zone de l'ADRAO

1981. 111 pages. 27.5 × 21.5 cm. Broché.

Une enquête régionale a été entreprise en septembre 1975 sur les techniques agricoles appliquées pendant et après la récolte du riz en vue de donner un tableau précis de la technologie utilisée par les pays membres de l'ADRAO. Le présent rapport a pour but de donner la vue d'ensemble adéquate pour déterminer où il est le plus nécessaire d'intervenir afin de réduire les pertes de riz en quantité et en qualité et pour augmenter l'efficacité. Les sujets couverts par l'enquête sont les suivants: récolte, battage, vannage, séchage, stockage, étuvage et usinage.

Egalement disponible en anglais.

J. L. Notteghem and P. Baudin

Main Rice Diseases in West Africa

1981. 33 pages. 21 × 14.5 cm. Paperback.

This manual does not aim to draw an exhaustive list of microorganic parasites on West African rice, but to enable the expert to recognize the main diseases. The chosen photographs and the text emphasize the main characteristics of the diseases described and the control measures.

Also available in French.

J. L. Notteghem et P. Baudin

Principales Maladies du Riz en Afrique de l'Ouest

1981. 33 pages. 21 × 14.5 cm. Broché.

Ce manuel n'a pas pour but de dresser une liste exhaustive des micro-organismes parasites du riz en Afrique de l'Ouest, mais de permettre au praticien de reconnaître les principales maladies. Les photos choisies et le texte mettent l'accent sur les principales caractéristiques des maladies décrites et les moyens de lutte.

Disponible également en anglais.

R. Vandevenne

Study of Seed Legislation and Proposals for Controlling and Organizing a Seed Service in the WARDA Region

1981. 204 pages. 29.5 × 21. Wrap around.

Undertaking a seed programme forms on of the legitimate preoccupations of any State that is aware of how important good quality seed is for increasing and improving agricultural output.

However, for a programme of this type to be successful and attain the desired result of providing certified seed for as many farmers as possible, it is necessary to set up a suitable organization and regulations adapted to the State's resources.

V. Vandenne

Etude D'une Législation Semencière et Propositions pour une Réglementation et une Organisation d'un Service Semencier Riz pour la Région de l'ADRAO

1981. 223 pages. 29.5 × 21. Broché.

Entreprendre un programme de production de semences fait partie des soucis légitimes de tout Etat conscient de l'importance que représente la disposition de semences de qualité pour augmenter et améliorer la production de son agriculture.

Cependant, un tel programme pour être mené à bien et pour aboutir au résultat souhaité qui est de fournir de la semence certifié au plus grand nombre possible de cultivateurs du pays, demande la mise en place d'une organisation et d'une reglementation appropriée et adaptée aux moyons do l'Etat.

Rice Varieties Recommended in the WARDA Region: 2nd Ed.

February 1982. 132 pages. 28 × 21.5 cm. Stapled.

This manual describes 128 rice varieties distributed in four ecologies. Varieties already recommended in the WARDA Region or in the process of being recommended are only included here. For simplication, only the major agronomic characters directly related to yield are described.

Also available in French.

Variétés de Riz Recommandées dans la Région de l'ADRAO; 2ème ed.

fevrier 1982. 132 pages. 28 × 21.5 cm.

Ce manuel décrit 128 variétés de riz appartenant à quatre écosystémes Seules les variétés déjà recommandées dans la région de l'ADRAO on sur le point de l'être ont été incluses. Pour simplifier, seules les principales caractéristiques agronomiques ayant un effet direct sur les rendements ont été décrites.

Disponible également en anglais.

J. Brenière

The Principal Insect Pests of Rice in West Africa and their Control: 2nd ed.

1983. 87 pages. 21 × 14.5 cm. Paperback.

The aim of this second edition, like the first, is to facilitate the identification of the principal pests of rice in West Africa by providing a minimum amount of information on their biology, their relationship with the plant and the protective measures required. Insects are the main object of this booklet. Nematodes, rodents and birds are briefly mentioned at the end.

Also available in French.

J. Brenière

Principaux Ennemis du Riz en Afrique de l'Ouest et leur Controle

1983. 87 pages. 21 × 14.5 cm. Broché.

Le but de cette seconde édition, tout comme la première, est de faciliter la reconnaissance des principaux ravageurs du riz en Afrique de l'Ouest en donnant un minimum d'informations sur leur biologie, leurs relations avec la plante et les moyens de protection de la riziculture qu'ils nécessitent. Les insectes constituent l'objet essentiel de cette brochure. Toutefois, les nématodes, les rongeurs et les oiseaux sont présentés plus sommairement à la fin de l'ouvrage.

Egalement disponible en anglais.

C. Van Hove, H. F. Diarra, and P. Goddard
Projet Azolla ADRAO/WARDA Azolla Project

Azolla en Afrique de l'Ouest = Azolla in West Africa: Bilingual Ed.

1983. 53 pages. 22 × 15.5 cm. Paperback.

Azolla est une fougère aquatique réalisant une symbiose systématique avec Anabaena azollae, cyanobactère diazotrophe (fixatrice d'azote), et apte, grâce à cette propriéte, à servir d'engrais vert. Le présent document fournit une information générale sur ces organismes, leur intérêt en riziculture, le project Azolla de l'ADRAO et sur les possibilités d'y participer.

Azolla is an aquatic fern which has a systematic symbiosis with anabaena azollae, a nitrogen fixing cyanobacteria, and which, by virtue of this property, is suitable for use as a green manure. This booklet intends to provide general information about these organisms, their importance in rice-growing, the WARDA-AZOLLA programme and possibilities for participation in it.

Conference, Workshop, and Symposia Proceedings

Seminar on Soil Fertility and Fertilizer Use. Monrovia, Liberia, 1, 22-27 January 1973. Seminar Proceedings

May 1979. 325 pages. 27.5 × 21.5 cm. Wrap around. (Seminar Proceedings, 2).

This volume contains 16 papers on rice soil fertility and fertilizer use problems in West Africa. Also included are recommendations on standardization of soil resources appraisal and evaluation for agricultural development in West Africa and WARDA programme.

Séminaire sur la Fertilité des Sols et l'utilisation des Engrais. Monrovia, Liberia, 22-27 janvier 1979. Délibérations du Séminaire

Juin 1979. 309 pages. 28 × 21.5 cm. Wrap around.

Ce volume renferme 16 documents sur les problèmes que pesant la fertilité des sols et l'utilisation des engrais an Afrique Occidentale présentés au séminaires. Des recommandations, notamment, sur la normalisation des méthodes de description et d'analyse des sols sont incluses.

Socioeconomic Aspects of Rice Cultivation in West Africa. Monrovia, Liberia, 22-25 April 1974. Seminar Proceedings

February 1975. 216 pages. 29 × 21 cm. Wrap around.

These proceedings contain 12 papers presented at the Seminar on Socioeconomic Aspects of Rice Cultivation in West Africa, such as production systems and policies, processing and marketing of rice.

Plant Protection for the Rice Crop. Monrovia, Liberia, 21-26 May 1973. Seminar Proceedings

March 1977. 443 pages. 27 × 21.5 cm. Wrap around (Seminar Proceedings; 4)

This volume contains 34 papers presented at the Seminar on various aspect of rice plant protection against pests in West Africa and recommendations resolutions concerning WARDA'S Plant Protection Programme.

Also available in French.

Protection des Plantes en Riziculture. Monrovia, Liberia, 21-26 mai 1973. Délibérations du séminaire

Mars 1977. 439 pages. 27.5 × 21.5 cm. Wrap around (Déliberation des séminaires; 4)

Ce volume contient 34 documents présentés au Séminaire sur les différents aspects de la protection des plantes en riziculture en Afrique de l'Ouest, et les recommandations et résolutions concernant le Programme de Protection des plantes de l'ADRAO.

Egalement disponible en anglais.

Rice Breeding and Varietal Improvement. Monrovia, Liberia, 14-19 January 1974. Seminar Proceedings

May 1974. 305 pages. 27.5 × 21.5 cm. Wrap around. (Seminar Proceeding; 1)

This volume contains various papers and summary of discussions on rice breeding and varietal improvement in West Africa, WARDA Varietal Programme and Coordinated Variety Trials, and Resolutions and recommendations approved by participants at the Seminar.

Also available in French.

Sélection et Amélioration Variétale du riz. Monrovia, Liberia, 14-19 janvier, 1974. Déliberation du Séminaire

janvier 1976. 235 pages. 33 × 21.5 cm. Ronéotypé et agrafé. (Déliberations des séminaires; 1).

Ce volume renferme différents documents ainsi que les résumés de discussions sur la sélection et l'amélioration variétale du riz en Afrique de l'Ouest, le programme d'amélioration variétale et les essais variétaux coordonnés de l'ADRAO, et les résolutions et recommandations adoptés par les participants à l'issue du séminaire.

Egalement disponible en anglais.

Rice Project Managers Meeting. Monrovia, Liberia, 4-8 February 1974. Seminar Proceeding

January 1976. 310 pages. 28 × 21.5 cm. Mimeographed and stapled. (Seminar Proceedings; 5)

This meeting was attended by Government officials and agricultural development project managers in particular concerned with rice development policies in their respective countries. This meeting was the first of its kind in pooling regional knowledge and experience for a better understanding of the problems relating to the development of rice cultivation in West Africa.

Varietal Improvement Seminar. 2nd. Monrovia, Liberia, 13-18 September 1976. Final Report

1976. 31 pages. 28 × 21.5 cm. Stapled. WARDA/S/76/38.

This report contains the summaries of discussions on 29 papers presented at the Seminar on Rice Varietal Improvement in West Africa organized by WARDA.

Séminaire sur l'Amélioration Variétale. 2ème. Monrovia, Liberia, 13-18 septembre 1976. Rapport final

1976. 30 pages. 28 × 21.5 cm. Agrafe. ADRAO/S/76/38.

Ce rapport fait l'économie des débats sur les 29 communications faites au cours du séminaire organisé par l'ADRAO sur l'amélioration variétale du riz en Afrique de l'Ouest.

Rice Breeding and Varietal Improvement. 2nd, Monrovia, Liberia, September 1976. Seminar Proceedings

January 1979. 499 pages. 28 × 21.5 cm. Wrap around. (Seminar Proceedings; 6).

These proceedings consist of 34 papers presented during the Second WARDA Variety Improvement Seminar held in Monrovia in September 1976.

Conférence-Atelier sur le Riz et Tournée des Projets Rizicoles en Sierra Leone, 8-15 octobre 1977. Rapport

octobre 1977. 73 pages. 28 × 21.5 cm. Reliure en spirale.

Rapport d'une conference - atelier sur le riz de mangrove et d'une tournée de visites aux divers projets rizicoles établis en Sierra Leone du 8 au 15 Octobre 1977 organisées par l'ADRAO à l'intention d'experts de ses stations de recherche et de délégues des pays membres.

Seminar on the Reinforcement of the Rice Growing Potentials of Member Countries: Adaptative Research, Training, Seed Multiplication. Monrovia, Liberia, 21-25 August 1978. Final report of the meeting

1978. 36 pages. 28 × 21.5 cm. Stapled. WARDA/78/S/20.

This seminar was intended to review and update the study by IRAT and BDPA on the reinforcement of the rice growing potential of WARDA member countries. It was attended by representatives of member countries which recommendations on and request for adaptative research, seed multiplication and training are included.

Also available in French.

Séminaire sur le Renforcement des Capacités Rizicoles des Pays Membres: Recherche d'Accompagnement, Formation, Multiplication des Semences. Monrovia, Liberia, 21-25 août 1978. Rapport final

1978. 39 pages. 28 × 21.5 cm. Agrafe ADRAO/78/S-20.

Ce séminaire avait pour but de passer en revue et de mettre à jour l'etude réalisée par l'IRAT et le BDPA sur le renforcement des capacités rizicoles des pays membres. Les recommandations et requêtes en matière de recherche d'accompagnement, multiplication des semences et formation des délégues des pays membres sont incluses.

Egalement disponible en anglais.

Seminar on Strategies for Rice Development in the WARDA Member Countries. Monrovia, Liberia, 15-20 October 1979

Papers presented and discussed at this Seminar addresses rice policies, rice production systems, incentives, comparative costs and marketing structures in the member countries of WARDA. Prospects and costs of self-sufficiency in rice in West Africa by 1980 was also presented recommendations for the improvement of rice production systems adopted.

Available papers:
* K. Craven and H. Tuluy
  Rice Policy in Senegal
  70 pages. WARDA/SD/79-6
* K. Prakah-Asante and V. K. Nyanteng
  Economic Study of Rice Production and Marketing in Ghana
  21 pages. WARDA/SD/70-13
* H. Tuluy
  Comparative Costs and Incentives in Senegalese Rice Production
  40 pages. WARDA/SD/79-14.
* J. C. Levasseur
  Rice Production Systems in Benin
  122 pages. WARDA/SD/79-16.
* Final Report
  29 pages. WARDA/SD/79-20.

Séminaire sur les Stratégies de Développement de la Riziculture dans les Pays Membres de l'ADRAO. Monrovia, Liberia, 15-20 octobre 1979.

1979. 20 papers. 27.5 × 21.5.

Au cours de ce Séminaire ont été présentés et débattus des rapports sur les politiques rizicoles, les systèmes de production rizicole, les incitations, les coûts comparatifs et sur les structures de commercialisation dans les pays membres de l'ADRAO, ainsi qu'un document sur les perspectives et les coûts de l'auto-suffisance en riz en Afrique de l'Ouest vers 1990.

Documents disponibles:
* C. P. Humphreys et P. L. Rader
  La politique rizicole en Côte d'Ivoire
  96 pages. ADRAO/SD/79-1

- E. A. Monke
  La politique du riz au Liberia
  59 pages. ADRAO/SD/79-2
- J. McIntire
  La politique rizicole du Mali
  61 pages. ADRAO/SD/79-3
- J. McIntire
  Coûts réels et incitations économiques dans la production de riz au
  Mali
  76 pages. ADRAO/SD/74-4
- K. Craven et H. Tuluy
  La politique du ruz au Sénégal
  71 pages. ADRAO/SD/79-6
- J. C. Levasseur
  Systèmes de production rizicole en Guinée-Bissau
  61 pages. ADRAO/SD/79-7
- D. S. C. Spencer
  La politique rizicole in Sierra Leone
  43 pages. ADRAO/SD/79-8
- D. S. C. Spencer
  Rentabilité sociale et finacière de la production du riz en Sierra
  Leone
  35 pages. ADRAO/SD/79-9
- J. C. Levasseur
  Analyse des techniques de production de riz au Togo
  46 pages. ADRAO/SD/79-11
- Guinée. Ministère de l'Agriculture, des Eaux et forêts et FAPA.
  Direction générále de l'agriculture
  Stratégie pour le développement de la riziculture en Guinée.
  24 pages. ADRAO/SD/79-12
- K. Prakah-Asante et V. K. Nyanteng
  Etude économique de a production et de la commercialisation du riz
  au Ghana
  23 pages. ADRAO/SD/79-13
- H. Tuluy
  Coûts et incitations de la production rizicole Sénégalaise
  50 pages. ADRAO/SD/79-14
- Perspectives et coûts de l'autosuffisance en riz en Afrique de l'Ouest
  en 1990.
  40 pages. ADRAO/SD/79-15
- J.C. Levasseur
  Systèmes de production rizicole en République populaire de Benin
  ____ pages. ADRAO/SD/79-16
- Stanford University. Food Research Institute et ADRAO
  L'économique politique du riz en Afrique de l'Ouest
  50 pages. ADRAO/SD/79-17
- Rapport final
  29 pages. ADRAO/SD/79-20

Workshop on Environmental Constraints on Rice Production in West Africa. Banjur, Gambia, 15-19 September 1980. Final Report of the Meeting

1980. 70 pages. 28 × 21.5 cm. Stapled.

This report contains the summaries of papers and discussions on the environmental constraints affecting rice production in the various ecologies of the WARDA region, such as: low yielding varieties, diseases and pest problems, weed infestations, human health problems, high and low temperature, limited rainfall and fresh water supplies, etc.

Integrated Pest Management in Rice in West Africa: Concepts, Techniques and Applications of Integrated Pest Management in Rice in West Africa. WARDA Regional Training Centre, Fendall, Liberia, 10-28 January 1982. Proceedings of a Course

1982. 505 pages. 27.5 × 21.5 cm. Wrap around.

These proceedings are the lecture papers emanating from the short course, "Integrated Pest Management in Rice in West Africa" held at the WARDA Training Centre and sponsored jointly by WARDA, CICP and USAID. The overall objective of the course was to create awareness on the concepts, techniques and applications of Integrated Pest Management in Rice in West Africa.

Also available in French.

Gestion Intégrée des Déprédateurs du Riz en Afrique de l'Ouest: Concepts, Techniques et Applications de la Gestion Intégrée des Déprédateurs du Riz en Afrique de l'Ouest. Centre Régional de Formation ADRAO, Fendall, Liberia, 10-28 janvier 1982. Compte rendu d'un stage

1982. 536 pages. 27.5 × 21.5 cm. Relié.

Ce compte rendu reprend les cours présentés au stage sur "la gestion intégrée des déprédateurs du riz en Afrique de l'Ouest" tenu au Centre de formation de l'ADRAO et parrainé conjointement par l'ADRAO, le CICI et l'USAID. L'objectif général de ce stage consistait à familiariser les participants avec les concepts, techniques et applications de la gestion intégrée des déprédateurs du riz en Afrique de l'Ouest.

Egalement disponible en anglais.

Seminar on Improved Rice Seed Production in West Africa. Freetown, Sierra Leone, 5-9 September 1983.

1983. Various papers. 28 × 21.5 cm. Stapled.

The main objectives of this Seminar was: to identify the main problems facing the rice industry in West Africa with particular reference to seed multiplications; to review the operations of seed production programmes in the WARDA region; and to recommend possible ways of improving seed rice production and supply in the sub-region.

Available papers:

- R. Vandevenne
  Introduction to Seed Legislation in the WARDA Region.
  6 pages.
- S. A. Botchey and A. Sandhu
  Extension Programmes to Promote the Use of Improved Rice Seed in WARDA Region
  17 pages.
- A. Joshua
  Rice Seed Production in Nigeria
  16 pages.
- J. C. Delouche and H. C. Potts
  The Importance of Seed in Agriculture and the Need for a Seed Programme
  20 pages.
- M. A. Larinde and M. A. Diallo
  West Africa Rice Development Association (WARDA) Regional Seed Activities
  32 pages.
- V. K. Nyanteng
  Rice (Paddy) Production in West Africa Strategies and Prospects in the 1980s
  46 pages.
- E. K. Sorgho
  Activity Report of the National Rice Seed Service in Upper Volta [Burkina Faso]
  6 pages.
- V. K. Nyanteng
  Improved Rice Seed Distribution and Pricing Policies in West Africa
  17 pages.
- A. O. Abifarin
  The Release of Variety, Production and Maintenance of Breeder Seed
  18 pages.
- B. K. Sarfo and L. Kone
  Economics of Alternative Seed Production Programmes in West Africa
  28 pages.

- G. E. Delhove
  Rice Seed Programmes in West Africa: Activities of the FAO in the Rice Seed Programmes in West Africa.
  16 pages.
- L. L. Delimini
  Improved Rice Seed Multiplication in Ghana
  14 pages.
- Final Report
  37 pages.

Séminaire sur la Production de Semences Améliorées en Afrique de l'Ouest. Freetown, Sierra Leone, 5-9 septembre 1983.

1983. Documents divers. 28 × 21.5 cm. Agrafés.

Les objectifs de ce séminaire étaient les suivants: identifier les principaux problemes qui se posent a la riziculture dans la région de l'ADRAO, et tout particulièrement ceux qui ont trait à la multiplication des semences; passer en revue les programmes de production de semences de riz dans la région et en examiner les modalités de fonctionnement; et formuler des recommandations en vue d'améliorer la production et la fourniture des semences de riz dans la région.

Documents disponibles:
- A. Joshua
  La production semencière au Nigeria
  21 pages.
- G. E. Delhove
  Le conditionnement des semences
  54 pages.
- L. Koné et M. A. Diallo
  Production de semences: cas du Mali
  15 pages.
- B. K. Sarfo et L. Koné
  Considérations économiques sur les différents systèmes de production semencière de l'Afrique de l'Ouest
  32 pages.
- A. O. Abifarin
  Homologation des variétés, production et conservation des semences de souche
  21 pages.
- E. K. Sorgho
  Rapport d'activité du Service national des semences en matieré de riz [Burkina Faso]
  6 pages.

- S. A. Botchey et A. Sandhu
  Programmes de vulgarisation destinés à promouvoir l'utilisation de semences de riz améliorées dans la région de l'ADRAO
  22 pages.
- V. K. Nyanteng
  La distribution des semences de riz améliorées et les politiques en matière de prix en Afrique de l'Ouest
  20 pages.
- E. T. O'Dowd
  Le stockage des semences et sa gestion dans les pays tropicaux
  46 pages.
- GTZ
  Agriculture et développement rural: production de semences et de plants
  12 pages.
- V. K. Nyanteng
  La riziculture en Afrique de l'Ouest: stratégies et perspectives pour les années 1980
  52 pages.
- J. C. Delouche et H. C. Potts
  Importance des semences en Agriculture et nécessité d'un programme semencier
  21 pages.
- M. A. Larinde et M. A. Diallo
  Le programme semencier régional de l'Association pour le développement de la riziculture en Afrique de l'Ouest (ADRAO)
  35 pages.
- A. Keita
  Rapport du Sénégal
  9 pages.
- Togo-Direction de la recherche agronomique: Rapport national du Togo
  7 pages.
- G. E. Delhove
  Programmes de production de semences de riz en Afrique de l'Ouest: activités de la FAO relatives à la production de semences de riz en Afrique de l'Ouest
  17 pages.
- Rapport final.
  36 pages.

Annual Rice Review Meetings

Various papers. 27.5 × 21.5 cm. Stapled.

In 1983 the decisions was taken to hold every two years the Rice Review Meeting that used to be organized yearly. They provide the opportunity for scientists from member countries programmes,

Cooperating institutions and WARDA to jointly, review and share their research experiences, discuss and plan research strategies and make specific recommendations on aspects relating to WARDA Research and development programme.

Available:

* Annual Rice Review Meeting Final Reports for 1974, 1975, 1982, 1983
* Annual Rice Review Meetings Papers for 1983.

Also available in French.

Revues annuelles de la riziculture de la riziculture

Documents divers: 27.5 × 21.5 cm. Agrafés.

La décision a été prise en 1983 que les revues qui avaient lieu chaque année se tiendraient désormais tous les deux ans. Elles offrent, aux scientifiques des programmes des pays membres, des Institutions coopérantes et de l'ADRAO, l'occasion de passer en revue et de partager leurs expériences en matière de recherche, de discuter et d'élaborer des stratégies de recherche et de faire des recommandations précises sur des aspects du Programme de recherche et de développement de l'ADRAO.

Disponibles:

* Rapports finaux des Revues annuelles de la riziculture de 1974, 1975, 1976, 1977, 1978, 1979, 1982 et 1983.
* Documents présentés aux revues annuelles de la riziculture de 1983.

Egalement disponible en anglais.

Miscellaneous Reports and Brochures

J. G. Vianen

Conflicts Between Farmer's Income and Cost of Living: a Quantitative Analysis of a Paddy Price Increase

September 1975. 27.5 × 21.5 cm. Stapled.

Conflict between interests of producer and consumer is analyzed and quantified for The Gambia. In addition, some other issues, which are closely resulted to this subject, such as production effects, inflationary effects and alternative policies are discussed.

COOP, BDPA, and IRAT

Reinforcement of the Rice-Growing Potential of Member Countries

December 1976. Various papers. 29.5 × 21 cm. Wrap around.

In support of efforts already undertaken by its Members States, WARDA has already initiated a series of regional programmes. These programmes however will not achieve their full effect unless they can be backed up and relayed by activities at national and local levels. The underlying intent of this study was therefore to identify in a sufficiently precise manner the problems likely to be encountered in this area and submit recommendations as to how these should be overcome.

Available papers on: Nigeria, Ghana and Sierra Leone.

COOP, BDPA et IRAT

Renforcement des capacités rizicoles des pays membres: recherche d'accompagnement, multiplication de semences, formation

Décembre 1976. Documents divers. 29.5 × 21 cm. Broché.

Pour appuyer les efforts de ses pays membres, l'ADRAO a déjà lancé une série de programmes régionaux. Ces différents programmes n'auront toutefois leur pleine efficacité que s'ils peuvent être relayés et démultiplies par des actions nationales et locales. C'est afin de cerner avec une precision suffisante les problèmes pouvant se poser à cet égard et formuler les recommandations utiles que la présente étude par pays a donc été décidée.

Documents disponibles: Bénin, Burkina Faso, Cote d'Ivoire, Mali, Mauritanie, Niger, Togo et Rapport general.

A Contribution to a Rice Development Strategy in Guinea Bissau

February 1977. 38 pages. 28 × 21.5 cm. Wrap around.

This report is a result of the study of a multidisciplinary team sent by WARDA to Guinea Bissau to help it identify the problems in rice cultivation and methods of solving them thus working out a rice development strategy. In conclusion, despite seasonal, regional and annual variations in climatic factors, conditions on the whole are reasonable for rice cultivation.

ADRAO et Stanford University Food Research Institute

Perspectives du Commerce Intra-Regional du Riz en Afrique de l'Ouest

Septembre 1977. 173 pages. 28 × 21.5 cm. Agrafé.

L'objectif de l'étude est d'examiner les perspectives de l'augmentation du commerce du riz entre les Etats membres de l'ADRAO. On prévoit que certains Etats membres augmenteront leur production de riz de manière plus rapide et plus efficace que les autres et que, par conséquent, l'attention doit se porter sur les perspectives d'un commerce intra-régional en Afrique de l'Ouest comme moyen de parvenir à l'autosuffisance regionale.

CCOP, BDPA and SCET INTER

Project Identification Study for the Development of the Rice-Growing Potential of Member Countries: General Report.

March 1979. 50 pages. 29.5 × 21 cm. Wrap around.

Project identification study pursuant to requests for reinforcing the rice growing capacities of WARDA member countries and more precisely in the fields of applied and accompanying research, seed multiplication and training.

COOP, BDPA et SCET INTER

Etude d'Identification de Projets pour le Renforcement des Capacités Rizicoles des Pays Membres

Mars 1978. Documents divers. 29.5 × 21 cm. Broché.

Etude d'identification des projets issus de requêtes visant au renforcement des capacités rizicoles des pays membres de l'ADRAO et precisement dans les domaines de la recherche appliquée et d'accompagnement, de la multiplication des semences et de la formation.

Documents disponibles: Bénin, Côte d'Ivoire, Gambie, Guinée-Bissau, Liberia, Mali, Sénégal, Sierrâ Leone. Rapport général.

J. C. Levasseur

Aspects Economiques du Développement de la Riziculture au Togo

Decembre 1979. 246 pages. 28 × 21.5 cm. Wrap around.

Cette étude constitue une analyse approfondie de l'ensemble des aspects économiques liés au développement de la riziculture au Togo. Elle comporte trois grandes parties: micro-économique principalement axée sur l'étude de la production, micro-économique axée sur l'etude de l'environnement du secteur de la production, et micro-économique axée sur l'étude de la politique économique.

J. C. Levasseur

La Commercialisation du Riz

Décembre 1980. 60 pages. 33 × 21.5 cm. Agrafé.

Ce manuel passe en revue la fonction et les phases de la commercialisation du riz, la normalisation et le classement, les prix et les marchés, et les circuits et l'organisation de la commercialisation dans les pays membres de l'ADRAO.

D. S. C. Spencer

Private and Social Profitability in Rice Production and Marketing in Sierra Leone

[1979]. 43 pages. 28 × 21.5 cm. Stapled.

There are many rice production and post-production systems in Sierra Leone. Government policy has had a differential impact on these various systems. The major objective of this analysis is to examine the structure of costs in order to estimate comparative advantage in domestic rice production and to determine the effects of government policy.

Rice Production in Liberia: its Problems and Prospects

January 1981. 39 pages. 28 × 21.5 cm. Stapled.

Traditional and improved existing methods of rice cultivation in Liberia with their problems, advantages and disadvantages are discussed. Projects and suggestions for increasing yields in each method are presented. Measures are recommended that need the full cooperation and coordination of all agents in rice production to be effective.

Basic Texts: Rev.

January 1981. 163 pages. 27.5 × 21.5 cm. Stapled.

Includes: Final Act and constitution; Financial Regulations; Staff Regulations, Rules of Procedure of the Governing Council; Rules of Procedure of the Scientific and Technical Committee; Appeals Rules and Procedure; Staff Provident Fund Rules; Agreement Establishing the West Africa Rice Development Association Fund; Cooperation Agreement; Special Project Agreement; and Sub-Regional Office Establishment Agreement.

Also available in French.

Textes fondamentaux: Rev.

Janvier 1981. 204 pages. 27.5 × 21.5 cm. Agrafé.

Comprend: Acte final et Acte constitutif; Réglement financier; Statut du personnel; Réglement intérieur du Conseil d'administration; Réglement intérieur du Comité scientifique et technique; Réglement sur le Recours; Réglement sur le Fonds de prévoyance, Accord portant creation d'un Fonds de l'association pour le développement de la riziculture en Afrique de l'Ouest; Accord de cooperation; Accords pour la création d'un Project spécial; Accord pour la création d'un bureau sous-régional.

Disponible également en anglais.

D. S. C. Spencer

Incentives for Increased Food Production in West Africa: the Role of Agricultural Economists and Extension Specialists

November 1982. 20 pages. 27.5 × 21.5 cm. Stapled.

The poor performance of food production in West Africa is shown to be partly due to lack of adequate incentives to farmers in the form of appropriate technology developed and adapted to local conditions and price guarantees. The role of agricultural economists in determining the right price level and extension specialists in transfer of technology and its adoption is highlighted.

Technology for Increased Rice Production in West Africa

April 1983. 31 pages. 27.5 × 21.5 cm. Stapled.

Paper prepared for the ECA Joint Expert Consultation on Survey of Existing Technology on Food and Agricultural Research for Food Self-Sufficiency in Africa, 23-27 May, 1983, Addis-Ababa, Ethiopia, and dealing with WARDA's role in technology generation assessment and transfer in West Africa.

A. S. Sandhu and D. S. C. Spencer

Farmer Participation in the Design and Modification of Production Technologies Meeting his Needs and Resources: Experiences with Rice in West Africa

July 1983. 20 pages. 28 × 21.5 cm. Stapled.

Paper prepared for the workshop by ICRISAT, SAFGRAD, and IRAT on Farmer Participation in the Development and Evaluation of Agricultural Technology, 20-24 September, 1983, Ouagadougou, Burkina Faso. WARDA's Technology Assessment and Transfer (TAT) model contains four phases which are discussed in this paper.

V. K. Nyanteng

Public Institutions in Rice Marketing in West Africa

September 1983. 22 pages. 28 × 21.5 cm. Stapled.

Public participation in rice marketing in West Africa has failed to achieve the stated objectives and has created more problems as well as distortions in the marketing systems. It appears that the last results in the marketing system could be achieved if public interventions are such that they supplement and complement the efforts of private traders instead of displacing them.

V. K. Nyanteng

Organismes Etatiques de Commercialisation du Riz en Afrique de l'Ouest

Septembre 1983. 27 pages. 28 × 21.5 cm. Agrafé.

L'intervention de l'Etat dans la commercialisation du riz en Afrique de l'Ouest n'a pas réussi à remplir les objectifs qui lui sont assignés. Une intervention de l'Etat, s'inscrivant en complément des opérations du secteur privé, donnerait de meilleurs résultats.

Reviews of WARDA's Member States' Rice Seed Production Programmes

To assist the member countries in their seed programmes, WARDA has completed studies on the seed multiplication programmes in all the member countries in order to determine the assistance they would need. The Association has through the studies identified the weaknesses in the national seed production programmes and made recommendations for improvement.

The following publications are available:

* A Review of the Gambian Rice Seed Multipication Programme. February 1983
* A Review of the Liberian Rice Seed Multiplication Programme. July 1983
* A Review of Nigeria's Seed Programmes. August 1984
* Evaluation of Phase II of the Seed Multiplication Project of the Ministry of Agriculture and Forestry, Sierra Leone. 1984

Projets d'Organisation ou de Réorganisation de la Production semencière rizicole des pays membres de l'ADRAO

Pour aider ses pays membres à mettre en oeuvre leurs programmes semenciers rizicoles, l'ADRAO a effectué des études sur leurs programmes de multiplication de semences en vue de déterminer l'assistance dont ils ont besoin. Sur la base de ces études, l'Association a identifié les faiblesses dont souffrent les programmes semenciers nationaux et formulé des recommendations pour y remédier.

Les documents suivants sont actuellement disponibles:

* Projet d'organisation de la production semencière rizicole du Burkina Faso. Septembre 1983
* Projet d'organisation de la production semencière rizicole en République populaire du Bénin. Septembre 1981
* Projet de reorganisation de la production semencière rizicole du Sénégal. Janvier 1983
* Projet d'organisation de la production semencière rizicole de la Guinée-Bissau. Avril 1983
* Projet d'organisation de la production semencière rizicole de la Mauritanie. Juin 1983
* Projet d'organisation de la production semencière rizicole au Mali. Jullet 1983
* Projet d'organisation de la production semencière rizicole de la République de Guinée. Décembre 1983

Rice Statistics Yearbook = Annuaire Statistique du Riz: Bilingual

Number of pages varies. 18.5 × 25 cm. Wrap around.

The editions of the WARDA Rice Statistics Yearbook are compilations of basic data on rice production, consumption, trade, prices, and other related statistics in West Africa.

Les éditions de l'Annuaire statistique du riz de l'ADRAO constituent une compilation de statistiques de base sur la production, la consomation, la commercialisation et les prix en Afrique de l'Ouest ainsi que de données connexes.

V. K. Nyanteng

Aspects of Rice Production, Marketing and Pricing in the Upper Volta (Burkina Faso)

April 1984. 63 pages. 28 × 21.5 cm. Stapled.

Following a brief presentation of rice situation in Burkina Faso as to production and consumption, imports and the level of self-sufficiency, the rice marketing with emphasis on Government policies regarding trading and pricing is analyzed in details. Recommendations for improving rice marketing in this country are also given.

V. K. Nyanteng

The Role of Rice in Food Self-Sufficiency in West Africa

April 1984. 38 pages. 27.5 × 21.5 cm. Stapled.

A paper presented at a Conference on Food Self-Sufficiency in West Africa, held at the University of Ghana, Legon-Accra, 10-13 April 1984, and dealing with the importance of rice in West Africa, food situation, rice production and problems, strategies for increasing rice production and prospects in the 1980's.

C. W. Winkel

Recherche sur les Problemes de Gestion de l'eau dans la Région de l'ADRAO (cas de la Vallée du Fleuve Sénégal)

Mai 1984. 39 pages. 28 × 21.5 cm.

Cette étude, phase préliminaire d'un projet de recherche ADRAO sur la gestion de l'eau dans la Vallée du Fleuve Sénégal, à été conduite par l'auteur á la Station ADRAO de recherche sur le riz irrigué de Richard-Toll (Sénégal). Elle a consisté en une évaluation de ce qui a été fait ou se fait en matière de gestion de l'eau en général et dans la Vallée du Fleuve Sénégal en particulier et en une sêne d'observations sur les pratiques de gestion de l'eau dans les petits périmètres irrigués villageois de la Vallée du Fleuve Sénégal.

West Africa Rice Development Association: Programme Achievement, Contribution to and Impact on Rice Development in West Africa

October 1984. 98 pages. 27.5 × 21.5 cm. Stapled. WARDA/84/STC-14/17.

The objectives of this study are threefold: to put in perspective the achievement on positive results so far obtained by WARDA in its programmes and activities; to analyse the contribution WARDA has made to rice development in West Africa; and to analyze the impact of the contribution on rice development in the respective member countries and in the West Africa region as a whole.

Also available in French.

Etat d'Exécution, Contribution et Impact du Programme de l'ADRAO sur le Developpement de la Riziculture en Afrique de l'Ouest

Octobre 1984. 126 pages. 27.5 × 21.5 cm. Agrafé. ADRAO/84/CST-14/17.

Cette étude se propose d'atteindre trois objectifs:

mettre en perspective les réalisations ou résultats positifs obtenus jusqu'ici par l'ADRAO dans l'exécution de ses programmes et activités; analyser la contribution de l'ADRAO au développement de la riziculture en Afrique de l'Ouest; et enfin analyser l'impact de la contribution de l'ADRAO au développement de la riziculture dans les pays membres respectifs ainsi que dans l'ensemble de l'Afrique de l'Ouest.

Egalement disponible en anglais.

V. K. Nyanteng

Women and Agricultural Technology in Rice Farming in West Africa

1984. 11 pages. 27.5 × 21.5 cm. Stapled.

This paper describes the role of women in agriculture particularly in rice production in West Africa. Women are actively participating in all the phases of rice cultivation but it varies from one region to the other. Constraints faced by women rice farmers and better technology are identified.

Bilan de la Formation 1973-1984

35 pages. 28 × 21.5 cm. Agrafé.

Ce document est une rétrospective des activités de formation de l'ADRAO, de 1973 à 1984. Il brosse, en particulier, un tableau des divers types de formation et des cours offerts et traite des perspectives en matière d'expansion et de mise en oeuvre des programmes de formation.

Bibliographies, Directories, and Abstracts

Documentation Unit Library Recent Accession List = Liste d'Acquisitions Récentes de la Bibliothèque de l'Unité de Documentation

Number of pages varies. 28 × 21.5 cm. Stapled.

This quarterly list gives the bibliographical references of publications added to the library collections.

Cette liste timestrielle donne les références bibliographiques des publications acquises par la bibliothèque.

Inventory of Rice Training, Research and Development Institutions in Member Countries

December 1984. 57 pages. 28 × 21.5 cm. Stapled. WARDA/84/ STC-14/16.

Inventory of agricultural institutions in the WARDA region involved directly or indirectly in rice development through training, research and development activities.

Inventaire des Institutions de Formation, Recherche et Développement Rizicoles des Pays Membres

Decembre 1984. 54 pages. 28 × 21.5 cm. Agrafé. ADRAO/84/CST-14/16.

Inventaires des institutions agricoles de la région de l'ADRAO ayant une action directe ou indirecte sur le développement de la production rizicole à travers des actions de formation, recherche et développement.

List of Available Publicatons = Liste des Publications Disponibles

1985. 104 pages. 28 × 21.5 cm. Stapled.

This list updated annually gives brief bibliographical references of WARDA publications available for distribution.

Cette liste, mise à jour annuellement, donne une description bibliographique succinte des publications de l'ADRAO disponibles pour distribution.

Rice Vocabulary: English, French, Latin: Draft Ed.

June 1975. 207 pages. 27.5 × 21.5 cm. Wrap around.

This edition of the Rice Vocabulary contains about 2,500 words listed in the English alphabetic order, each word being followed by its French equivalent. For insects, weeds and other predators, the scientific name in Latin is also given, and for diseases the causal organism and the vector of the disease. The vocabulary also includes a French index and a Latin index.

Periodicals

WARDA Technical Newsletter

Number of pages varies. 28 × 21.5 cm. Saddle stitched.

Articles submitted for publication in the Technical Newsletter should be short, technical and related to rice work done in the WARDA region.

Articles from outside should have a bearing on rice in West Africa. Articles should in no case exceed three (3) pages (22 × 28 cm) double spaced typing. One table or figure per article will be accepted. Units of measures should be metric and abbreviations are to conform to international standards. Citations or references are not to be included. Authors are identified by name, title, and research organization.

Six issues have been released.

Bulletin Technique de l'ADRAO

Nombre de pages variable. 28 × 21.5 cm. Broché.

Les articles soumis pour publications dans le Bulletin doivent être courts, de nature technique et avoir traít aux activités entreprises dans le domaine de la riziculture dans la région de l'ADRAO. Les articles provenant d'autres parties du monde doivent présenter un intérêt spécial pour la riziculture en Afrique de l'Ouest. Ils doivent être long de trois (3) pages (22 × 28 cm) au maximum dactylographiées en double interligne et ne pas comporter plus d'un tableau. Le système métrique est adopté pour les unités de mesure et les abréviations doivent être conformes aux normes internationales. Les citations et références ne sont pas incluses. Les auteurs sont identifiés par leur nom, leur titre et leur institut de recherche.

Six numéros ont été édités.

Case Studies

Number of pages varies. 27 × 21.5 cm. Stapled.

The issues report case studies of encouraging methods of rice production in West Africa for Dissemination to Member States who might have similar conditions and might therefore wish to try out the systems.

Three titles have been published:

No. 1 The Irrigated Rice Area of the Kou Valley Upper Volta. June 1975.
No. 2 The Mac Carthy Islands Irrigation Project in The Gambia Economy. February 1976.
No. 3 Large Scale Mechanized Rice Production in Northern Ghana. September 1975.

Etudes de Cas

Nombre de pages variables. 27 × 21.5 cm. Agrafés.

Les études de cas analysent les expériences accumulées par un projet rizicole d'un Etat membre de l'ADRAO à la fois prometteur et riche d'enseignement pour les autres Etats membres.

Deux titres ont été publiés:

No. 1 Le périmètre de riziculture irriguée de la Valée du Kou (Haute Volta). Juin 1975.
No. 4 Projet pilote d'aménagement de petits bas-fonds en région forestière de Côte d'Ivoire. Mars 1977.

WARDA Occasional Papers

Number of pages varies. 28 × 21.5 cm. Stapled or Wrap around.

The WARDA Occasional Papers reports research and other findings by WARDA staff members and their collaborators. Emphasis is on technical, social and economics aspects of rice production, trade and consumption in West Africa. Articles of more than journal article length are preferred and are published in the two official languages of WARDA, i.e. English and French.

Six titles have been published:

No. 1 Prospects of Self-Sufficiency in Rice in West Africa. July 1980
No. 2 Types of Rice Cultivation in West Africa. July 1980
No. 3 S. A. Oni and A. E. Ikpi
Rice Production, Marketing and Policy in Nigeria. July 1980
No. 4 K. Prakah-Asante and V. K. Nyanteng
Economic Study of Rice Production, Marketing and Policy in Ghana. January 1981
No. 5 M. Kamanga and D. S. C. Spencer
Losses of Rice in West Africa: The Case of the Office du Niger in Mali. January 1981
No. 6 V. K. Nyanteng
Some Recent Policies and Programmes Related to Rice Production, Consumption and Trade in West Africa. September 1983

Publications Spéciales de l'ADRAO

Nombre de page variable. 28 × 21.5 cm. Agrafé ou relié.

Les publications spéciales de l'ADRAO rendent compte de resultats de recherche et autres travaux des membres du personnel de l'ADRAO et de leurs collaborateurs. L'accent y est sur les aspects techniques, sociaux et économiques de la production de riz, de sa commercialisation et de sa consommation en Afrique de l'Ouest. Ces publications spéciales, contrairement aux articles de revue, comportent des textes plus longs et plus élaborés publiés dans les deux langues officielles de l'ADRAO, à savoir le français et l'anglais.

Cing titres ont été publiés:

No. 1 Perspectives de l'autosuffisance en riz en Afrique de l'Ouest
       Juillet 1980
No. 2 Classification des types de riziculture en Afrique de l'Ouest
No. 3 S. A. Oni et A. E. Ikpi
       Production, commercialisation et politique rizicoles au
       Nigeria, Juillet 1980
No. 4 K. Prakah-Asante et V. K. Nyanteng
       Etude économique sur la production, la commercialisation et la
       politique rizicoles au Ghana, Janvier 1981
No. 5 M. Kamanga et D. S. C. Spencer
       Les pertes de riz en Afrique de l'Ouest: le cas de l'Office du
       Niger, Janvier 1981
No. 6 V. K. Nyanteng
       Politiques et programmes récents en matière de production, de
       consommation et de commercialisation du riz en Afrique de
       l'Ouest.

A paraître courant 1985.

AVRDC

The Asian Vegetable Research and Development Center

P.O. Box 42, Shanhua, Tainan, 741,
Taiwan, Republic of China
Cable: ASVEG Shanhua - Telex: 73580 AVRDC
Telephone: (06)5837801

About AVRDC

AVRDC is the international research and training center charged with improving vegetable crop production in the tropics. Its multinational scientific staff conducts research aimed at increasing the production potential of tomato, Chinese cabbage and sweet potato as well as soybean and mungbean. The Center's research agenda also includes the study of the nutritional, environmental and management factors that influence vegetable production in the tropics. Emphasis is similarly directed to scientific and training programs conducted in collaboration with national and international research organizations, agricultural development agencies and institutions of higher education.

How to Order

The publications listed below can be ordered from the AVRDC Office of Information Services (OIS). Most are available free of charge and will be posted upon request via surface mail. If airmail delivery or multiple copies are required, please check the price list located below. All charges should be paid with a US dollar check made out to AVRDC.

Notice to Librarians

Single copies of all AVRDC publications are available to libraries free of charge. If more than one copy is required, or if our records show that your library received a free copy in the past, you will be charged for the publication at the "additional copies" price listed below. All library copies will be sent by surface mail unless otherwise indicated in your request. Please consult the price list for airmail charges.

Publication	Standard Price*	Additional Copies	Airmail Postage (per copy) Asia Oceania	Airmail Postage (per copy) Africa Americas Europe
	Includes surface postage			
Pre- & Post-harvest Vegetable Technology Proceedings	$ 6.00	$ 5.00	$2.00	$ 3.00
Mungbean Proceedings	$10.00	$ 8.00	$3.75	$ 6.00
Tomato Proceedings	$10.00	$ 8.00	$4.50	$ 7.25
Chinese Cabbage Proceedings	$15.00	$12.00	$7.50	$13.00
Sweet Potato Proceedings	$15.00	$12.00	$8.20	$13.00
Soybean, Tropical & Subtropical Cropping Systems	$18.00	$14.00	$8.20	$13.00
Progress Report (before '83)	Free	$ 3.00	$2.00	$ 3.00
Progress Report (after '83)	$10.00	$ 8.00	$7.50	$11.50
Monographs	Free	$ 2.00	$1.00	$ 1.00

*Library copies free; libraries listed on the AVRDC mailing list receive all AVRDC publications automatically.

Research Highlights

AVRDC Progress Report Summaries 1982

Sept. 1983. 77 pages. 19.5 × 21 cm. Soft cover. Gratis (includes surface postage), plus air postage US$3.00.

AVRDC Progress Report Summaries 1983

Sept. 1984. 90 pages. 18 × 25.5 cm. Soft cover. Gratis (includes surface postage), plus air postage US$3.00. ISBN 95-9058-0009-9.

AVRDC Progress Report Summaries 1984

Sept. 1985. ___ pages. 18 × 25.5 cm. Soft cover. Gratis (includes surface postage), plus air postage US$3.00.

Annual Reports

AVRDC Progress Report 1975

1976. 77 pages. 17.5 × 21.5 cm. Soft cover. Gratis (includes surface postage), plus air postage US$3.00.

AVRDC Progress Report 1977

1978. 77 pages. 17.5 × 21.5 cm. Soft cover. Gratis (includes surface postage), plus air postage US$3.00.

AVRDC Progress Report 1979

1981. 107 pages. 17.5 × 21.5 cm. Soft cover. Gratis (includes surface postage), plus air postage US$3.00.

AVRDC Progress Report 1981

1982. 85 pages. 19.5 × 21.5 cm. Soft cover. Gratis (includes surface postage), plus air postage US$3.00.

AVRDC Progress Report 1982

February 1984. 338 pages. 19 × 26 cm. Soft cover. Gratis (includes surface postage), plus air postage US$9.00.

AVRDC Progress Report 1983

February 1985. 460 pages. 19 × 26 cm. Soft cover. US$10.00 (includes surface postage), plus air postage US$11.50. ISBN 92-9058-0013-7.

AVRDC Progress Report 1984

In press. ___ pages. 19 × 26 cm. Soft cover. US$10.00 (includes surface postage), plus air postage US$11.50.

Monographs

Robert F. Chandler, Jr.

The Potential for Breeding Heat Tolerant Vegetables for the Tropics

March 1983. 10 pages. 17.5 × 25 cm. Saddle stitched. Gratis (includes surface postage), plus air postage US$1.00.

A review of selected research concerning heat tolerance of vegetable crops in the humid and subhumid tropics. Discusses specific advances made to date with tomato, Chinese cabbage, and white potato.

Ricardo Bressani

World Needs for Improved Nutrition and the Role of Vegetables and Legumes

May 1982. 22 pages. 17.5 × 25 cm. Saddle stitched. Gratis (includes surface postage), plus air postage US$1.00.

An analysis of the potential for using vegetable and legume crops as a means of improving nutrition in the tropics. Dr. Bressani, one of Latin America's leading nutrition scientists, outlines a number of potential useful systems that combine vegetables and legumes with both cereals and root crops.

Sylvan H. Wittwer

Frontiers for Science and Technology in Vegetable Production

April 1983. 19 pages. 17.5 × 25 cm. Saddle stitched. Gratis (includes surface postage), plus air postage US$1.00.

A comprehensive review of current vegetable production technology and prospects for the future. Includes sections on genetic improvement, environmental stress, seed viability, biological stress, hormonal mechanism, photosynthetic enhancement, and the use of plastics in vegetable production.

N. L. Innes

Breeding Field Vegetables

July 1983. 34 pages. 17.5 × 25 cm. Saddle stitched. Gratis (includes surface postage), plus air postage US$1.00.

A summary of vegetable breeding research worldwide. The author, one of the United Kingdom's most eminent horticultural researchers, provides insights on breeding for pest resistance, environmental stress, and the efforts made to improve the nutritional content of vegetables.

M. S. Liu and Paul C. Ma

Postharvest Problems of Vegetables and Fruits in the Tropics and Subtropics

September 1983. 14 pages. 17.5 × 25 cm. Saddle stitched. Gratis (includes surface postage), plus air postage US$1.00.

The authors, researchers at the Food Industry Research Institute in Taiwan, describe the general problems associated with handling perishable fruits and vegetables in the tropics. Various protection methods are highlighted including temperature control, drying, pickling, and salting.

B. Choudhury

Vegetables and the Quality of Life in the Year 2000

1984. 7 pages. 17.5 × 25 cm. Saddle stitched. Gratis (includes surface postage), plus air postage US$1.00. ISBN 92-9058-0005-6.

The author, an internationally acclaimed vegetable breeder, discusses various issues facing vegetable research and production in the years ahead. Specific attention is drawn to the need for integrated research at the national and international levels and the expansion of extension and training services.

S. Tsou and S. Y. Huang

The Economics of Vegetable Production in the Year 2000

In press. ___ pages. 17.5 × 25 cm. Saddle stitched. Gratis (includes surface postage), plus air postage US$1.00.

The authors, consider the desirability of reducing dependency on rice and other cereals while increasing nutritional and economic benefits from fruits and vegetables in Asian LDCs.

Conference, Symposia and Workshop Proceedings

Proceedings of the Workshop on Pre- & Post-Harvest Vegetable Technology in Asia

1977. 156 pages. 17.5 × 25 cm. Soft cover. US$6.00 (includes surface postage) plus air postage. US$3.00. ISBN 92-9058-0011-0.

Robert Cowell (Editor)

Mungbean, Proceedings of the First International Symposium

February 1978. 278 pages. 20 × 25.5 cm. Soft cover. US$10.00 (includes surface postage), plus air postage US$6.00.

The first comprehensive review of mungbean research and production in the tropics. Contains more than 50 papers presented at the First International Symposium on Mungbean held at the University of the Philippines, August 1977. General subjects include mungbean production, utilization and economics, management, plant protection, and varietal improvement.

Robert Cowell (Editor)

Tomato, Proceedings of the First International Symposium

February 1979. 290 pages. 17.5 × 24.5 cm. Soft cover. US$10.00 (includes surface postage), plus air postage US$7.25.

An authoritative summary of tomato production in the subhumid and humid tropics. Contains 27 papers presented at the First International

Symposium on Tropical Tomato held at the Asian Vegetable Research and Development Center, October 1978. Subject matter includes tomato physiology and plant nutrition, pest and disease control, management, and varietal improvement.

N. S. Talekar and T. D. Griggs (Editors)

Chinese Cabbage, Proceedings of the First International Symposium

1981. 489 pages. 17.5 × 25 cm. Hard cover. US$15.00 (includes surface postage), plus air postage US$13.00.

The most up-to-date review presently available on Chinese cabbage research. Includes 45 papers presented at the First International Symposium on Chinese cabbage held in 1980 at Tsukuba, Japan. Subject matter encompasses plant protection, general physiology, reproduction physiology and plant breeding.

R. L. Villareal and T. D. Griggs (Editors)

Sweet Potato, Proceedings of the First International Symposium

October 1982. 481 pages. 17.5 × 25 cm. Hard cover. US$15.00 (includes surface postage), plus air postage US$13.00.

An in-depth examination of the current status of sweet potato research in the world's tropical and temperate zones. Includes 44 papers presented at the First International Symposium on Sweet Potato in Taiwan, March 1981. Subject matter includes general production characteristics, plant physiology, pest control, crop management and varietal improvement.

S. Shanmugasundaram and Edward Sulzberger (Editors)

Soybean in Tropical and Subtropical Cropping Systems

June 1985. 400+ pages. 17.5 × 25 cm. Hard cover. US$18.00 (includes surface postage), plus air postage US$13.00. ISBN 92-9058-0015-3.

Miscellaneous

Merle R. Menegay

Taiwan's Specialized Vegetable Production Areas: An Integrated Approach

December 1975. 17 pages. 17.5 × 25 cm. Saddle stitched. Gratis (includes surface postage), plus air postage US$1.00. ISBN 92-9058-0004-8.

A description of government efforts in the mid-1970's to deal with vegetable production and marketing problems in Taiwan through the establishment of special production zones.

Merle R. Menegay

Farm Management Research on Cropping Systems

November 1976. 19 pages. 17.5 × 25 cm. Saddle stitched. Gratis (includes surface postage), plus air postage US$1.00. ISBN 92-9058-0001-3.

A guide to the "CII" analytical system of determining how new crop technologies benefit specific farmer groups, especially those using cropping systems.

Peter H. Calkins

Four Approaches to Risk and Uncertainty for Use in Farm Management Extension

1976. 18 pages. 17.5× 25 cm. Saddle stitched. Gratis (includes surface postage), plus air postage US$1.00. ISBN 92-9058-0008-10.

A discussion of the role of farm management researchers and extension personnel in providing the cultivator with useful information on new varieties, cropping practices and farm organization.

P. H. Calkins, S. Y. Huang, and J. F. Hong

Farmers' Viewpoint of Sweet Potato Production in Taiwan

December 1977. 44 pages. 17.5 × 25 cm. Saddle stitched. Gratis (includes surface postage), plus air postage US$1.00.

An analysis of why farmers in Taiwan grow less sweet potato than in the past. Includes evaluations on the availability of irrigation, the influence of soil type fertilizer and cropping patterns, production costs and profits.

Peter H. Calkins

Vegetable Consumption Patterns in Five Cities of Taiwan

September 1978. 24 pages. 17.5 × 25 cm. Saddle stitched. Gratis (includes surface postage), plus air postage US$1.00.

A survey of vegetable consumption patterns in Taiwan urban centers. Includes analyses of consumption trends as influenced by geographic location, income and consumer preference.

S. Y. Huang and P. H. Calkins

Summer Tomato Production in Taiwan

1978. 29 pages. 17.5 × 25 cm. Saddle stitched. Gratis (includes surface postage), plus air postage US$1.00. ISBN 92-9058-0007-2.

Although tomatoes are a popular vegetable in Taiwan, this study shows that consumption drops significantly in the summer while prices rise substantially because of low production levels deriving from increased risk of crop failure due to high temperatures and rainfall.

K. R. Huang and P. H. Calkins

Vegetable Production in Taiwan: A survey of 300 Farmers

September 1978. 36 pages. 17.5 × 25 cm. Saddle stitched. Gratis (includes surface postage), plus air postage US$1.00.

This report provides detailed information in the area planted to vegetable crops in Taiwan, the species selected for production, and the yield of major vegetable crops. It also focuses on the way farmers make decisions about the crops they plant, choice of cropping system, and how they cope with production problems.

Peter H. Calkins

Why Farmers Plant What They Do

1978. 92 pages. 17.5 × 25 cm. Saddle stitched. Gratis (includes surface postage), plus air postage US$1.00. ISBN 92-9058-0010-2.

This study attempts to analyze the decision-making processes of the farmer.

Ruben L. Villareal and S. H. Lai

Procedures for Tomato Evaluation Trials

1978. 4 pages. 20.5 × 26 cm. Single fold. Gratis (includes surface postage), plus air postage US$1.00.

Peter H. Calkins and Hui-Mei Wang

Improved Marketing of Perishable Commodities: A Case Study of Selected Vegetables in Taiwan

February 1978. 65 pages. 17.5 × 25 cm. Saddle stitched. Gratis (includes surface postage), plus air postage US$1.00. ISBN 92-9058-0012-9.

Seven target vegetables are examined with the aim of developing practical analytical techniques that can be used by government planners and private sector entrepreneurs in developing marketing schemes for perishable crops.

Peter H. Calkins and Su-hua Tu

White Potato Production in Taiwan: A Farm Survey

September 1978. 29 pages. 17.5 × 25 cm. Saddle stitched. Gratis (includes surface postage), plus air postage US$1.00.

The study investigates the productivity and profitability of white potato production in Taiwan, one area where the introduction of heat-tolerant varieties of potato could have a great impact on both producers and consumers.

Peter H. Calkins and Kuang-rong Hwang

Soybean Production in Taiwan: A Farm Survey

1978. 28 pages. 17.5 × 25 cm. Saddle stitched. Gratis (includes surface postage), plus air postage US$1.00. ISBN 92-9058-0017-10.

The study describes profitability and yield from soybean, the motivations and problems of farmers, and the reaction of farmers in the past to measures such as the extension of improved varieties.

Ruben L. Villareal and S. H. Lai

Pollen Collector — An Aid to Plant Breeders

1978. 2 pages. 20.5 × 26 cm. Gratis (includes surface postage), plus air postage US$1.00.

H. G. Park

Suggested Cultural Practices for Mungbean

1978. 2 pages. 20.5 × 26 cm. Gratis (includes surface postage), plus air postage US$1.00.

James J. Riley

AVRDC Crop Environment

1978. 2 pages. 20.5 × 26 cm. Gratis (includes surface postage), plus air postage US$1.00.

S. Shanmugasundaram

Varietal Development and Germplasm Utilization in Soybeans

1979. 36 pages. 17.5 × 25 cm. Side stitched. Gratis (includes surface postage), plus air postage US$1.00. ISBN 92-9058-0016-1.

This study provides an overview of soybean varietal development and germplasm utilization.

James R. Riley

Evaluation of Environmental Parameters in the Humid Tropics for Crop Scheduling Purposes

December 1979. 31 pages. 17.5 × 25 cm. Saddle stitched. Gratis (includes surface postage), plus air postage US$1.00.

How can researchers analyze environmental variation for crop planning purposes using reports from small, unsophisticated weather stations? In this report the author describes a simple weather analysis system developed for researchers at AVRDC.

C. G. Kuo and S. H. Lai

Suggested Cultural Practices for Tomato

1979. 4 pages. 20.5 × 26 cm. Single fold. Gratis (includes surface postage), plus air postage US$1.00.

S. Shanmugasundaram

Evaluating AVRDC Soybeans

1979. 4 pages. 20.5 × 26 cm. Single fold. Gratis (includes surface postage), plus air postage US$1.00.

John N. Hubbell

Suggested Cultural Practices for Sweet Potato Experiments

1979. 2 pages. 20.5 × 26 cm. Gratis (includes surface postage), plus air postage US$1.00.

S. Shanmugasundaram

Suggested Cultural Practices for Soybean

1979. 2 pages. 20.5 × 26 cm. Gratis (includes surface postage), plus air postage US$1.00.

Romeo T. Opeña and S. H. Lo

Procedures for Chinese Cabbage Evaluation Trials

1980. 4 pages. 20.5 × 26 cm. Single fold. Gratis (includes surface postage), plus air postage US$1.00.

Ruben L. Villareal and H. F. Lo

Procedures for Sweet Potato Evaluation Trials

1980. 4 pages. 20.5 × 26 cm. Single fold. Gratis (includes surface postage), plus air postage US$1.00.

Romeo T. Opeña

Cultural Practices for Chinese Cabbage at AVRDC

1981. 4 pages. 20.5 × 26 cm. Single fold. Gratis (includes surface postage), plus air postage US$1.00.

T. H. Menegay

The AVRDC Vegetable Preparation Manual

March 1977. Reprinted December 1982. 105 pages. 28 × 21.5 cm. Side stitched. Gratis (includes surface postage), plus air postage US$1.00.

A guide to preparing simple, nutritious, low cost meals using AVRDC vegetable crops — soybean, mungbean, Chinese cabbage, tomato, and sweet potato.

John S. Caldwell and Donald W. Newsom

Vegetable Consumption and Production in Two Municipalities in Ilocos Norte, Philippines

April 1984. 24 pages. 17.5 × 25.5 cm. Saddle stitched. Gratis (includes surface postage), plus air postage US$1.00.

This socioeconomic study describes vegetable farming trends in two villages in the northern Philippines. Specific attention is drawn to farm size and tenancy and consumption and production patterns for tomato, mungbean, sweet potato, common cabbage, pepper, soybean and Chinese cabbage.

S. K. Green

Guidelines for Diagnostic Work in Plant Virology

1974. 43 pages, 17.5 × 25 cm. Side stitched. Gratis (includes surface postage), plus air postage US$1.00. ISBN 92-9058-0003-10.

This handbook is a general guide to diagnostic work in plant virology and is addressed primarily to the extension worker.

Annotated Bibliography of Diamondback Moth

Feb. 1985. 475 pages. 17 × 25 cm. Soft cover. Limited quantity. Gratis to entomologists (includes postage). ISBN 92-9058-0014-5.

This annotated bibliography was published for the International Workshop on Diamondback Moth Management. Its purpose is to bring about all available information on one of the most destructive pests for cruciferous crops in the world.

C. J. Franssen

Insect Pests of Sweet Potato Crop in Java

In press. Approx. 20 pages. 17.5 × 25 cm. Saddle stitched. Gratis (includes surface postage), plus air postage US$1.00.

This booklet was translated from the Dutch "Insecten shadelisk aan het Batatengewas," Archipel drukkerii-Buitenzovo (Bogor), 1934.

Compiled by N. S. Talekar, Hui-chin Yang, Song-tay Lee, Bor-shyan Chen, and Li-yi Sun

Jack Gershon and Pura Lastimosa (Editors)

International Gardening Workshop — Thailand/Taiwan (April 21-27)

In press. ___ pages. Soft cover.

Planners, policy-makers and decision-makers met to examine first hand the transfer of home garden technology developed at AVRDC to a tropical country (Thailand).

A Strategy for Progress in Tropical Vegetable Research (1986-1990)

In press. ___ pages. Soft cover.

This strategy report analyzes the AVRDC's recent achievements, sets future goals and outlines the strategy for their achievement.

Newsletters/Bulletins

TVIS News

Approx. 26 pages/issue. 21 × 27.5 cm. Side stitched. Gratis (includes postage).

This newsletter is prepared by the Tropical Vegetable Information Services through AVRDC and provides a platform for scientists working with mungbean, soybean and Chinese cabbage to exchange information.

Soybean Rust Newsletter

Approx. 50 pages/issue. 21 × 27.5 cm. Saddle stitched. Gratis (includes postage).

This annual newsletter is published by the International Working Group on Soybean Rust through AVRDC.

Centerpoint

Approx. 4 pages/issue. 21.5 × 27.5 cm. Gratis (includes postage).

This quarterly bulletin provides up-to-date information on the AVRDC's activities on research, training, publications, etc.

Audiovisual Materials

A. Perlis and E. W. Sulzberger

The Asian Vegetable Research and Development Center

1983. 12.5 minutes. US$25.00 (includes surface postage), plus air postage US$10.00.

An introduction to the research, training, and development work of the Asian Vegetable Research and Development Center.

ICIPE

The International Centre of Insect Physiology and Ecology

P. O. Box 30772, Nairobi, Kenya
Cable: ICIPE, Nairobi
Telex: 22053
Telephone: Nairobi 43049/43081/43719

Introduction

The International Centre of Insect Physiology and Ecology (ICIPE) was established in 1970 and became functional in 1972. The idea was inspired by the realization that long-term solutions to pest problems in the tropics could not rely on chemical insecticides alone, and that even the effective use of these depended on a thorough understanding of insect biology, ecology and related disciplines.

Research Programmes

ICIPE's Research Programme is primarily concerned with tropical insect pests of crops and vectors of livestock diseases. The crop pests under investigation include the crop borers of maize, sorghum, and cowpeas. The livestock pests being studied are tsetse, the vector of trypanosomiasis, and ticks, the vector of several debilitating cattle diseases including East Coast Fever. In earlier years, there was also substantial emphasis on insect vectors of human diseases. All that remains of this medical vectors research is a single project on leishmaniases, which is concerned with the sandfly vector of the disease, which is important in the semi-arid rural areas. The ultimate impact of ICIPE's research will be the development of integrated pest and vector management systems for these economically important pests and disease vectors.

ICIPE's research programmes are supported by four research units which provide assistance in chemistry, fine structure, sensory physiology and biostatistics.

Training

Training at the ICIPE is conducted in three major areas: (a) Professional scientific training is accomplished at both the postdoctoral and postgraduate levels. The latter is done in close collaboration with several African universities. (b) Short-term training courses in pest and vector management are offered to practising insect scientists and technologists from the developing world. (c) There is programme for the upgrading of special skills and techniques of the staff of the ICIPE itself.

Research Highlights

ICIPE Research Highlights

Forthcoming - Late 1985.

This will be the first time ICIPE is producing Research Highlights. This first report will therefore be fairly comprehensive giving a summary of work and achievements over the last 14 years, in easy to read language. Thereafter, this will be a yearly report.

Annual Reports

ICIPE Annual Reports 1970-1984

Annual. Pages vary. 29 × 21 cm. Paperbound. Free.

ICIPE Annual Reports give an overview of the centre's research activities and results for each preceding year. It contains a list of scientific publications and seminar presented in the year of reporting.

Conferences/Symposia/Workshop Proceedings

F. D. Obenchain and R. Galun (Editors)

Physiology of Ticks

1982. 509 pages. 23 × 15 cm. Hardbound. ISBN 0-08-024937-X. US$125.00 plus air postage ($30.00) or surface postage ($3.00).

Concentrates on various aspects of Tick Physiology. Based on papers contributed in a workshop on Physiology Significance of Tick Behavior. Represents an international effort by 25 experts each giving an up-to-date review of the subject.

D. L. Whitehead and W. S. Bowers (Editors)

Natural Products for Innovative Pest Management

1983. 586 pages. 23 × 15 cm. Hardbound. ISBN 0-08-028893-6. US$125.00 plus air postage ($30.00) or surface postage ($3.00).

This volume contains a collection of papers, presented by Internationally recognized authors, concerned with current methods of developing new pesticides. The chapters focus on the use of natural products and describe methods for evaluating chemicals from a wide range of biological materials.

Proceedings of the International Study Workshop on Host Plant Resistance and its Significance in Pest Manangement

1984. 15 pages. 21 × 30 cm. Paperbound.

Report of a 1984 workshop organized; to review the status of research on host plant resistance and its potential role in pest management; to review advances made in recent years on this research; to promote increased awareness and stimulate effort in practical host resistance research and to explore ways of collaborating in this area of research.

Recommendations from this workshop are included in this report.

International Study Workshop on Leishmaniasis Epidemiology

1984. 12 pages. 21 × 30 cm.

Report of a 1984 workshop on Leishmaniasis Epidemiology convened to review current knowledge and exchange experiences on various aspects of the disease in order to hasten the design and testing of effective control measures; to provide a forum for discussing key problem areas encountered by researchers in this field; to expose scientists to the magnitude of leishmaniasis and its epidemiology in the world and how scientists are approaching the survey, surveillance and research on this disease complex and to discuss international research collaboration.

Recommendations formulated took into account recommendations of the WHO Expert Committee on Leishmaniases (WHO Technical Report No. 701,1984).

Miscellaneous

African Regional Postgraduate Programme in Insect Science (ARPPIS)

2nd Annual Report - 1984 - 26 pages. Free
Brochure 1984 - 30 pages. Free
Calendar 1984 - 22 pages. Free

These 3 documents give information on the abovementioned post-graduate training undertaken by ICIPE in collaborative with 12 African Universities.

J. A. L. Watson, B. M. Okot-Kotber, Ch. Noirot (Editors)

Caste Differentiation in Social Insects

1985. 405 pages. 23 × 15 cm. Hardbound. ISBN 0-08-030783-3. US$125.00 plus air postage ($30.00) or surface postage ($3.00).

In this book, for the first time, a panoramic presentation is made of the many facets of caste differentiation in termites together with comparative discussion of other social insects, including bees, ants, and wasps.

Mbita Point Field Station Brochure

1984. 18 pages. 15 × 21 cm. Paperback.

This brochure gives an illustrated summary of Mbita Point Field Station, ICIPE's Main field facility located on the shores of Lake Victoria, about 500 km. from Nairobi in Western region of Kenya. It describes ICIPE's Crop Pest Research Programme which is accomodated in this station and other facilities used for the development, testing and packaging of pest management technologies.

ICIPE Brochure

Forthcoming - Late 1985.

This brochure gives a broad overview of the structure and work of ICIPE. It explains the establishment and evolution of the Centre, the research of the institute and a highlight of achievement in the various programme.

J. M. Ojal

Directory of Insect Scientists in Africa

1979. 50 pages. 21 × 15.5 cm. Saddle stitched. US$5.00 surface or US$8.00 airmail

A directory compiled at ICIPE inspired by a meeting in Ibadan, in 1974 where some eighty or so insect scientists from East, West, North, and Central Africa met to discuss the status of insect science in Africa.

The Directory is intended to facilitate greater interaction among insect scientists of Africa.

The Status of Insect Science in the Tropical World

A series of ICIPE Annual Public Lectures delivered by the ICIPE Director, Pofessor Thomas R. Odhiambo, which examine the problems and progress of insect science in all its many manifestations, but especially in the way it contributes to national development in Tropical Africa.

1. This is a DUDU World 1975. 17 pages 14.5 × 20.5 cm paper back. Free upon request.
2. National Scientific Capabilities 1976. 20 pages 14.5 × 20.5 cm paper back. Free upon request.
3. Science and Technology for the Rural Farmer 1977. 21 papers. 14.5 × 20.5 cm paper back. Free upon request.
4. The Use and Non-Use of Insects.1978. 17 pages. 14.5 × 20.5 cm paper back. Free upon request.
5. Administration of Scientific Research in Africa. 1979. 17 pages. 14.5 × 20.5 cm paper back. Free upon request.

Periodicals

Thomas R. Odhiambo (Editor)

Insects Science and Its Application
(The International Journal of Tropical Insect Science)

Bimonthly. Pages vary. 27 × 19 cm. Paperbound. ISSN 0191-9040.

Subscription can be purchased at US$45. Single copies available at US$8.00 plus airmail postage ($2.00) or surface postage ($1.00).

Deals comprehensively with all aspects of research on tropical insects (and related arthropods) and the application of new discoveries to such diverse fields as pest and vector management and the use of insects for human welfare.

Special Issues of the Journal

D. A. T. Baldry and M. F. B. Chaudhury (Editors)

Epidemiology of African Trypanosomiasis (Vol. 1 No. 1)

1980. 121 pages. 27 × 19 cm. Paperbound. ISSN 0191-9040.

This issue contains the proceedings of a study workshop on the Epidemiology of African Trypanosomiasis held at the international

Centre of Insect Physiology and Ecology (ICIPE) in Nairobi, Kenya, which reviewed major advances and development in all aspects of the epidemiology of African trypanosomiasis up to 1979.

Thomas R. Odhiambo (Editor)

The Biology, Ecology, and Control of the Sorghum Shootfly *Atherigona Soccata Rondanl* (Volume 2, No. 1/2)

1981. 127 pages. 27 × 19 cm. Paperbound. ISSN 0191-9040.

This special issue consists of 20 papers presented by practitioners in the field of pest management, and in particular sorghum shootfly, during an international study workshop on the sorghum shootfly in Nairobi, 1981. The workshop reviewed the global progress in sorghum shootfly research and examined its relevance to the practical problems of shootfly control.

Thomas R. Odhiambo (Editor)

Crop-Borers and Emerging Strategies for their Control (Volume 4 No. 1/2)

1983. 222 pages. 27 × 19 cm. Paperbound. ISSN 0191-9040.

This issue contains the proceedings of an international study workshop on crop-borers and strategies for their control held at ICIPE's Mbita Point Field Station in June 1982. It includes some 32 papers presented by participants from all over the world which go some way to reviewing a great deal of recent information on crop-borers which might lead to new crop-borer management strategies.

Dudu

8 pages. 29 × 21. Saddle stitched. Free.

This newsletter is published 4 times a year and contains news of ICIPE's activities and usually one or two science features. Announcements of meetings, courses, and appointments are also made in the newsletter.

ICLARM

International Center for Living Aquatic Resources Management

MCC P.O. Box 1501, Makati, Metro Manila
Cable Address: ICLARM Manila
Telex Address: ITT 45658 ICLARM PM
Telephone: 818-04-66; 818-92-8

Introduction

ICLARM was organized in 1975 in response to a broadly perceive need to strengthen research on tropical aquatic resources for the benefit of developing countries. It was recognized that development of aquaculture and of viable fisheries management practices is often hindered by the inadequacy of technical information pertinent to problems encountered in the tropics. ICLARM's research has been oriented toward the goal of improving the condition of the rural poor in developing countries by improving their incomes, employment opportunities and productivity through the wise use of aquatic resources ICLARM was incorporated in the Philippines in 1977. Since then ICLARM has played an important role in fisheries development and has been widely recognized for its research contributions.

Objectives

The objectives of ICLARM as stated in its Articles of Incorportation may be summarized as follows:
- To conduct and assist with research on fish production, management, presentation, distribution, and utilization to assist peoples of the world in meeting their nutritional and economic needs.
- To improve the efficiency of culture and capture fisheries through coordinated research, education and training, linked with appropriate development and extension programs.
- To upgrade the social, economic, and nutritional status of people in less-developed areas through improvement of small-scale fisheries.
- To encourage labor intensive and low-energy input systems where appropriate.
- To publish and disseminate research findings in support of the Center's objectives.
- To organize and conduct conferences, forums, and workshops for discussion of current problems and for exchange of research results.

Annual Reports

J. L. Maclean (Editor)

ICLARM Report 1977-80

1980. 168 pages. 18 × 25.5 cm. Perfect. ISSN 0115-4494. Available free on request by surface mail. US$6.00 airmail.

J. L. Maclean (Editor)

ICLARM Report 1981

1982. 112 pages. 18 × 25.5 cm. Perfect. ISSN 0115-4494. Available free on request by surface mail. US$6.00 airmail.

J. L. Maclean and L. B. Dizon (Editors)

ICLARM Report 1982

1983. 95 pages. 18 × 25.5 cm. Perfect. ISSN 0115-4494. Available free on request by surface mail. US$6.00 airmail.

J. L. Maclean and L. B. Dizon (Editors)

ICLARM Report 1983

1984. 115 pages. 18 × 25.5 cm. Perfect. ISSN 0115-4494, ISBN 971-1022-05-2. Available free on request by surface mail. US$6.00 airmail.

J. L. Maclean and L. B. Dizon (Editors)

ICLARM Report 1984

1985. 108 pages. 18 × 25.5 cm. Perfect. ISSN 0115-4494, ISBN 971-1022-15-X. Available free on request by surface mail. US$6.00 airmail.

Monographs

D. Pauly

Theory and Management of Tropical Multispecies Stocks: a Review, with Emphasis on the Southeast Asian Demersal Fisheries

1979. Reprinted 1983. 35 pages. 28 × 21.5 cm. Saddle stitched. ISSN 0115-4389. US$2.50 surface or US$7.00 airmail.

This is a critical review of the demersal fisheries of Southeast Asia and the models used for managing them. Many are overcapitalized; they are always extremely difficult to monitor; and they are beset with problems related to effective enforcement of any selected management scheme.

The theory behind the stock assessment models and the derived rules of thumb used in the region have been notably neglected, the result being that models which now appear unrealistic have been used for years.

I. R. Smith

A Research Framework for Traditional Fisheries

1979. Reprinted 1983. 40 pages. 28 × 21.5 cm. Saddle stitched. ISSN 0115-4389. US$2.50 surface or US$7.00 airmail.

Identifies areas of traditional fisheries research that have the greatest potential for contributing to the solution of problems facing traditional fishermen and their communities; traces the changing emphasis of past development programs; discusses alternative development strategies; analyzes the relevant theoretical predictions and research issues associated with each; and concludes that long-term solutions to problems of low standards of living lie in reducing rather than in increasing fishing effort.

C. E. Nash and Z. H. Shehadeh (Editors)

Review of Breeding and Propagation Techniques for Grey Mullet, *Mugil cephalus* L.

1980. Reprinted 1984. 87 pages. 28 × 21.5 cm. Perfect. ISSN 0115-4389. US$6.00 surface or US$12.50 airmail.

This review is a compendium of most of the available biological and engineering knowledge relevant to the breeding and mass propagation of the Mugilidae species, particularly the grey mullet *Mugil cephalus* Linn. This knowledge has been accumulated over 20 years with most major advances occurring the last 10 years.

I. R. Smith, M. Y. Puzon and C. N. Vidal-Libunao

Philippine Municipal Fisheries: a Review of Resources, Technology and Socioeconomics

1980. Reprinted 1981, 1983. 87 pages. 28 × 21.5 cm. Perfect. ISSN 0115-4389. US$6.00 surface or US$12.50 airmail.

Synthesizes publicly available research studies and secondary data; provides evidence of a trend toward overfishing of Philippine coastal waters, and of a willingness among fishermen to consider alternative activities to capture fishing; the shift in government programs from a resource "development" to "management"; concludes with a discussion of the implications of these research findings to fisheries management and research.

P. Edwards

Food Potential of Aquatic Macrophytes

1980. Reprinted 1984. 51 pages. 28 × 21.5 cm. Saddle stitched. ISSN 0115-4389. US$4.00 surface or US$8.00 airmail.

A review is presented of the pathways in which aquatic macrophytes may be involved in the food production process, directly as human food, as livestock fodder, as fertilizer (mulch and manure, ash, green manure, compost, biogas slurry), and as food for aquatic herbivores, such as fish, turtles, rodents, and manatees.

G. Wohlfarth and G. Hulata

Applied Genetics of Tilapias

1981. Revised 1983. 26 pages. 28 × 21.5 cm. Saddle stitched. ISSN 0115-4389. US$3.75 surface or US$7.50 airmail.

This paper reviews the successes and failures of genetic experiments on tilapias and points to various characteristics of tilapias which require improvement for increased yields.

J. L. Munro (Editor)

Caribbean Coral Reef Fishery Resources

1983. 276 pages. 28 × 21.5 cm. Smythe-sewn (cloth), Perfect (paper). ISSN 0115-4389, ISBN 971-1022-00-1 (cloth), ISBN 971-1022-01-X (paper). US$20.00 surface or US$41.25 airmail (paper); US$24.40 surface or US$46.25 airmail (cloth).

An edited republication of a series of 10 mimeoed reports of the 1969-1973 phase of the Fisheries Ecology Research Project at the University of the West Indies, Jamaica, with a concluding chapter detailing advances in knowledge of reef fish biology and stock assessment to the present time; some data have been reanalyzed using new techniques.

D. Pauly

Fish Population Dynamics in Tropical Waters: a Manual for Use with Programmable Calculators

1984. 325 pages. 28 × 21.5 cm. Smythe-sewn. ISSN 0115-4389, ISBN 971-1022-03-6 (cloth). ISBN 971-1022-04-4 (paper). US$17.50 surface, US$31.25 airmail (paper); US$21.85 surface, US$37.00 airmail (cloth); 30 preprogrammed HP 67/97 cards, US$12.50.

A selection of methods applicable to tropical fish and fisheries which can be implemented with the help of programmable calculators. Methods cover length/weight relationships, mesh selection, growth, mortality, population size estimation, yield-per-recruit assessments, stock-recruitment relationships, surplus-yield-models, rate of increase of populations and aspects of multispecies stocks and fisheries.

Includes user instructions for 30 programs for use with HP 67/97 programmable calculators, 60 examples and complete keystroke sequences. Preprogrammed cards available.

A supplement will be available in 1985 for users of HP 41 C calculators.

C. Bailey, A. Dwiponggo and F. Marahudin

Review of Indonesian Marine Capture Fisheries

1985. 200 pages. 28 × 21.5 cm. Perfect. ISSN 0115-4389. US$12.50 surface or US$25.00 airmail.

A review of the biological, technical and socioeconomic aspects of Indonesian capture fisheries based on a large volume of "grey" literature obtained from regional universities and government agencies. A result of the cooperative effort between researchers from the Directorate-General of Fisheries, the Marine Fisheries Research Institute and ICLARM.

V. G. Jingran and R. S. V. Pullin

A Hatchery Manual for the Common, Chinese and Indian Major Carps

1985. 180 pages. 28 × 21.5 cm. Perfect. ISSN 0115-4389, ISBN 971-1022-17-6. US$10.50 surface or US$24.00 airmail.

This major work on carp hatchery and nursery methods was written as part of an Asian Development Bank (ADB) project aimed at improving carp seed production technology in ADB-member countries notably Bangladesh, Burma, Indonesia, Nepal, Pakistan and Sri Lanka. Includes chapters on all aspects of the cultured carps. Designed to be a reference source for all working on carp seed production and to act as a mini-library for those stationed at seed production centers remote from scientific information channels.

Conferences, Symposia, and Workshop Proceedings

J. M. Reinhart (Editor)

Small Boat Design

1979. 79 pages. 28 × 21.5 cm. Perfect. ISSN 0115-4435. US$5.00 surface or US$13.00 airmail.

Papers from a 1975 conference on Small Boat Design in Noumea, New Caledonia. Its goal was to develop small-scale fishing craft for future fisheries programs.

Contains 16 papers on boats and engines for small-scale fisheries in the South Pacific.

F. T. Christy, Jr. (Editor)

Law of the Sea: Problems of Conflict and Management of Fisheries in Southeast Asia

1980. 68 pages. 28 × 21.5 cm. Perfect. ISSN 0115-4435. US$5.00 surface or US$13.00 airmail.

Proceedings of a 1978 workshop held jointly by ICLARM and the Institute of Southeast Asian Studies in Manila.

The workshop brought together experts from foreign ministries, departments of fisheries, universities, and the private sector. The meeting focused on two fundamentally important issues — allocation of fisheries resources and implementation of fisheries agreements.

F. T. Christy, Jr. (Editor)

Summary Report of the ICLARM/ISEAS Workshop on the Law of the Seas: Problems of Conflict and Management of Fisheries in Southeast Asia

1980. Reprinted 1984. 11 pages. 18 × 25.5 cm. Saddle stitched. ISSN 0115-4435. Available free on request.

Contains the summary report only of the ICLARM/ISEAS Workshop on the Law of the Sea.

R. S. V. Pullin and Z. H. Shehadeh (Editors)

Integrated Agriculture-Aquaculture Farming Systems

1980. 259 pages. 28 × 21.5 cm. Perfect. ISSN 0115-4389. US$12.50 surface or US29.00 airmail.

Proceedings of a 1979 conference held jointly by ICLARM and the Southeast Asian Regional Center for Graduate Study and Research in Agriculture in Manila.

Provides an overview of integrated agriculture-aquaculture farming systems as currently practiced in a number of Southeast Asian countries; reviews available experience and technology; discusses the social and economic aspects of these systems and identifies research and development requirements. 24 papers.

J. E. Bardach, J. J. Magnuson, R. C. May, and J. M. Reinhart (Editors)

Fish Behavior and Its Use in the Capture and Culture of Fishes

1980. 512 pages. 23 × 15.5 cm. Perfect. ISSN 0115-4389. US$19.00 surface or US$29.00 airmail.

Papers and discussions of the Conference on Physiological and Behavioral Manipulation of Food Fish as Production and Management Tools, held in Bellagio, Italy, 3-8 November 1977.

The papers are grouped into three categories: 1) manipulation of fish behavior essentially through the animals' senses, 2) controlling or predicting the reproduction or recruitment of fishes, and 3) predicting the distribution of fishes and their responses to fishing gear.

R. S. V. Pullin

Summary Report of the ICLARM Conference on the Biology and Culture of Tilapias

1981. 13 pages. 18 × 25.5 cm. Saddle stitched. ISSN 0115-4435. US$1.50 surface or US$3.50 airmail.

A summary of findings and research recommendations of a 1980 international conference. Topics covered are biology, physiology and culture of tilapias and research requirements.

R. S. V. Pullin and R. H. Lowe-McConnell (Editors)

The Biology and Culture of Tilapias

1982. Reprinted 1983, 1984. 432 pages. 18 × 25.5 cm. Smythe-sewn (cloth); Perfect (paper). ISSN 0115-4389, ISBN 971-04-0003-7 (cloth), ISBN 971-04-0004-5 (paper). US$16.25 surface, US$31.25 airmail (paper); US$21.85 surface, US$36.90 airmail (cloth).

Papers and discussions of an international conference, 2-5 September 1980, Bellagio, Italy.

The conference was built around 15 major review papers on tilapias tabled by leading researchers in their fields. Summaries of discussions on the papers as well as consensus statements on research and information needs are included in the proceedings.

R. C. May, I. R. Smith, and D. B. Thomson (Editors)

Appropriate Technology for Alternative Energy Sources in Fisheries

1982. Reprinted 1984. 215 pages. 28 × 21.5 cm. Perfect. ISSN 0115-4389. ISBN 971-04-0008-8. US$12.50 surface or US$25.00 airmail.

The proceedings of a joint Asian Development Bank-ICLARM workshop, held in Manila, February 1981.

Includes 5 major state-of-the-art papers and 15 contributions dealing with specific technologies. There were five major subject areas — wind energy alternatives for onshore activities; solar power alternatives for onshore activities; biofuels in fisheries; low energy fishing boats and methods and low cost fish processing; integrated systems of fishing villages, incorporating appropriate technology and alternative energy sources.

D. Pauly and G. I. Murphy (Editors)

Theory and Management of Tropical Fisheries

1982. 360 pages. 18 × 25.5 cm. Smythe-sewn (cloth); Perfect (paper). ISSN 0115-4389, ISBN 971-04-0021-5 (cloth), ISBN 971-04-0022-3 (paper). US$21.90 surface, US$35.60 airmail (paper); US$26.90 surface, US$40.60 airmail (cloth).

Proceedings of a workshop in Sydney, January 1981; reviews current research and methodology used in tropical stock assessment; formulates an action plan to overcome present restraints to stock assessment and management of tropical fisheries; includes review of the Southeast Asian data base on tropical multispecies stocks.

I. R. Smith, E. B. Torres, and E. O. Tan (Editors)

Summary Report of the PCARRD-ICLARM Workshop on Philippine Tilapia Economics

1983. 45 pages. 18 × 25.5 cm. Saddle stitched. ISSN 0115-4435, ISBN 971-1022-02-8. Available free on request by surface mail or US$2.50 airmail.

Contains abstracts of 17 presented papers and reports of four working groups of this workshop, held on 10-13 August 1983 at Laguna, Philippines, which brought together Philippine researchers who conducted an economic analysis of tilapia operations.

R. C. May, R. S. V. Pullin and V. G. Jhingran (Editors)

Summary Report of the Asian Workshop on Carp Hatchery and Nursery Technology

1984. 38 pages. 18 × 25.5 cm. Saddle stitched. ISSN 0115-4435, ISBN 971-1022-09-5. Available free on request by surface mail or US$6.25 airmail.

Contains abstracts of the papers presented and summaries of the discussion sessions at a workshop on carp hatchery/nursery technology in Asia. Scope includes the common carp, Chinese carps, Indian major carps and some minor species of local importance. Recommendations for further development of carp culture in Asia are also included.

I. R. Smith, E. B. Torres and E. O. Tan (Editors)

The PCARRD-ICLARM Workshop on Philippine Tilapia Economics

1985. 320 pages. 18 × 25.5 cm. Perfect. ISSN 0115-4435, ISBN 971-1022-18-4. US$20.00 surface or US$35.00 airmail.

Full proceedings of this conference held on 10-13 August 1983 at Laguna, Philippines.

Miscellaneous

Technical Reports

I. R. Smith

The Economics of the Milkfish Fry and Fingerling Industry of the Philippines

1981. 148 pages. 28 × 21.5 cm. Perfect. ISSN 0115-5547. US$10.00 surface or US$23.50 airmail.

An in-depth report on all aspects of this important industry; gives details of the complicated distribution system for fish fry and how the whole industry works.

K-C. Chong, M. S. Lizarondo, V. F. Holzao, and I. R. Smith

Inputs as Related to Output in Milkfish Production in the Philippines

1982. Reprinted 1984. 82 pages. 28 × 21.5 cm. Perfect. ISSN 0115-5547. US$9.35 surface or US$12.50 airmail.

Recall and record-keeping surveys of 324 farms were used to demonstrate the potential and means for improving yields in the majority of Philippine milkfish farms. Implications of observed economies of scale are also discussed.

T. Panayotou, S. Wallanutchariya, S. Isvilanonda, and R. Tokrisna

The Economics of Catfish Farming in Central Thailand

1982. 60 pages. 28 × 21.5 cm. Saddle stitched. ISSN 0115-5547. US$8.00 surface or US$12.40 airmail.

A survey of 41 catfish farms was undertaken to ascertain why production has been falling since 1974. The major factors influencing production were found, and guidelines to increase profitability provided.

K. D. Hopkins and E. M. Cruz

The ICLARM-CLSU Integrated Animal-Fish Farming Project: Final Report

1982. 96 pages. 28 × 21.5 cm. Perfect. ISSN 0115-5547. US$9.70 surface or US$18.50 airmail.

Full results of 18 major livestock-fish farming experiments over a three-year period. Includes economic analyses and forecasts and 26 pages of raw and summary data.

D. Pauly and A. N. Mines (Editors)

Small-Scale Fisheries of San Miguel Bay, Philippines: Biology and Stock Assessment

1982. 124 pages. 28 × 21.5 cm. Perfect. ISSN 0115-5547. US$11.25 surface or US$20.00 airmail.

The first of a series of five reports on the first truly multidisciplinary study of any fishery in Asia. Contains eight separate papers dealing with biology and ecology, catch and effort, stock assessment and status of the various sectors of the fisheries in a heavily exploited 840-km^2 bay in the Philippines. The study has relevance to all inshore tropical fisheries.

I. R. Smith and A. N. Mines (Editors)

Small-Scale Fisheries of San Miguel Bay, Philippines: Economics of Production and Marketing

1982. 143 pages. 28 × 21.5 cm. Perfect. ISSN 0115-5547. US$15.00 surface or US$31.25 airmail.

Second of a series of five reports on the first truly multidisciplinary study of any fishery in Asia.

C. Bailey (Editor)

Small-Scale Fisheries of San Miguel Bay, Philippines: Social Aspects of Production and Marketing

1982. 57 pages. 28 × 21.5 cm. Saddle stitched. ISSN 0115-5547. US$10.00 surface or US$17.50 airmail.

Third of a series of five reports on the firtst truly multidisciplinary study of any fishery in Asia.

C. Bailey

Small-Scale Fisheries of San Miguel Bay, Philippines: Occupational and Geographic Mobility

1982. 57 pages. 28 × 21.5 cm. Saddle stitched. ISSN 0115-5547. US$8.75 surface or US$14.40 airmail.

Fourth of a series of five reports on the first truly multidisciplinary study of any fishery in Asia.

C. S. Lee

Production and Marketing of Milkfish in Taiwan: an Economic Analysis

1983. 41 pages. 28 × 21.5 cm. Saddle stitched. ISSN 0115-5547. US$7.85 surface or US$11.85 airmail.

A study of the entire Taiwanese milkfish industry from fry gathering to marketing. Based on a survey of over 200 respondents.

I. R. Smith, D. Pauly, and A. N. Mines

Small-Scale Fisheries of San Miguel Bay, Philippines: Options for Management and Research

1983. 72 pages. 28 × 21.5 cm. Saddle stitched. ISSN 0115-5547, ISBN 971-1022-08-7. US$7.85 surface or US$11.85 airmail.

A synthesis of the findings of the San Miguel Bay Project from Technical Reports 7-10. Management policy options are outlined. The discussions are relevant to many tropical fisheries.

S. K. Meltzoff and E. S. LiPuma

A Japanese Fishing Joint Venture: Worker Experience and National Development in the Solomon Islands

1983. 63 pages. 28 × 21.5 cm. Saddle stitched. ISSN 0115-5547. US$8.20 surface or US$13.60 airmail.

A penetrating report on the problems of industrialization of a fishery. Based on a 2-year study, the report examines the early problems and conflicts of a joint-venture tuna fishery, and suggests guidelines for future avoidance of conflict areas.

J. Ingles and D. Pauly

An Atlas of the Growth, Mortality and Recruitment of Philippine Fishes

1984. 127 pages. 28 × 21.5 cm. Perfect. ISSN 0115-5547, ISBN 971-1022-12-5. US$7.00 surface or US$12.50 airmail.

Presents growth and catch curves as well as recruitment and selection patterns for 112 sets of length-frequency data collected between 1957 and 1981 from all areas of the Philippines on 23 families, 34 genera and 56 species of commercially important fishes. Illustrates the use of stock assessment methodologies based on length data. Valuable for single- or multispecies stock assessments in Asia and elsewhere in the tropical Indo-Pacific.

W. O. Watanabe, C-M. Kuo and M-C. Huang

Experimental Rearing of Nile Tilapia Fry (*Oreochromis niloticus*) for Saltwater Culture

1984. 28 pages. 28 × 21.5 cm. Saddle stitched. ISSN 0115-5547, ISBN 971-1022-11-7. US$3.75 surface or US$7.50 airmail.

Represents a preliminary evaluation of various approaches of early salinity exposure for saltwater culture of tilapias. Studies the reproductive performance of Nile Tilapia under laboratory conditions at various salinities; salinity tolerance of progeny; survivorship of fertilized eggs, spawned in freshwater but removed from the mouth of the parent female and artificially incubated at various salinities.

K-C. Chong, M. S. Lizarondo, Z. S. dela Cruz, C. V. Guerrero and I. R. Smith

Milkfish Production Dualism in the Philippines: a Multidisciplinary Perspective on Continuous Low Yields and Constraints to Aquaculture Development

1984. 70 pages. 28 × 21.5 cm. Saddle stitched. ISSN 0115-5547, ISBN 971-1022-10-9. US$8.75 surface or US$13.50 airmail.

Determines and measures the constraints to the adoption of more intensive fertilizer application rates among Philippine milkfish farmers. Fifty-six explanatory variables, categorized into socioeconomic, institutional, physical and bio-technical parameters, were hypothesized to explain variations in fertilizer use. Focused on farmers' perceptions of constraints. Data from 447 milkfish farmers in seven provinces and from a previous survey of 324 farmers in seven provinces.

W. O. Watanabe, C-M. Kuo and M-C. Huang

Salinity Tolerance of the Tilapias *Oreochromis aureus, O. niloticus* and an *O. mossambicus* $\times$ *O. niloticus* Hybrid

1985. 22 pages. 28 $\times$ 21.5 cm. Saddle stitched. US$3.00 surface or US$6.50 airmail.

Studies onto genetic changes in salinity tolerance in tilapias spawned and reared in freshwater using the indices of median lethal salinity, mean survival time and median survival time. Discusses implications of findings for brackish and seawater culture of tilapias.

Posters

Cultured Fish of Southeast Asia

1983. 92 $\times$ 60 cm. A single poster costs US$7.00; two, US$10.00; three, US$13.50; four, US$16.50; five, US$19.50 (all airmail prices). Prices are for orders mailed to one address; if multiple orders are for different addresses, use the single rate. Posters will be airmailed in a tube.

Some of Asia's most popular cultured food fish are featured on this new colorful poster. Nine species are included in large paintings showing taxonomic details of seabass, gourami, catfish, Nile tilapia, common carp, snakehead, greasy grouper, milkfish and grass carp. The names are given in seven languages and synopses of relevant culture biology are provided below the paintings.

Translations

P. Chonchuenchob, K. Chalayondeja, and K. Muttarasin

Hanging Culture of the Green Mussel (*Mytilus smaragdinus* Chemnitz) in Thailand

1980. 12 pages. 28 $\times$ 21.5 cm. Stitched. ISSN 0115-4141. US$1.25 surface or US$2.25 airmail.

Contains results of recent Thai experiments. Manila rope was used for the hanging method of green mussel (*Mytilus smaragdinus* Chemnitz) culture. The monthly density of spat on rope, their rate of growth and environmental factors in the experimental area were determined.

H. M. Peters. Translated and edited by D. Pauly

Fecundity, Egg Weight and Oocyte Development In Tilapias (Cichlidae, Teleostei)

1983. 28 pages. 28 × 21.5 cm. Stitched. ISSN 0115-4141. US$3.75 surface or US$6.70 airmail.

A detailed study on the fecundity and related aspects of the biology of tilapiine fishes by Prof. H. M. Peters of the University of Tübingen based on data from seven species of substrate-spawning and mouth-brooding tilapias.

F. Brouard, R. Grandperrin, M. Kulbicki, and J. Rivaton

Note on Observations of Daily Rings on Otoliths of Deepwater Snappers

1984. 8 pages. 28 × 21.5 cm. Saddle stitched. ISSN 0115-4141, ISBN 971-1022-07-9. US$1.70 surface or US$2.50 airmail.

M. Boonyubol and S. Pramokchutima
Translated by T. Bhukaswan

Trawl Fisheries in the Gulf of Thailand

1984. 12 pages. 28 × 21.5 cm. Side stitched. ISSN 0115-4141, ISBN 971-1022-13-3. Available free on request by surface mail or US$3.00 airmail.

Discusses the status of demersal fishery in the Gulf of Thailand. Translated from Thai.

I. R. Smith and D. Pauly
Translated by J. M. F. Bartolome

Paglutas ng Tunggalian ng Iba't Ibang Pamamaraan ng Pangingisda sa Baybay-Dagat

1985. 6 pages. 28 × 21.5 cm. Side stitched. ISSN 0115-4141, ISBN 971-1022-16-8. Available free on request by surface mail or US$2.00 airmail.

Translation into Tagalog of a summary report on a research project in San Miguel Bay, Philippines. Originally in English — "Small-scale fisheries of San Miguel Bay, Philippines: resolving multigear competition in nearshore fisheries."

Bibliographies

R. M. Temprosa and Z. H. Shehadeh

Preliminary Bibliography of Rice-Fish Culture

1980. 20 pages. 28 × 21.5 cm. Stitched. ISSN 0115-5997. US$2.50 surface or US$7.00 airmail.

A listing of over 300 references to research and development of the combined farming of rice and fish. All forms of aquatic animals that have been grown with rice are included.

D. Pauly and S. Wade-Pauly

An Annotated Bibliography of Slipmouths (Pisces: Leiognathidae)

1981. 62 pages. 28 × 21.5 cm. Saddle stitched. ISSN 0115-5997. US$3.75 surface or US$7.75 airmail.

A bibliography of the fishes of the Leiognathidae family (slipmouths, ponyfishes, silver-bellies, etc.) is presented. A total of 941 references, most of which are annotated, are listed, covering all aspects of their biology, including fishery related aspects, physiology and anatomy (e.g., their bioluminescence) as well as their use as human food. Comprehensive subject and geographic indices are provided.

P. Schoenen

A Bibliography of Important Tilapias (Pisces: Cichlidae) for Aquaculture

1982. 336 pages. 28 × 21.5 cm. Perfect. ISSN 0115-5997. US$22.50 surface or US$40.60 airmail.

Eight separate, complete and indexed bibliographies on the commercially important tilapias are contained in this volume. Species include: *Oreochromis macrochir, O. aureus, O. hornorum, O. mossambicus, O. niloticus, Sarotherodon galilaeus, Tilapia rendalli* and *T. zillii.*

P. Schoenen

A Bibliography of Important Tilapias (Pisces: Cichlidae) for Aquaculture. Supplement 1

1984. 193 pages. 28 × 21.5 cm. Perfect. ISSN 0115-5997, ISBN 971-1022-06-0. US$15.00 surface or US$31.25 airmail.

A supplement to the 1982 edition.

P. Schoenen

A Bibliography of Important Tilapias (Pisces: Cichlidae) for Aquaculture.

1985. 97 pages. 28 × 21.5 cm. perfect. ISSN 0115-5997, ISBN 971-1022-19-2. US$9.75 surface or US$18.50 airmail.

Indexed bibliographies on the following tilapia species: *Oreochromis variabilis, O. andersonii, O. esculentus, O. leucostictus, O. mortimeri, O. spilurus niger, Sarotherodon melanotheron* and *Tilapia sparrmanii.*

Periodicals

J. L. Maclean (Editor)

ICLARM Newsletter

Quarterly from July 1978. 24-32 pages. 28 × 21.5 cm. Saddle stitched. US$7.50 or US$17.50/year (four issues). ISSN 0015-4575.

Articles on all aspects of fisheries and aquaculture in developing countries; research summaries; news reports; notices of new publications, and upcoming meetings of interest.

ICRAF

International Council for Research in Agroforestry

P. O. Box 30677, Nairobi, Kenya
Cable: ICRAF-Telex: 22048
Tel: 29867

General Information

ICRAF is an autonomous, nonprofit international council governed by a board of trustees with equal representation from developed and developing countries. With the exception of a representative of the host country, Kenya, trustees do not represent countries or organization but are elected on individual merit. ICRAF derives its financial support from governments and international, private, and public organizations and agencies.

ICRAF's objective is to improve the nutritional, economic, and social well-being of the peoples of developing countries by promoting agroforestry systems for better land use without harming the environment. ICRAF acts as an an international catalyst in agroforestry research.

ICRAF's program of work includes

- the development of methodologies to identify social, economic, and ecological constraints in land-use systems and to asses the potential of agroforestry technologies to overcome such constraints;
- the systematic collation and assessment of agroforestry knowledge and the development of methods of studying and evaluating agroforestry technologies; and
- the efficient dissemination of methodologies and knowledge to scientists and development planners in the tropical and subtropical developing world.

In this catalog ICRAF publications are listed in the following categories:

> Books, proceedings, and reviews
> Science and practice of agroforestry
> ICRAF communications
> ICRAF reprints
> ICRAF working papers
> Agroforestry systems
> About ICRAF
> Miscellaneous papers

Books, Proceedings, and Reviews

H. O. Mongi and P. A. Huxley (Editors)

Soils Research in Agroforestry: Proceedings of an Expert Consultation

1979. XXXIX + 584 pages. 14.3 × 20.6 cm. Paperback.

These papers review soil research related to agroforestry systems. They summarize relevant work in agronomy and soils under monoculture, mixed and intercropping conditions, and the research methods used as they relate to soil management for agroforestry. A common outlook on current soil research problems is developed and some proposals are made as to how best the problems could be solved.

Trevor Chandler and David Spurgeon (Editors)

International Cooperation in Agroforestry: Proceedings of an International Conference

1980. XVIII + 469 pages. 14.5 × 20.8 cm. Paperback.

The conference on International Cooperation in Agroforestry provided the first worldwide overview of tropical agroforestry. Conference papers and discussions reveal wide support for agroforestry principles among foresters. Recommendations for international action were made.

Louise Buck (Editor)

Proceedings of Kenya National Seminar on Agroforestry

1981. XI + 638 pages. 20.5 × 29.2 cm. Paperback.

The seminar was the first major meeting in Kenya for exchange of views and experiences on agroforestry land use among professionals from a wide variety of backgrounds. Papers presented confirmed that there is wide interest in agroforestry in Kenya. The recommendations underline the need to elaborate and support a wide range of further activites in agroforestry in that country.

D. A. Hoekstra and F. M. Kuguru (Editors)

Agroforestry Systems for Small-Scale Farmers

1983. XXI + 283 pages. 15 × 25 cm. Paperback.

These papers presented at a workshop in September 1982 describe the potential role of agroforestry on small farms where fuelwood is used domestically and to cure tobacco. Papers give a general introduction to agroforestry, discuss selected topics in agroforestry and case studies of three tobacco-growing areas in Kenya, and describe agroforestry in Kenya, Nigeria, Sierra Leone, Zimbabwe, and Zaire.

P. A. Huxley (Editor)

Plant Research and Agroforestry: Proceedings of a Consultative Meeting

1983. XXIV + 619 pages. 15 × 24.4 cm. Limp.

This volume brings together a broad range of information about the potential contribution of plant science to agroforestry. It includes plants and agroforestry (12 papers), understanding agroforestry systems (9 papers), applications of plant science to agroforestry (13 papers), and working group reports.

Dan M. Etherington and Peter J. Matthews

MULBUD User's Manual

1982. XII + 77 pages. 17.6 × 24.0 cm. Flexibound.

MULBUD is an interactive computer package designed to assist in the economic appraisal of perennial and complex agroforestry systems where several tree species are grown together.

The manual describes MULBUD's budgeting applications.

ICRAF

Proceedings of the International Workshop on Professional Education in Agroforestry

1984. 320 pages. 17.6 × 25 cm. Paperback.

This volume summarizes activities and plans for teaching of agroforestry in the major geographical regions of the world. It includes workshop recommendations for program, course content, teaching methods, and teaching materials.

P. K. R. Nair

Agroforestry Species: a Crop Sheets Manual

1980. IX + 336 pages. 14.9 × 20.1 cm. Paperback.

As interest in agroforestry as a sustainable land use system increases, so does interest in forestry species, their cultivation, and management. This manual gives concise descriptions of forest species with potential for agroforestry.

J. F. Burley and P. von Carlowitz (Editors)

Multipurpose Tree Germplasm

1984. XVI + 298 pages. 15.5 × 24.2 cm. Limp.

This volume brings together the proceedings, recommendations and documents of a planning workshop that brought together leading scientists, experts and donor representatives from around the world to identify the problems of collection, storage and distribution of multipurpose tree germplasm, and to discuss and propose appropriate strategies and recommendations for their possible solution.

Louise Fortmann and James Riddell

Trees and Tenure: an Annotated Bibliography for Agroforesters and Others

1985. VII + 135 pages. 13.5 × 21.2 cm. Paperback.

This bibliography represents the first fruits of collaboration — initiated in 1982 — between ICRAF and the Land Tenure Center of the University of Wisconsin at Madison in an effort to understand the traditional tenure systems under which peasant cultivators — the target recipients of agroforestry technologies — operate.

Science and Practice of Agroforestry

Peter Huxley (General editor) and Richard C. Ntiru (Technical editor)

This is a series of small booklets on a range of scientific and practical topics in agroforestry. The series is designed for high school and university students; resource planners, administrators, and scientists; and development personnel in agroforestry and related disciplines.

P. K. R. Nair

Soil Productivity Aspects of Agroforestry

1983. 85 pages. 17.6 × 25 cm. Paperback.

That agroforestry depends upon self-maintenance, soil productivity is of utmost concern. This booklet examines the likely effect of agro-forestry practices on the long-term soil productivity and suggests appropriate soil management practices.

J. F. Burley

Global Needs and Problems of the Collection, Storage and Distribution and Multipurpose Tree Germplasm

1985. IX + 179 pages. 17.6 × 25 cm. Paperback.

In this booklet, Dr. Burley describes the special characteristics of multipurpose tree germplasm and the major problems and issues in taxonomy, exploration, collection, evaluation and conservation of species and populations.

P. O'Keefe (Editor)

Agroforestry in Kenya: an Outline

1983. 100 pages. 17.6 × 25 cm. Paperback.

This is a collection of short papers on important aspects of agro-forestry development and research in Kenya. Contributors include a land lawyer, foresters, an agronomist, and a commercial organization.

B. Lundgren

Agroforestry for Improved Productivity of Tropical Lands

ICRAF Communications

ICRAF Communications

ICRAF Communications is a series of original or previously published article-length material in definitive form.

ICRAF Reprints

The ICRAF Reprints series consists of articles by ICRAF or papers by other research scientists first published elsewhere but presented at ICRAF conferences.

IR 1 P. K. R. Nair. Multiple land use and agroforestry. Reprinted from *Better Crops for Food*-CIBA Foundation Symposium 97. London: Pitman Books Ltd., 1983. pp. 101-115.
IR 2 B. Lundgren and L. Lundgren. Socioeconomic effects and constraints in forest management: Tanzania. Reprinted from

Socioeconomic Effects and Constraints in Tropical Forest Management. New York: John Wiley and Sons Ltd., 1983. pp. 42-49.

IR 3 B. Lundgren and J. B. Raintree. Sustained agroforestry. Reprinted from *Agricultural Research for Development: Potentials and Challenges in Asia.* The Hague: ISNAR, 1983. pp. 37-49

IR 4 F. Torres. Role of woody perennials in animal forestry. Reprinted from *Agroforestry Systems* 1(2): pp. 131-163, 1983.

IR 5 P. K. R. Nair. Tree integration on farmlands for sustained productivity on smallholdings. Reprinted from *Environmentally Sound Agriculture.* New York: Praeger Publishers. 1983.

IR 6 J. B. Raintree. Strategies for enhancing the adoptability of agroforestry innovations. Reprinted from *Agroforestry Systems.* The Hague: Martinus Nijhoff/Dr. W. Junk Publishers, 1983. pp. 173-187.

IR 7 P. A. Huxley. Some characteristics of trees to be considered in agroforestry. Reprinted from *Plant Research and Agroforestry.* Nairobi: ICRAF, 1983. pp. 3-12.

IR 8 P. K. R. Nair. Agroforestry with coconuts and other tropical plantation crops. Reprinted from *Plant Research and Agroforestry.* Nairobi: ICRAF, 1983. pp. 79-102.

IR 9 P. A. Huxley. Comments on agroforestry classifications: with special reference to plant aspects. Reprinted from *Plant Research and Agroforestry.* Nairobi: ICRAF, 1983. pp. 161-171.

IR 10 P. A. Huxley. The role of trees in agroforestry: some comments. Reprinted from *Plant Research and Agroforestry.* Nairobi: ICRAF, 1983. pp. 257-270.

IR 11 J. B. Raintree. Bioeconomic considerations in the design of agroforestry cropping systems. Reprinted from *Plant Research and Agroforestry.* Nairobi: ICRAF, 1983. pp. 271-289.

IR 12 H. A. Steppler and J. B. Raintree. The ICRAF research strategy in relation to plant science research in agroforestry. Reprinted from *Plant Research and Agroforestry.* Nairobi. ICRAF, 1983. pp. 297-305.

IR 13 P. A. Huxley. Phonology of tropical woody perrenials and seasonal crop plants with reference to their management and agroforestry systems. Reprinted from *Plant Research and Agroforestry.* Nairobi: ICRAF, 1983. pp. 503-525.

IR 14 T. T. Koslowski and P. A. Huxley. The role of controlled environment in agroforestry research. Reprinted from *Plant Research and Agroforestry.* Nairobi: ICRAF, 1983. pp. 551-567.

IR 15 F. Torres. Potential contribution of Leucaena hedgerows intercropped with maize to the production of organic nitrogen and fuelwood in the lowland tropics. Reprinted from *Agroforestry Systems* 1(4):323-333. The Hague: Martinus Nijhoff/Dr. W. Junk Publishers, 1983.

IR 16 D. A. Hoekstra. An economic analysis of a simulated alley cropping system for semiarid conditions, using microcomputers. Reprinted from Agroforestry Systems 1(4):335-345. The Hague: Martinus Nijhoff/Dr. W. Junk Publishers, 1983.

IR 17 P. K. R. Nair and E. Fernandes. Agroforestry as an alternative to shifting cultivation. Reprinted from FAO Soils Bulletin 53. Rome: FAO, 1984.

IR 18 P. K. R. Nair, E. Fernandes and P. N. Wambugu. Multipurpose leguminous trees and shrubs for agroforestry. Reprinted from Agroforestry Systems 2:145-163. The Haque: Martinus Nijhoff/ Dr. W. Junk Publishers, 1984.

IR 19 Louise Fortmann and Dianne Rocheleau. Women and agroforestry: four myths and three case studies. Reprinted from Agroforestry Systems 2:253-272. The Haque: Martinus Nijhoff/ Dr. W. Junk Publishers, 1985.

IR 20 D. A. Hoekstra. Choosing the discount rate for analyzing agroforestry systems/technologies from a private economic viewpoint. Reprinted from Forest Ecology and Management 10(1985): 177-183. Amsterdam: Elsevier Science Publishers B. V., 1985.

IR 21 M. Baumer. Les terres a pâturage en zone aride: réflexions pur un aménagement de milieu. Reprinted from Forêt Méditerranéenne, t. v. No. 2. Marseille: Louis-Jean, 1983.

ICRAF Working Papers

ICRAF Working Papers are available in limited numbers for comment and discussion and to inform interested colleagues about work in progress. Comments and suggestions are invited and should be directed to the authors. Material may be cited from Working Papers but Working Papers may not be reproduced without permission.

IWP 1 J. B. Raintree. Preliminary diagnosis of land use problems and agroforestry potentials in Northern Mbere Division, Embu District, Kenya. Nairobi: ICRAF. 16 pp.

IWP 2 D. A. Hoekstra. The use of economics in agroforestry. Nairobi: ICRAF: 43 pp. + refs.

IWP 3 D. A. Hoekstra. *Leucaena leucocephala* hedgerows intercropped with maize beans: an ex ante analysis of a candidate agroforestry land use system for the semiarid areas in Machakos District. Nairobi: ICRAF. 7 pp. + refs. + printouts.

IWP 4 L. E. Buck. Kenya agroforestry tree seed project report. With assistance from W. Teel, Mennonite Central Committee. Nairobi: ICRAF. 61 pp. + appendices.

IWP 5 A. Young. An environmental data base for agroforestry. Nairobi: ICRAF. 60 pp.

IWP 6 ICRAF. Draft guidelines for agroforestry diagnosis and design. Nairobi: ICRAF. 25 pp.

IWP 7 ICRAF. Draft resources for agroforestry diagnosis and design. Nairobi: ICRAF. 383 pp.

IWP 8 R. Labelle. A preliminary agroforestry word list, with definitions. Nairobi: ICRAF. 30 pp.

IWP 9 D. A. Hoekstra. Choosing the discount rate for analyzing agroforestry systems/technologies from farmer's point of view. Nairobi: ICRAF. 9 pp.

IWP 10 D. A. Hoekstra. An annotated bibliography of economic analysis of agroforestry systems/technologies. Nairobi: ICRAF. 44 pp.

IWP 11 D. Rocheleau. The application of ecosystems and landscape analysis in agroforestry and design: a case study from Kathama sublocation, Machakos District, Kenya. Nairobi: ICRAF.

IWP 12 P. A. Huxley. Systematic design for field experimentation with multipurpose trees. Nairobi: ICRAF. 6 pp. + annexure.

IWP 13 P. A. Huxley. Investigations into tree-crop interface or simplifying the biological environment study of mixed cropping agroforestry systems. Nairobi: ICRAF. 20 pp. + annexure.

IWP 14 T. Darnhofer. Meteorological elements and their observation. Nairobi: ICRAF. 30 pp.

IWP 15 P. A. Huxley. Considerations when experimenting with changes in plant spacing. Nairobi: ICRAF. 32 pp. + appendix.

IWP 16 D. A. Hoekstra. An ex-ante economic analysis of proposed mixed and zonal agroforestry systems for Batu Arang Forest Reserve, Malaysia. Nairobi: ICRAF. 16 pp. + annexure.

IWP 17 P. G. von Carlowitz. Multipurpose trees and shrubs: opportunities and limitations. The establishment of an MPT data bank. Nairobi: ICRAF. 28 pp.

IWP 18 F. Torres and J. B. Raintree. Agroforestry for smallholder upland farmers in a land reform area of the Philippines: the Tabango case study. Nairobi: ICRAF. 25 pp.

IWP 19 D. A. Hoekstra. Agroforestry systems for the semiarid areas of Machakos District, Kenya. Nairobi: ICRAF. 28 pp.

IWP 20 P. J. Wood. Mixed systems of plant production in Africa, past, present, and future. Nairobi: ICRAF. 18 pp.

IWP 21 W. C. Beets. Aspects of traditional farming systems in relation to integrated pest management. Nairobi: ICRAF. 12 pp.

IWP 22 W. C. Beets. Agroforestry in African farming systems. Nairobi: ICRAF. 79 pp.

IWP 23 A. Young. Site selection for multipurpose trees. Nairobi: ICRAF. 30 pp.

IWP 24 A. Young. Land evaluation for agroforestry: the tasks ahead. Nairobi: ICRAF. 54 pp.

IWP 25 P. A. Huxley. The basis for selection, management and evaluation of multipurpose trees: an overview. Nairobi: ICRAF. 85 pp. + 23-page appendix.

IWP 26 P. A. Huxley and P. J. Wood. Technology and research considerations in ICRAF's "Diagnosis and Design" procedures. Nairobi: ICRAF. 49 pp.

IWP 27 A. Young. Evaluation of agroforestry potential in sloping areas. Nairobi: ICRAF. 33 pp.
IWP 28 P. K. R. Nair. Classification of agroforestry systems. Nairobi: ICRAF. 52 pp.
IWP 29 D. A. Hoekstra. The use of economics in diagnosis and design of agroforestry systems. Nairobi: ICRAF. 85 pp.
IWP 30 D. A. Hoekstra. Economic concepts of agroforestry. Nairobi: ICRAF. 12 pp.
IWP 31 F. Torres. Networking for the generation of agroforestry technologies in Africa. Nairobi: ICRAF. 24 pp.
IWP 32 P. K. R. Nair. Fruit trees in tropical agroforestry systems. Nairobi: ICRAF/EWC. 89 pp.

H. J. von Maydell (Chairman of the editorial board)

Agroforestry Systems

Length varies. 14.7 × 21 cm. Perfect bound. Published quarterly in cooperation with Martinus Nijhoff/Dr. W. Junk Publishers.

Agroforestry Systems is an international, multidisciplinary journal that publishes research on various aspects of agroforestry systems and critical reviews of all sustainable land management systems that combine agronomy/agriculture, animal husbandry, and trees.

Editorial Office Agroforestry Systems, P. O. Box 566, 2501 CN The Hague, The Netherlands.

Orders and subscription information:

> Klurver Academic Publishers Group,
> Distribution Centre, P.O. Box 322, 3300 AH
> Dordrecht, The Netherlands

This series of agroforestry system descriptions is edited by Dr. P. K. R. Nair and published by Martinus Nijhoff/Dr. W. Junk Publishers, of The Hague, in the journal *Agroforestry Systems.*
AFSD 1 E. Fernandes, A. O'Kting'ati and J. Maghembe. The Chagga home gardens: a multistoried agroforestry cropping system on Mt. Kilimanjaro (N. Tanzania). Reprinted from "Agroforestry Systems" 2:73-86. 1984.
AFSD 2 S.-A. Boonkird, E. C. M. Fernandes and P. K. R. Nair. Forest villages: an agroforestry approach to rehabilitating forest land degraded by shifting cultivation in Thailand. Reprinted from "Agroforestry Systems" 2:87-102. 1984.
AFSD 3 A. O'Kting'ati, J. A. Maghembe, E. C. M. Fernandes and G. H. Weaver. Plant species in the Kilimanjaro agroforestry system. Reprinted from "Agroforestry Systems" 2:177-186. 1984.

AFSD 4 P. F. Fonzen and O. Oberholzer. Use of multipurpose trees in hill farming systems in Western Nepal. Reprinted from "Agroforestry Systems" 2:187-197. 1984.

AFSD 5 P. T. Evans and J. S. Rombold. Paraiso (*Melia azedarach* var. "Gigante") woodlots: an agroforestry alternative for the small farmer in Paraguay. Reprinted from "Agroforestry Systems" 2:199-214. 1985.

AFSD 6 R. M. Bourke. Food, coffee and casuarina: an agroforestry system from the Papua New Guinea highlands. Reprinted from "Agroforestry Systems" 2:273-279. 1984.

AFSD 7 M. de S. Liyanage, K. G. Tejwani and P. K. R. Nair. Intercropping under coconuts in Sri Lanka. Reprinted from "Agroforestry Systems" 2:215-228. 1984.

AFSD 8 D. V. Johnson and P. K. R. Nair. Perennial crop-based agroforestry systems in Northeast Brazil. Reprinted from "Agroforestry Systems" 2:281-292. 1984.

AFSD 9 B. J. Allen. Dynamics of fallow successions and introduction of Robusta coffee in shifting cultivations areas in lowlands of Papua New Guinea. Reprinted from "Agroforestry Systems" 3: (in press). 1985.

AFSD 10 E. Escalante. Promising agroforestry systems in Venezuela. Reprinted from "Agroforestry Systems" 3: (in press). 1985.

AFSD 11 P. H. May, A. B. Anderson, J. M. Frazao and M. J. Ballick. Babassu palm in the agroforestry systems in Brazil's Mid-North Region. Reprinted from "Agroforestry Systems". 1985.

About ICRAF

ICRAF Newsletter

8 pages. 21 × 29.6 cm. Saddle-stitched.

Published quarterly in English, French, and Spanish, and distributed free of charge. (To become a subscriber, complete and return our Mailing List and Reader Survey Form.)

An Account of the Activities of the International Council for Research and Agroforestry

1983. 36 pages. 17.6 × 25 cm. Color photos. Paperback.

This account of the activities concisely describes ICRAF's Programme of Work for 1983-86 to improve the nutritional, economic, and social well-being of the peoples of developing countries by promoting agroforestry systems designed for better land use without harming the environment.

Compte Rendu des Activites du Conseil International pour la Recherche en Agroforesterie

1984. 40 pages. 17.6 × 25 cm. Colour photos. Paperback.

Las Actividades de Consejo Internacional para Investigación en Agrosilvicultura

1984. 36 pages. 17.6 × 25 cm. Colour photos. Paperback.

Report of the External Review Panel of the International Council for Research in Agroforestry

1985. 119 pages. 20.7 × 29 cm. Paperback.

Annual Report 1983

1983. 40 pages. 21 × 29.6 cm. Paperback.

Among the achievements and developments reported are the establishment of ICRAF's Agroforestry Advisory Unit, the commercial dissemination of MULBUD 3, the multipurpose tree germplasm workshop, the publication of two workshop proceedings, the organization of the first agroforestry training course, and the installation of an agrometeorological unit at the field station.

Annual Report 1984

1985. 48 pages. 21 × 29.6 cm. Paperback.

Among the major developments reported on are the Board's agreement in principle to the construction of a permanent headquarters' building and the creation of a formal donor group, and the Board's decision to appoint an independent evaluation team to evaluate ICRAF's performance; to review critically ICRAF's mandate and strategy; and specifically to evaluate and make recommendations on the future development options of the Council.

Miscellaneous Papers

A list of miscellaneous documents and papers produced by ICRAF staff on agroforestry and/or ICRAF is available on request. A reasonable number of copies of those publications are free to persons from nonprofit institutions in developing countries. For developed countries and profit-motivated institutions there is a nominal charge.

WINROCK

WINROCK International Institute for Agricultural Development

Route 3, Petit Jean Mountain, Arkansas 72110, U.S.A.
Telex: 910-720-6616
Telephone: (501)-727-5435
Electronic mail: Dialcom system 57 — AGC017

Introduction

The Winrock International Institute for Agricultural Development was established in 1985, and incorporates more than 50 years of combined organizational experience in providing technical assistance and professional expertise to international agricultural development. One of the largest independent entities engaged in international agricultural development, Winrock International was created through the merging of the Agricultural Development Council (A/D/C), the International Agricultural Development Service (IADS), and the Winrock International Livestock Research and Training Center. Winrock International is an autonomous, nonprofit corporation.

Mission

Winrock International's mission is to help people around the world by increasing agricultural productivity and improving nutrition. Since agricultural development is by its nature a team effort, cooperative relationships are the key to its success. Accordingly, Winrock staff members work with host governments, private voluntary organizations, producer groups, and others to provide research and analysis, education and training, technical assistance, and communications services. Areas of emphasis are human resources, renewable resources, food policy, animal agriculture and farming systems, and agricultural research and extension.

Annual Reports

Winrock International

1984 Annual Report

1984. 32 pages. 22 × 28 cm. Saddle stitched.

This publication provides an overview of the Winrock International Livestock Research and Training Center's activities during the 1983-1984 fiscal year. The publication presents a description of the organization as well as highlights of its national, international, and public policy programs.

Monographs

A. J. De Boer (Editor)

Ruminant Livestock in Intensive Agricultural Areas of Sichuan Province, China: Current Status and Development Prospects. Report of Joint Study Team on Livestock Development Prospects in Sichuan Province, China.

1984. 137 pages. 21 × 27 cm. Perfect. $5.00.

This technical report was written for the Chinese government by a livestock-sector study team. It outlines for the Chinese what the current situation is in Sichuan, and what might be done to improve livestock production.

R. D. Child, H. F. Heady, W. C. Hickey, R. A. Peterson, and R. D. Pieper

Arid and Semi-arid Lands: Sustainable Use and Management in Developing Countries

1984. 205 pages. 16 × 23 cm. Perfect. $10.00.

This document reviews the functions of arid-land ecosystems and their responses to development-imposed stresses. It also describes methods of economic land use that sustain a wide range of options, such as livestock grazing, wildlife, native plants, water, and traditional land uses. It stresses that development projects will be more successful if they are planned to take into account all cultural, economic, and natural resource elements.

A. D. Tillman

Animal Agriculture in Indonesia

1981. 80 pages. 22 × 28 cm. Perfect. $6.50.

This is a report of the status of the livestock industry in Indonesia and the potential for its further development. It examines the factors that affect animal agriculture, taking into consideration those that serve as constraints to future development as well as those that provide opportunities.

H. A. Fitzhugh and G. E. Bradford (Editors)

Ovejas de Pelo del Africa Occidental y de las Americas. Un Recurso Genetico para los Tropicos

1983. 58 pages. 15 × 23 cm. Perfect.

A Spanish-language translation of the first two chapters, and four-color illustrations, from Hair Sheep of Western Africa and the Americas.

M. W. Demment and P. J. Van Soest

Body Size, Digestive Capacity, and Feeding Strategies of Herbivores

1983. 66 pages. 22 × 28 cm. Perfect. $5.00.

This study looked at body size as a mechanism for differentiating the feeding requirements of herbivores. The report describes the physiological and qualitative phenomena related to the consumption and use of lignocellulosic materials by herbivores.

T. C. Byerly, H. A. Fitzhugh, H. J. Hodgson, T. D. Nguyen, and O. J. Scoville

Role of Ruminants in Support of Man

1978. 136 pages. 22 × 28 cm. Perfect. $5.00.

"Increased world population and consumer buying power will create a demand for 74% more milk, 82% more beef, and 90% more sheep and goat meat in the year 2000 from what was consumed in 1970."

This is one of the many findings presented in the publication that resulted from a 2-year worldwide study on ruminants. The publication assesses the world population of ruminants, and their productivity and feed requirements — with projections to the year 2000. The publication also identifies the opportunities and constraints to improving ruminant efficiency and productivity.

R. O. Wheeler, G. L. Cramer, K. B. Young, and E. Ospina

The World Livestock Product, Feedstuff, and Food Grain System

1981. 85 pages. 22 × 28 cm. Comb bound. $25.00.

The task of documenting and evaluating the animal-agriculture components of the world food system has been a central focus of Winrock International. Thus, in 1979, Winrock called upon a wide range of professional expertise to assist the U.S Department of Agriculture with a global study of the world livestock product, feedstuff, and food grain system.

This study charts interaction within the current food system and develops procedures for spelling out the specific energy and protein inputs required for selected food energy and protein outputs.

R. (Dick) Newton

Sheep Pens for Ewe and Me

1983. 116 pages. 13 × 18 cm. Stapled and taped. $4.85.

A very practical, down-to-earth handbook for sheep producers or anyone else who is going to build, rebuild, or remodel sheep pens. It tells the reader how to plan pens for his or her particular set of needs and circumstances — in terms of principles and guidelines.

Winrock International

World Agriculture: Review and Prospects into the 1990s

1983. 607 pages. 22 × 28 cm. Comb bound. $150.

The complete report of this baseline study, as described in World Agriculture: Review and Prospects into the 1990s: A Summary.

R. D. Child and E. K. Byington

Potential of the World's Forages for Ruminant Animal Production

1981. 110 pages. 22 × 28 cm. Perfect. $6.50.

Nearly 3 billion hectares of the world's land is permanent grassland — twice the area used for cultivated crops. It is becoming increasingly apparent that this major renewable natural resource must be conserved and enhanced. The papers in this publication highlight the role and importance of forage-crop, pasture, and range resources in meeting human food needs in a complex world.

H. A. Fitzhugh and G. E. Bradford (Editors)

Hair Sheep of Western Africa and the Americas

1983. 319 pages. 16 × 24 cm. Perfect.

Most of the world's 1 billion sheep are found in temperate or arid/semiarid tropical environments. However, an estimated 1 million sheep are favorably adapted to the humid tropics because of their hair coat. Unfortunately, relatively little has been documented about the performance of hair sheep. This report, through papers submitted by a number of contributors, attempts to survey hair sheep resources and their native environment, emphasizing those hair sheep in the western hemisphere and western Africa, and to document the performance traits, including size, growth, fertility, and adaptive ability.

BOSTID

The Board on Science and Technology for International Development

Office of International Affairs
National Academy of Sciences/National Research Council

2101 Constitution Avenue,
NW . JH-210 Washington, D.C. 20418, USA
Cable: NARECO
Telex: 248664 NASW UR
Telephone: 202/334-2639

What are the National Academy of Sciences and the National Research Council?

The National Academy of Sciences (NAS) is a private, honorary society of scientists and engineers, dedicated to the furtherance of science and its uses for the general welfare. Although the Academy is not a federal agency, it is called upon by the terms of its 1983 charter to examine and report on any subject of science or technology upon request of any department of the federal government.

The National Research Council (NRC), serves as the operating arm of the National Academy of Sciences (NAS) and the National Academy of Engineering. The Institute of Medicine, a sister organization, also participates in the activities of the NRC.

The NRC is composed of eight major units, called assemblies and commissions. The Office of International Affairs has the broad function of conducting the international activities of the National Research Council.

What is the Board on Science and Technology for International Development (BOSTID)?

BOSTID is a division of the Office of International Affairs and is responsible for programs with developing countries. Established in 1969, BOSTID examines ways to apply science and technology to problems of economic and social development, through overseas programs, studies, advisory committees, and other mechanisms.

Participants in BOSTID activities work with counterpart groups in developing countries. This joint effort is directed toward strengthening local scientific and technological capabilities related to agriculture, environmental planning, energy, forestry, health, natural resource management and conservation, nutrition, water supply and quality, and other areas. Overseas activities also address the national organization and planning capabilities needed in applying science and technology to development. Studies examine specific development problems and suggest possible scientific and technological solution.

BOSTID's work relies on scientist and engineers who are selected for their expertise and who contribute their time and services as members of study panels and participants in overseas activities. BOSTID's permanent staff provides professioal support and continuity and plans future programs.

What are the objectives of BOSTID's program?

- BOSTID seeks to help developing countries strengthen their own capabilities for dealing with important development-related problems and for moving toward greater scientific and technological self-reliance.
- BOSTID seeks to stimulate and support R&D within and among the developing countries on problems of high priority for development and human welfare.
- BOSTID seeks to provide developing countries with greater access to the scientific and technological know-how and expertise found in the United States and other countries.
- BOSTID seeks to provide a focal point within the U.S. scientific and technical community for assistance to the developing countries and to encourage greater cooperation between U.S. scientists and engineers and their colleagues in the Third World.

Monographs

National Research Council

More Water for Arid Lands: Promising Technologies and Research Opportunities

1974. 153 pages. 23 cm. Paperback. Free.

This report outlines little-known but promising technologies to supply and conserve water in arid areas. French language edition is also available from BOSTID.

National Research Council

Underexploited Tropical Plants with Promising Economic Value

1975. 187 pages. 23 cm. Paperback. Free.

Describes 36 little-known tropical plants that, with research, could become important cash and food crops in the future. Includes cereals, roots and tubers, vegetables, fruits, oilseeds, forage plants and others.

National Research Council

Making Aquatic Weeds Useful: Some Perspectives for Developing Countries

1976. 175 pages. 23 cm. Paperback. Free.

This report describes ways to exploit aquatic weeds for grazing, and by harvesting and processing for use as compost, animal feed, pulp, paper, and fuel. Also describes utilization for sewage and industrial wastewater treatment. Examines certain plants with potential for aquaculture.

National Research Council

Postharvest Food Losses in Developing Countries

1978. 202 pages. 23 cm. Paperback. Free.

Assesses potential and limitations of food loss reduction efforts; summarizes existing work and information about losses of major food crops and fish; discusses economic and social factors involved; identifies major areas of need; and suggests policy and program options for developing countries and technical assistance agencies.

National Research Council

Tropical Legumes: Resources for the Future

1979. 331 pages. 23 cm. Paperback. Free.

Describes plants of the family Leguminosae, including root crops, pulses, fruits, forages, timber and wood products, ornamentals, and others.

National Research Council

Firewood Crops: Shrub and Tree Species for Energy Production

1980. 237 pages. 25 cm. Paperback. Free.

Examines the selection of species suitable for deliberate cultivation as firewood crops in developing countries.

National Research Council

Food, Fuel and Fertilizer from Organic Wastes

1981. 150 pages. 23 cm. Paperback. Free.

Examines some of the opportunities for the productive utilization of organic wastes and residues commonly found in the poorer rural areas of the world.

National Research Council

Sowing Forests from the Air

1981. 61 pages. 23 cm. Paperback. Free.

Describes experiences with establishing forests by sowing tree seed from aircraft. Suggests testing and development of the techniques for possible use where forest destruction now outpaces reforestation.

National Research Council

Winged Bean: a High Protein Crop for the Tropics (2nd Edition)

1981. 59 pages. 23 cm. Paperback. Free.

An update of BOSTID's 1975 report of this neglected tropical legume. Describes current knowledge of winged bean and its promise.

National Research Council

The Water Buffalo: New Prospects for an Underutilized Animal

1981. 118 pages. 23 cm. Paperback. Free.

The water buffalo is performing notably well in recent trials in such unexpected places as the United States, Australia, and Brazil. This report discusses the animal's promise, particularly emphasizing its potential for use outside Asia.

National Research Council

Mangium and Other Fast-Growing Acacias for the Humid Tropics

1983. 62 pages. 23 cm. Paperback. Free.

Highlights ten acacia species that are native to the tropical rain forest of Australasia. That they could become valuable forestry resources elsewhere is suggested by the exceptional performance of Acacia mangium in Malaysia.

National Research Council

Calliandra: a Versatile Small Tree for the Humid Tropics

1983. 52 pages. 23 cm. Paperback. Free.

This Latin American shrub is being widely planted by villagers and government agencies in Indonesia to provide firewood, prevent erosion, provide honey and feed livestock.

National Research Council

Crocodiles as a Resource for the Tropics

1983. 59 pages. 23 cm. Paperback. Free.

In most parts of the tropics crocodilian populations are being decimated but programs in Papua New Guinea and a few other countries demonstrate that, with care, the animals can be raised for profit while protecting the wild populations.

National Research Council

Butterfly Farming in Papua Ne./ Guinea

1983. 34 pages. 23 cm. Paperback. Free.

Indigenous butterflies are being reared in Papua New Guinea villages in a formal government program that both provides a cash income in remote rural areas and contributes to the conservation of wildlife and tropical forests.

National Research Council

Little-Known Asian Animals with a Promising Economic Future

1983. 133 pages. 23 cm. Paperback. Free.

Describes banteng, madura, mithan, yak, kouprey, babirusa, Javan warty pit, and other obscure but possibly globally useful wild and domesticated animals that are indigenous to Asia.

National Research Council

Alcohol Fuels: Options for Developing Countries

1983. 128 pages. 23 cm. Paperback. HDC $8.95. LDC Free.

Examines the potential for the production and utilization of alcohol fuels in developing countries. Includes information on various tropical crops and their conversion to alcohols through both traditional and novel processes.

National Research Council

Producer Gas: Another Fuel for Motor Transport

1983. 101 pages. 23 cm. Paperback. Free.

During World War II, Europe and Asia used wood, charcoal, and coal to fuel over a million gasoline and diesel vehicles. However, the technology has since been virtually forgotten. This report reviews producer gas and its modern potential.

National Research Council

Firewood Crops: Shrub and Tree Species for Energy Production. Volume II

1983. 92 pages. 25 cm. Paperback. Free.

A continuation of the 1980 BOSTID report. Describes 27 species of woody plants that seem suitable candidates for fuelwood plantations in developing countries.

National Research Council

Diffusion of Biomass Energy Technologies for Developing Countries

1984. 123 pages. Paperback. HDC $9.25. LDC Free.

This report examines the technical, economic, cultural, and political factors that affect the introduction and diffusion of biomass-based energy technologies in developing countries. It includes information on the opportunities and limitations for these technologies as well as conclusions and recommendations for their application.

The report was prepared by an international panel under the sponsorship of the International Relations Division of the Rockefeller Foundation.

National Research Council

Amaranth: Modern Prospects for an Ancient Crop

1984. 80 pages. 23 cm. Paperback. HDC $6.95. LDC Free.

Before the time of Cortes, grain amaranths were staple foods of the Aztec and Inca. Today this extremely nutritious food has a bright future. The report discusses vegetable amaranths also.

National Research Council

Casuarinas: Nitrogen-Fixing Trees for Adverse Sites

1984. 118 pages. 23 cm. Paperback. Free.

These robust nitrogen-fixing Australasian trees could become valuable resources for planting on harsh eroding land to provide fuel and other products. Eighteen species for tropical lowlands and highlands, temperate zones, and semiarid regions are highlighted.

National Research Council

Jojoba: New Crop for Arid Lands

1985. Paperback. HDC $10.00. LDC Free.

In the past 10 years the domestication of jojoba, a little-known North American desert shrub, has been all but completed. This report describes the plant and its promise to provide a unique vegetable oil with many likely industrial uses.

National Research Council

Leucaena: Promising Forage and Tree Crop for the Tropics (2nd Edition)

1984. 100 pages. 23 cm. Paperback. Free.

This update of the 1977 edition describes *Leucaena leucocephala*, a little known Mexican plant. Vigorously growing, bushy types of *Leucaena* produce nutritious forage and organic fertilizer. Tree types produce timber, firewood, and pulp and paper. The plant is also useful for revegetating hillslopes and providing firebreaks, shade, and city beautification.

Conferences, Workshops, and Symposia Proceedings

National Research Council

Proceedings, International Workshop on Survey Methodologies for Developing Countries

1980. 220 pages. 23 cm. Paperback. Free.

Report of a 1980 workshop organized to examine past and ongoing energy survey efforts in developing countries. Includes reports from rural, urban, industry, and transportation working groups, excerpts from 12 background papers, and a directory of energy surveys for developing countries.

National Research Council

Priorities in Biotechnology Research for International Development: Report of a Workshop

1982. 261 pages. 23 cm. Paperback. Free.

Report of a 1982 workshop organized to examine opportunities for biotechnology research in developing countries. Includes general background papers and specific recommendations in six areas: (1) vaccines, (2) animal production, (3) monoclonal antibodies, (4) energy, (5) biological nitrogen fixation, and (6) plant cell and tissue culture.

Periodicals

BOSTID Developments

Three times a year. Free.

Issued 3 times/year. Available free to researchers and officials in developing countries and to members of the national and international development communities.

GTZ

Deutsche Gesellschaft für
Technische Zusammenarbeit (GTZ) GmbH

Dag Hammarskjöld Weg 1, Postfach 5180, D-6236 Eschborn.
Cable: Germatec Eschborn Taunus. Telex: 417405 gtz d
Telefon: 06196-790.

What is GTZ?

The Deutsche Gesellschaft für Technische Zusammenarbeit (GTZ) is a non-profit enterprise owned by the Federal Republic of Germany, represented by the Federal Minister for Economic Cooperation (BMZ) and the Federal Minister of Finance (BMF).

The GTZ has existed in its present form since January 1, 1975, when it took over the activities of two previous development agencies — the Bundesstelle für Entwicklungshilfe (BfE) and the Deutsche Förderungsgesellschaft für Entwicklungsländer (GAWI), to streamline operations.

The basis of the GTZ's work in official technical operation is a General Agreement concluded on December 12, 1974. This agreement clearly delineates the tasks between the Federal Ministry for Economic Cooperation (BMZ) which is responsible for development policy and the GTZ as implementing agency:

- The BMZ, the negotiating partner of the governments of the developing countries, defines the general development policy aims and the objectives of individual activities and supervises their implementation; it decides what projects are to be supported and sets the financial framework in its commissions to the GTZ.

- The GTZ carries out BMZ commissions under its own responsibility, either itself or by recalling in other state or private enterprises; it supports the BMZ in the further development of principles and instruments of development policy.

What does GTZ?

GTZ puts development policy into practice. Most of its work is commissioned by the Government of the Federal Republic of Germany whose July 1980 "Policy Paper on German Cooperation with Developing Countries" defines technical cooperation as follows:

> "The purpose of technical cooperation is to enhance the performance of manpower and institutions in the developing countries. It involves the assignment of technical personnel and the supply of technical equipment. As a rule, technical cooperation are services granted on a non-repayable basis. In the more advanced developing countries, this cooperation is either partly or wholly against payment."

In addition to implementing Federal German development policy, the GTZ also takes direct commissions against payment from developing countries or from international organizations and it operates projects with its own funds on a limited scale.

On average in any given year the GTZ is involved in 1,700 projects in 90 countries. They are implemented by the GTZ's own field staff or by consultant firms and include the elaboration of studies and evaluations, handling of financial contributions, food aid, promotion of trade fair participation, etc. More than 10,000 supply consignments are made to projects every year.

From 1975 to 1982 annual turnover rose from DM 460 to DM 945 million. Commissions in hand developed similarly over this period from DM 1.379 to DM 3.126 million. At the end of 1982, 1445 GTZ fieldstaff and 740 experts employed by consultant firms were on assignment abroad.

Main tasks

Technical cooperation projects span almost the entire economic and social spectrum of the developing countries. GTZ tasks and services are correspondingly broad:

- Assignment and funding of experts for agreed projects as well as advisers, instructors, specialists, appraisers and short-term experts;

- supply of equipment and materials for the facilities supported, supply of industrial and agricultural production inputs, provision of services and works;

- support of the reintegration in their home countries of expert manpower from developing countries who trained in the Federal Republic of Germany and wish to return home;

- salary topping up subsidies for German experts under direct contract to a developing country;

- coordination of the initial and further training of specialists and managerial personnel from developing countries either in their own countries, in other developing countries, in the Federal Republic of Germany or in other industrial countries;

- subsidization of the cost of training programmes carried out by companies in the developing countries;

- financial contributions to projects and programmes of efficient implementing agencies in developing countries.

To achieve these tasks the development projects are planned and implemented by three departments:

Department 1: Agriculture, Health and Rural Development

Department 2: Science and Technology, Education, Vocational Training, Industry and Trade

Department 3: Infrastructure

Agriculture, Health and Rural Development

The promotion and development of rural areas has always been a top priority in German development cooperation with developing countries. The activities of the Department of Agriculture, Health and Rural Development are therefore determined by the following goals:

- to raise and diversify production in agriculture, forestry and fisheries,

- to create more jobs by promoting the agricultural structure,

- to improve storage, distribution, marketing and processing of agricultural products,

- to establish appropriate production, credit and marketing organizations,

- to intensify agricultural research and technology,

- to improve living conditions by supporting health, nutrition and family planning programmes.

The department cooperates closely with national and international research institutions and organizations in its research and development activities.

Publications

Experience gained in gtz projects is available to a wide circle of readers in gtz publications.

In this catalog-focussing on agriculture, health and rural development — these are listed in the following categories:

	Page
Schriftenreihe der gtz (gtz series)	461
Special issues	497
Appropriate technology	524

The GTZ series of publications can be obtained directly from the TZ-Verlagsgesellschaft mbH, Bruchwiesenweg 19, D-6101 Roßdorf 1, Telefon: Your code for Germany, then 6154-81119, agains payment of the stipulated price, plus postage.

Schriftenreihe der gtz

gtz Series

Nr. 1
Gachet, Paul und Jaritz, Günter

Situation und Perspektiven der Futterproduktion im Trockenanbau in Nordtunesien

1973. 30 Seiten. Deutsch. Dm 10.-.

Auch heute noch fehlen wichtige Grundlagen für eine Entwicklung der Viehhaltung in Nordtunesien. Für eine Reihe von Gebieten ist es jedoch schon jetzt möglich, neue Produktionssysteme anzuwenden, die einige Probleme der nordtunesischen Landwirtschaft lösen können, so zum Beispiel durch Schutz vor Bodenerosion und Verbesserung der Bodenfruchtbarkeit, Intensivierung der Fruchtfolgen und Integration der Viehhaltung in den Ackerbau. In dieser Publikation wurden die in Nordtunesien gesammelten Erfahrungen festgehalten.

Nr. 3
Jaritz, Günter

Die Weidewirtschaft im australischen Winterregenklima und ihre Bedeutung für die Entwicklung der Landwirtschaft in den nordafrikanischen Maghrebländern

1973. 40 Seiten, Deutsch. DM 10,-

In der Weidewirtschaft und Futterproduktion verfügt die nordafrikanische Landwirtschaft nur über einen spärlichen Kenntnisstand und

Erfahrungsschatz. Unter den entwickelten Ländern mit ähnlichem Klima kann Australien auf dem Gebiet der mediterranean Weidewirtschaft als führend angesehen werden.

Der Autor hatte Gelegenheit, die Weidewirtschaft im australischen Winterregenklima kennenzulernen. Seine dabei gewonnenen Eindrücke und Informationen hat er hier zusammengefaßt und am Beispiel Tunesiens die Bedeutung der australischen Leistungen auf dem Gebiet de Weidewirtschaft für den sehr ähnlichen nordafrikanischen Raum erörtert.

Nr. 4
Wienberg, Dieter, Weiler, Norbert und Seidel, Helmut

Der Erdbeeranbau in Südspanien

1972. 92 Seiten. 22 Abbildungen. Deutsch. DM 15.-.

Deutsche Agrarfachleute haben ihre mehrjährigen Erfahrungen in Finca La Mayora/Malaga, Spanien, in dieser Broschüre festgehalten. Mit dieser zweiten, verbesserten Aufliage wird dem Erwerbsanbauer auf der Grundlage bisheriger Erkenntnisse eine Anbauanleitung an die Hand gegeben, die ihn allerdings nicht davon entbindet, seine eigenen bzw. zusätzlichen Erfahrungen zu sammeln.

Nr. 5
Neumaier, Thomas (Redaktion)

Beiträge deutscher Forschungsstätten zur Agrarentwicklung in der Dritten Welt

1973. 568 Seiten. Deutsch. DM 15.-.

Die Agrarentwicklung in den Ländern Asiens, Afrikas und Lateinamerikas gehörte bisher nicht zu den Schwerpunkbereichen deutscher Forschung. Andererseits sind viele wissenschaftliche Arbeiten unbekannt und dadurch unberücksichtigt geblieben, weil ihr Auffinden mit zu großem Zeit- und Arbeitsaufwand — vor allem für im Ausland praxisorientiert tätige Fachkräfta — verbunden war. Diesem Tatbestand will die Publikation Rechnung tragen.

Nr. 7
Seidel, Helmut und Wienberg, Dieter

Gemüsesortenversuche in Südspanien

1973. 104 Seiten. Deutsch. DM 10,-.

Im südspanischen Küstengebiet der Provinzen Málaga, Granada und Almeria hat sich in den letzten Jahrzehnten ein intensiver Gemüseanbau entwickelt, der das früher kultivierte Zuckerrohr immer mehr verdrängt.

Um die Qualität des angebotenen Gemüses auf dem lokalen Markt zu verbessern, die Flächenerträge zu erhöhen und durch den Export hochwertigen Gemüses neue Absatzmöglichkeiten zu schaffen, wurden auf der Versuchsstation "La Mayora" Sortenversuche mit den wichtigsten Gemüsearten durchgeführt. Die Ergebnisse dieser Versuche geben den Landwirten in Südspanien die Möglichkeit, durch die Verwendung besserer Varietäten zu Ertrags- und Qualitätssteigerungen und damit zu einer Einkommenserhöhung zu gelangen. Auch für die Gemüseproduzenten in anderen Ländern mit ähnlichen Klimabedingungen können die Versuchsergebnisse eine Hilfe bei der Sortenwahl sein.

Nr. 11
Neumaier, Thomas (Redaktion)

Internationale Agrarentwicklung zwischen Theorie und Praxis (Bericht über die vierte landwirtschaftliche Projektleitertagung, Bonn 1973)

1974. 390 Seiten. Deutsch. ISBN 3-9800030-1-9. DM 18,-.

Der zehntägige Dialog zwischen der deutschen Entwicklungshilfe-Administration und den Führungskräften in Agrarprojekten ist in dieser Publikation enthalten. In der Schrift wurden die Referate, Diskussionsergebnisse, Randbemerkungen und Vorschläge zusammengestellt und für jene Interessenten aufbereitet, die an der Tagung nicht teilnehmen konnten. Der Bericht dient auch dazu, den Gedankenaustausch zwischen den Teilnehmern fortzusetzen, und zu einer permanenten kritischen Fortschreibung der entwicklungspolitischen Vorstellungen, Möglichkeiten und vor allem auch Grenzen beizutragen. Die Publikation vermittelt auch einen aktuellen Einblick in die Themen "Allgemeine Entwicklungspolitik", "Mitarbeiter", "Projekt" und "Projektmodelle".

Nr. 13
Neumaier, Thomas (editor)

Mokwa Cattle Ranch (Model of a Beef Production Unit for West Africa — Feedlot)

1974. 46 pages. English, French and German. 35 ills. ISBN 3-9800030-3-5. DM 5,-.

A summary of the establishment and development of an agricultural project handed over to the sole responsibility of the Nigerian partners after an operating period of 10 years. This illustrated booklet deals with the Mokwa Cattle Ranch, planning for which began in 1962 and which has meanwhile been handed over. The publication is intended to contribute to the discussion of the beef production problem in West Africa and to promote the interchange of experience between all concerned with this problem.

Nr. 15
Neumaier, Thomas (Editor)

Mandi — Projekt in einer indischen Bergregion
(Mandi — A Project in a Mountainous Region of India)

1974. 76 pages. 1 map, 41 ills. English and German. ISBN 3-9800030-5-1. DM 5,-.

Texts: Tim Zeuner and H. Stiegler.

This publication presents an agricultural extension project implemented in the sixties in the interior of the pre-Himalayan Highlands. The illustrated text gives an account of the project establishment as well as data indicating the results achieved.

Nr. 16
Rüchel, Werner-Michael

Chemoprophylaxe der bovinen Trypanosomiasis

1974. 252 Seiten. Deutsch. ISBN 3-9800030-6-X. DM 15,-.

In dieser Publikation wurden die Kenntnisse über die Bekämpfungsmöglichkeiten der bovinen Trypanosomiasis zusammengefaßt. Besonderes Gewicht wurde dabei auf die Möglichkeiten der Chemotherapie und Chemoprophylaxe gelegt. Dazu ist die verfügbare relevante internationale Literatur gesichtet und erörtert worden.

Also available in English (No. 30).

Nr. 18
Kopp, Erwin

Das Produktionspotential des semiariden tunesischen Oberen Medjerdatales bei Beregnung.
Hydrotechnische und agronomische Folgerungen und die Entwicklung einer neuen Produktionskonzeption.

1975. 332 Seiten. 28 Abbildungen. Deutsch. ISBN 3-88085-000-3. DM 15,-.

Diese Veröffentlichung ist der Versuch einer umfassenden Auswertung der Beobachtungen, Analysen, Messungen und Versuchsergebnisse, die von 1969 bis Anfang 1974 im Rahmen der Technischen Zusammenarbeit der Bundesrepublik Deutschland im Institut National de Recherche Agronomique en Tunisie (INRAT) gemacht wurden.

Nr. 21
Burgemeister, Rainer

Elevage de Chameaux en Afrique du Nord (Camel Breeding in North Africa)

85 pages. French. ISBN 3-88085-010. DM 5,-.

The author gives an account of camel breeding and management in Southern Tunisia. In Tunisia camel breeding still represents an important source of income for a large proportion of the population. Also important is the work performed by the animals on the large plantations in the Sahelian zone.

Nr. 22
Agpaoa, A., Endanga, D., Festin, S., Gumayagay, J., Hoenninger, Th., Seeber, G., Unkel, K. and Weidelt H. J. (Compiled by H. J. Weidelt)

Manual of Reforestation and Erosion Control for the Philippines

1975. 569 pages. English. ISBN 3-88085-020-8. DM 18,-.

This work is based on a training program in the Philippine-German Training Center for Reforestation and Erosion Control in Baguio City. It is intended as a textbook for reforestation and erosion control.

Nr. 23
Jürgens, Gerhard (editor)

Curso Basico sobre Control de Malezas en la Republica Dominica (Basic Course for Weed Control in the Dominican Republic)

Spanish. ISBN 3-88085-010-0. DM 5,-.

This publication contains the papers given at a basic course in weed control held in the Dominican Republic in 1975. In addition to general contributions contains papers on weed control in maize and sorghum, rice, legumes, vegetables, sugar cane and cocoa.

Nr. 25
Rohrmoser, Klaus

Ölpflanzenzüchtung in Marókko — Sélection des Oléagineux au Maroc (Oil Plant Breeding in Morocco)

1975. 278 pages. 8 coloured photos, 1 general map. German and French. ISBN 3-88085-035-6, DM 5,-.

This publication describes the tasks, methods used and results obtained in nine years of work in Moroccan oil plant breeding. The book is intended to serve as a basis for the continuation, in Moroccan plant breeding, of the techniques introduced, to give impeti for new projects with similar objectives and to inform interested specialists about a successful project of German Technical Cooperation with developing countries.

Nr. 26
Bonarius, Helmut

Physical Properties of Soils in the Kilombero Valley (Tanzania)

1975. 34 pages. English. DM 5,-.

This publication provides information on the physical properties of the soils in the Kilombero Valley (Tanzania).

Nr. 29

Landwirtschaftliche Entwicklung West-Sumatra

1976. 30 Seiten. 13 Schwarzweißfotos. 1 farbige Standortkarte. Deutsch. ISBN 3-88085-007-0. DM 10,-.

In dieser Publikation wird ein indonesisch-deutsches Vorhaben der Technischen Zusammenarbeit vorgestellt. Nach einem kurzen Abriß der drei Aufbauphasen werden die einzelnen Sektionen behandelt.

Also available in English (No. 37).

Nr. 33

Journées Agrostologie — Elevage des Ruminants (Exchange of Experience on Pasture Improvement)

1976. 183 pages. French. ISBN 3-88085-009-7. DM 5,-.

Report on a pasture improvement symposium held in Fianarantsoa, Madagascar, from May 21 to 23, 1975.

Nr. 35

Colheitas melhores para Minas Gerais — Bessere Ernten für Minas Gerais (Better Harvests for Minas Gerais. Five Years of Brazilian-German Cooperation in Minas Gerais)

54 pages. 52 ills. Portuguese and German. ISBN 3-88085-018-6. DM 7,50.

Compiled by Ernst Lamster and Thomas Neumaier.

An account of five years of Brazilian-German cooperation in the Federal State of Minas Gerais, Brazil.

Nr. 36
Kassebeer, von Keyserlingk, Lange, Pink, Pollehn, Zehrer and Bohlen

La Défense des Cultures en Afrique du Nord — en considérant particulierement la Tunisie et le Maroc (Crop Protection in North Africa with Special Reference to Tunisia and Morocco)

1976. 272 pages. 375 colour ills. French. DM 41, 20.

This publication is divided into seven sections, namely cereals and maize, sugar beets, pulses, Solanaceae, Cucurbitaceae, fruit trees and olives and deals with major diseases and pests.

Nr. 37

Agricultural Development in West Sumatra

1976. 30 pages. 13 black-and-white photographs. 1 coloured locality map. English. ISBN 3-88085-007. DM 5,-.

The publication reports on an Indonesian-German Technical Cooperation project. Short history of the three establishment phases is followed by the project sections.

Nr. 38
Kopp, Erwin

Le Potential de Production dans la Région semiaride de la Haute Vallée de la Medjerda tunisienne sous irrigation par aspersion
(The Production Potential Under Spray Irrigation of the Semi-Arid Upper Medjerda Valley of Tunisia)

1977. 360 pages. French. ISBN 3-88085-021. DM 26,-.

This publication is an attempt at comprehensive evaluation of observations, analyses, measurements and trial results made within the framework of Technical Cooperation with the Federal Republic of Germany at the Institut National de Recherche Agronomique en Tunisíe (INRAT).

Nr. 39
Schmutterer, Heinz

Plagas e Enfermedades de Algodon en Centro America
(Cotton Diseases and Pests in Central America)

1977. 104 pages. 50 color ills. Spanish. DM 22,-.

Advisory brochure about integrated cotton pest control in Central America.

Nr. 40

Dritte externe Veterinärtagung
(Berichte und Arbeitsergebnisse)

1977. 370 Seiten. Deutsch. ISBN 3-88085-022-4. DM 24,50.

Die Publikation enthält alle Referate, die anläßlich der dritten externen Vetennärtagung vom 11. bis. 19. November 1975 in Ouagadougou/ Obervolta gehalten wurden.

Nr. 41
Becker, Günther

Holzzerstörung durch Termiten im Zentralafrikanischen Kaiserreich — Destruction du bois par les termites dans l'Empire Centralafricain
(Destruction of Wood by Termites in the Central African Empire)

1977. 96 pages. 16 ills. German and French. DM 12, 60.

Termite destruction of wood is particularly high in the Central African Republic. This publication is to help promote wood protection against termites.

Nr. 42
Furtmayr, Ludwig

Besamungsstationen an tropischen und subtropischen Standorten

1977. 64 Seiten. Deutsch. ISBN 3-88085-031-3. DM 10,80.

Der Verfasser behandelt die Problematik und Methoden der künstlichen Besamung. Er liefert außerdem detaillierte Angaben über den Bau von Besamungsstationen in tropischen und subtropischen Gebieten.

Nr. 43
Wirth, Frigga

Culture de plants à parfum en Tunisie
(Perfume Plant Growing in Tunisia)

1977. 196 pages. French and German. ISBN 3-88085-060-7. DM 18,40.

The publication was produced as a practical guide for plant growing and for extension work. It is based on several year's experience gained in private and statal agricultural sectors in Tunisia.

Nr. 44
Hofmann, Rudolf and Otte, Kai-Christian

Nutzung der Vikunjas in Peru
(Utilization of Vicugnas in Peru)

1978. 48 pages. 39 ills. German, English, Spanish. DM 7,-.

Multi-lingual information booklet about a vicuna raising project in Peru.

Nr. 45
Grove, Dietrich

Diagnostico Andrológico Ambulante en el Bovino en Paises Calidos
(Ambulatory Andrological Diagnosis of Cattle in Warm Countries)

1977. 280 pages. Spanish. ISBN 3-88085-038-0. DM 24,50.

The publication was produced within the framework of the university partnership between the Veterinary University of Hannover and San Carlos University of Guatemala.

In view of the growing importance of beef production in the increasingly export-oriented Central American countries the procedures of testing and assessing the suitability of breeding bulls from the health and sexual points of view have become particularly important for veterinarians. The collated working instructions and evaluation guidelines are to serve as practical guidelines for students. The publication is also to serve as a bridge between the rapidly expanding veterinary-scientific and zoological-technical body of knowledge on the one hand the practical demands and possibilities in tropical and sub-tropical areas on the other hand. It is also to be on incentive and guide to veterinarians working in Technical Cooperation projects in warm climatic zones to examine male beasts.

Nr. 46
Nägel, Ludwig

Aquakultur in der Dritten Welt

1977. 110 Seiten. 21 Abbildungen. Deutsch. ISBN 3-88085-029-1. DM 18,50.

Während weltweit gesehen die Produktion aus der Aquakultur bis vor kurzer Zeit nur marginal war, gewinnt sie neuerdings an Bedeutung für die Versorgung der Bevölkerung mit tierischem Eiweiß. Unter Aquakultur versteht man die kontrollierte Produktion und die dazugehörenden Methoden der Vermehrung, Aufzucht und Haltung vor allem von Fischen, Krebstieren, Mollusken und Algen. Der Autor gibt einen Überblick über verschiedene Formen der Aquakultur wie Teichwirtschaft, Fischkultur in Netzgehegen, in Bewässerungskanälen etc. Voraussetzungen für den Erfolg von Aquakulturprojekten werden ebenso behandelt wie die einzelnen heirfür geeigneten Tiere. Abschließend geht der Autor noch auf Ausbildungsmöglichkeiten für Fischwirte und anhand von Projekten auf die Wirtschaftlichkeit von Aquakulturanlagen ein.

Nr. 47
Wagener, Wilhelm E.

Model for Practical-Educational Counterpart Training

1977. 106 pages. English. ISBN 3-88085-006-02. DM 18,50.

The author deals with the learning target-oriented planning and evaluation of instruction in technical subjects. This "assembly kit" has been tested under conditions of lacking linguistic breadth of understanding in the training of technical teachers at the Faculty of Technical Education, King Mongkut's Institute of Technology in Bangkok, a Thai-German Technical Cooperation project. The "assembly kit" is variable for practical-pedagogy counterpart training at various levels.

Nr. 49
Bischof, Friedrich

Common Weeds from Iran, Turkey, the Near East and North Africa

1979. 234 pages. 204 colour ills. English. ISBN 3-88085-061-5. DM 56,-.

This is a weed definition book about the Near East, Iran, Turkey and North Africa. The publication is to contribute to the provision of understandable information on weed recognition and control to indigenous experts in developing countries.

Nr. 50
Count Ballestrem, Conrad and Holler, Hans-Joachim

Potato Production in Kenya. Experiences and Recommendations for Improvement

1977. 88 pages. 69 ills. English. ISBN 3-88085-026-7. DM 19,60.

The authors give an overview of the present status of potato production in Kenya, dealing also with problems and possibilities of improvement. Although potato growing is fairly new in Kenya, the problems which arose could be overcome quickly. Kenya already produces certified seed potatoes. The brochure gives a brief summary of the botanic characteristics of the potato and its demands on the growing-location. Growing and multiplication techniques are described as are the economics of potato production. The remaining chapters deal with fertilization, plant protection, storage and marketing.

Nr. 51
Neumaier, Thomas, Richter, Wolfang and Weers, Egge

Savar Farm — The Central Breeding-Station of Bangladesh

1977. 44 pages. 41 ills. English and German. ISBN 3-88085-039-9. DM 7,50.

Nr. 52

**Progress on Lake Malawi — the Central Region
Lakeshore Development Project 1976-1977**

1978. 54 pages. 37 ills. English. ISBN 3-88085-036-4. DM 7,50.

The various activities and results of a Malawian-German project are described. On behalf of the Federal German Ministry for Economic Cooperation the GTZ promoted the Salima central region with DM 30 million from 1967 to 1977. The aim of the German support was to boost the production of cotton, groundnuts, rice and maize as well as breeding, dairy, slaughter and draught livestock. The family incomes of the population were to be raised. Equality of opportunity was to be created for all families by using all socioeconomic resources and improving infrastructure and health care.

Nr. 53
Kisselmann, Erwin (Redaktion)

**Gutachten — Studien — Berichte (Beiträge aus 20 Jahren
Internationaler Zusammenarbeit
im ländlichen Raum)**

1977. 540 Seiten. Deutsch. ISBN 3-88085-037-2. DM 28,-.

Dokumentation von 1350 Gutachten und Studien des Agrarbereichs über Problembereiche in 88 Ländern.

Nr. 54

**Tierärztliche diagnostische Labors in Malaysia — Beispiel
malaysisch-deutscher Zusammenarbeit
Veterinary Diagnostic Laboratories in Malaysia**

1978. 32 Seiten. 17 Abbildungen. Deutsch, Englisch. ISBN 3-88085-51-8. DM 10,60.

Informationsschrift über den Aufbau eines veterinärmedizinischen Labors in Malaysia.

Nr. 56
Leppack, Eberhardt and Rosskamp, Robert

**Almacenamiento de Papas en Panama — un Ejemplo para Zonas Tropicales y Subtropicales
(Potato Storage in Panama — an Example for Tropical and Subtropical Zones)**

1978. 102 pages. 28 coloured ills. 19 sketches. Spanish. ISBN 3-88085-059-3. DM 26,50. (See publication No. 77.)

This techno-scientific publication deals with potato production, storage and marketing in Panama. Improved storage has reduced annual losses through shrinkage; producers can obtain better prices for their qualitatively faultless produce.

Nr. 57
Walker, Jane B., Mehlitz, Dieter and Jones, Garroth, E.

Notes on the Ticks of Botswana

1978. 83 pages. 32 ills. English. ISBN 3-88085-052-6. DM 18,40.

The tick plays an important role in the transmission of various animal diseases. Even for man, ticks represent a permanent source of infection. This publication describes the ticks occurring in Botswana, their distribution, life cycle and the diseases they cause.

Nr. 58
Hueck, Kurt

**Los Bosques de Sudamérica — Ecologí, composicióne e importancia economica
(The Forests of South America)**

1978. 476 pages. With vegetation map of South America. Spanish. ISBN 3-88085-053-4. DM 49,50.

Nr. 59
Dorow, Eberhard

Hubschrauber in der Feldheuschreckenbekämpfung — Helicopter in Grasshoppers Control — L'hélicoptére dans la lutte contre les criquets

1978. 66 pages. 13 ills. German, English, French. ISBN 3-88085-047-X. DM 8,20.

The damage caused by grasshoppers often leads to complete defoliation of the infested areas. In the north-east of Nigeria infestation reached such dimensions that an efficient and highly effective method of control became necessary which provided large-area coverage, was independent of local water resources and did not have to rely on inexperienced local workers. This publication provides information on the experience gathered in the use of helicopters for grasshopper control in northern Nigeria.

Nr. 60
Hubert, Klemens and Sander, Friedrich

The Rehabilitation of Rural Roads in Handeni District (Tanzania). Project Description and Assessment of Experiences

1978. 81 pages. 3 maps. English. ISBN 3-88085-046-1. DM 8,60.

The Handeni District in north-eastern Tanzania is one of the least developed regions of the country. Extensive-type farming, partly still at subsistence level, a population density below the country's average and an only slightly developed infrastructure characterize the area. The existing road network is entirely inadequate. Its improvement was a task tackled since 1972 by the GTZ together with the Office of the Regional Development Director, Tanga. The experience gathered in the project forms the subject of this publication.

Nr. 61
German Agricultural Team (GAT) in Kenya

Passion Fruit Growing in Kenya — a Recommendation for Smallholders

1978. 46 pages. 57 ills. English. ISBN 3-88085-056-9. DM 11,50.

Passion fruit is growing eminently suitable for Kenya's smallholding. The crop can be grown with profit even in small acreages. This extension booklet describes the requirements of the plant in terms of soil, climate and cultural measures. Other aspects dealt with include propagation, fertilization and crop protection; hints are given on the so-called interplanting.

Nr. 62
Lippmann, Dieter

Cultivation of Passiflora edulis S. (General Information on Passion Fruit Growing in Kenya)

1978. 88 pages. 88 ills. English. ISBN 3-88085-065-8. DM 18,-.

Whilst publication No. 61 deals quite specifically only with the cultivation techniques in passion fruit growing, we are here given a comprehensive survey of the subject. Information is provided on the origin of passion fruits, on the different varieties and their distribution in the individual cultivation areas. Using the Kenyan Kisii District cultivation area as an example, the author explains the reasons for the expansion of this crop; details are also given on yield developments, the marketing system and extension service. This chapter is followed by a section on different cultivation techniques and their application in the Kisii District. Subsequently the major plant diseases and pests affecting passion fruits are described together with the respective control methods. The last chapter lists the material and labour input required per hectare in passion fruit growing, the data being based on the working conditions in smallholdings.

Nr. 63
Kopisch-Obuch, Friedrich-W., Ulsperger, Walter, Werner, Hans-J., Wieland, Thomas

Rückstandsprobleme im Pflanzenschutz in der Dritten Welt (Der Beitrag des GTZ-Rückstandslabors und seine Außenaktivitäten)

1978. 67 Seiten. 35 Abbildungen. Deutsch. ISBN 3-88085-057-7. DM 14,50.

Die Einführung intensiver Anbaumethoden in den Entwicklungsländern hat zwar zu erheblichen Steigerungen der Flächenerträge geführt, doch war im Rahmen einer intensiven Bodennutzung die zunehmende Anwendung von Pflanzenschutzmitteln unerläßlich. Eine Überwachung der ausgebrachtten Pflanzenschultzmittelmengen und Kontrolle der Rückstände finden aber nur in unzureichendem Maße statt. Die Bundesrepublik Deutschland hat diese Situation schon früh erkannt und schuf im Rahmen der Technischen Zusammenarbeit ein Projekt, das sich weltweit mit den Rückstandsproblemen befaßt. Die vorliegende Publikation gibt einen Einblick in die Arbeit, die das GTZ-Zentrallabor in Darmstadt und die mit ihm in enger Verbindung stehenden Rückstandslaboratorien in zahlreichen Ländern der Dritten Welt leisten.

Also available in English (No. 82).

Nr. 64
Schmutterer, Heinz

Cotton Pests in the Philippines

1978. 110 pages. 47 ills. English. ISBN 3-88085-064-X. DM 21,-.

Although the cotton industry in the Philippines has experienced a considerable upturn in the past several years, the numerous cotton pests represent a constant threat and suitable control measures consequently have to be undertaken. For this purpose, an exact knowledge of the major pests is indispensable. This textbook gives a survey of the most important cotton pests followed by a description of the plant parts most frequently attacked. The main section contains a description of the pests and the damage symptoms caused and gives details of their distribution, life cycle, natural enemies and the relevant control measures.

Nr. 65

Dune Stabilization — a Survey of Literature on Dune Formation and Dune Stabilization

1977. 407 pages. English. ISBN 3-88085-032-1. DM 18,50.

Ever growing destruction of cultivable land through intensified exploitation and interference with the ecological balance are leading to increasing desertification and trasformation into steppeland of great parts of the earth; even now dry areas make up a third of the globe. The battle against increasing formation of steppeland is worldwide task in which the Federal Republic of Germany is also participating together with countries of the Third World within the framework of Technical Cooperation. This publication on dune stabilization was prepared by the GTZ in cooperation with the Geographical Institute of the Würzburg university. The work is based on the experience and findings of an Egyptian-German project and is divided into four sections. The first chapter describes interalia the different types of dunes, the morphogenetic processes of dune formation and the transport of drifting sand. Using the New Valley, Egypt, as an example, the second chapter deals with methods of stabilizing dunes and drifting sand. In the final section a model is presented, giving once more a clear and concise description of the absolutely necessary steps of a dune stabilization programme. A detailed bibliography forms the second part of this work.

Nr. 67
Becker, Silke

La Propagación de la Roya del Cafeto
(The Spread of Coffee Rust)

1979. 70 pages. 13 ills. Spanish, English, German. ISBN 3-88085-075-05. DM 7,50.

Translation of a work about the course of infestation and spore catching in Hemileia vastatrix in Kenya.

Nr. 69
Marouani, A., Kouki, M., Ghanmi, M. and Grell, P.-H.

Gutes Saatgut — eine Voraussetzung für hohe Erträge (Good seed — a Precondition for High Yields)

1979. 56 pages. 61 ills. Trilingual edition (German, French, Arabic). ISBN 3-88085-066-6. DM 15,50.

A basic precondition for stepping up crop yields is without any doubt the use of high quality seed of locally adapted varieties. Already at the beginning of this century, the first seed legislation was passed in Tunisia to promote this sector. Later the Institut National de Recherches Agronomiques Tunisie (INRAT) was established and charged with the production of good varieties and the control of seed multiplication. When Tunisia initiated the establishment of the Tunis seed testing station in 1970 within the framework of Tunisian-German cooperation, her object was to satisfy the domestic requirements of good seed of the major crops, over and above this, to participate in international seed trade. In many years of cooperation, seed legislation was drawn up and a seed testing center was established with its own laboratory and field control agency which fully meet present requirements.

Nr. 70
Michels, Thomas

Medical Laboratory Development in Tanzania

1078. 30 pagoo. 22 ills. English. ISDN 3-88085-003-1. DM 12,50.

This publication provides information on the work of the Laboratory Service Station in Dar es Salaam. This medical laboratory, the largest in Tanzania, is a Tanzanian-German Technical Cooperation Project and was set up on "historic" ground: it was here that Robert Koch studied malaria and the sleeping sickness.

The station carries out laboratory tests for various hospitals and the public health service, supervises and assists the District laboratories and trains laboratory technicians and medical students.

Nr. 74
Heidemann, C. und Ries, H.O.

Raumordnung, Regional - und Stadtentwicklung. Ein methodisches Konzept

1979. 52 Seiten. Deutsch, ISBN 3-88085-067-4. DM 11,25

Ausgehend von den Erfahrungen mit bislang abgewickelten Projekten der Technischen Zusammenarbeit kamen die Autoren zu der Auffassung, daß punktuelle und nur auf einen Sektor der Volkswirtschaft ausgerichtete Entwicklungsförderungsmaßnahmen nur einen geringen Beitrag zur Lösung der gesellschaftlichen Probleme der Länder der Dritten Welt leisten können.

Das in der vorliegenden Publikation vorgestellte Konzept will die Besonderheiten problemezogener Entwicklungsprojekte aufzeigen, das diesem Projekttyp angemessene methodische Konzept aufschlüsseln, die bislang entwickelten Projektansätze diskutieren, einige aus der praktischen Durchführung ableitbare Erfahrungen festhalten und in handlungsbezogene Vorschläge für die Vorbereitung, Steuerung und Auswertung dieser Projekte umsetzen.

Hierzu liegen auch eine englische (Nr. 48) und eine spanische (Nr. 98) Übersetzung vor.

Nr. 75

Landtechnische Ausbildungs - und Beratungszentren als Mittel zur Förderung der Landwirtschaft in Entwicklungsländern

1979. 96 Seiten. Deutsch. ISBN 3-88085-069-0. DM 16,50.

Aufgrund der sehr unterschiedlichen Bedingungen in genossenschaftlicher, betriebswirtschaftlicher, ökologischer und sozialer Hinsicht, gibt es bisher noch keine allgemeingültige Entwicklungsstrategie für die Mechanisierung der Landwirtschaft in Entwicklungsländern. In der vorliegenden Studie wird anhand der Länder Thailand, Sri Lanka, Türkei und Marokko die Agrarsituation der einzelnen Länder auf die Möglichkeiten hin untersucht, die der Landtechnik und ihrer Verbesserung jeweils eingeräumt werden können. Dabei wird auch der Stellenwert analysiert, den die deutschen Ausbildungs- und Trainingszentren bei der Entwicklung der Landwirtschaft im jeweiligen Land einnehmen. Im Anschluß daran werden Strategien zur Förderung der Technisierung der Landwirtschaft vorgestellt, denen Empfehlungen zur Auslegung landtechnischer Entwicklungsprojekte folgen.

Also available in English (No. 86).

Nr. 76
Gassert, Werner C.

Research on Coffee Berry Disease in Ethiopia

1979. 56 pages. English. 17 ills. ISBN 3-88085-070-4. DM 12,60.

This publication presents the findings of a research project on the coffee berry disease Colletotrichum coffeanum Noack in Ethiopia.

Nr. 77
Rosskamp, Robert, Leppsack, Eberhardt

Potato Storage in Panama

1979. English. ISBN 3-88085-073-9. DM 26,50.

This book is the translation of publication No. 56.

Nr. 79
Krause, R. und Lorenz, F.

Bodenbearbeitung in den Tropen und Subtroben Einige Grundlagen, Geräte und Verfahren

1979. 252 Seiten. Deutsch. ISBN 3-88085-079-8. DM 24,50.

Die vorliegende Schrift — im Rahmen des Arbeitskreises "Internationale Agrartechnische Zusammenarbeit" der Max-Eyth-Gesellschaft (MEG) entstanden — behandelt Geräte und Verfahren zur Primär -und Sekundärbodenbearbeitung. Sie ist das erste Werk, einer lockeren Reihe von Handbüchern zur Mechanisierung der Landwirtschaft in den Entwicklungsländern. Die Schrift richtet sich an einen weiten Kreis von Interessenten, ist jedoch in erster Linie ein Nachschlagewerk für Berater, Ausbildor und Praktiker.

Nr. 80

Die Ökologie und Bekämpfung des Blutschnabelwebervogels (Quelea quelea [L.]) in Nordostnigeria

1983. 274 Seiten. Deutsch. ISBN 3-88085-167-0. DM 76,-.

Die Bekämpfung von Schadvögeln ist seit nahezu zwei Jahrzehnten ein Schwerpunkt der deutschen Technischen Zusammenarbeit. In dieser Veröffentlichung versuchen die Autoren nicht nur Antworten auf noch viele offene wissenschaftliche Fragen über Art, Verbreitung und Lebensraum der sich erhelich vermehrenden Webervögel zu geben. Von ganz besonderer Bedeutung ist die Darstellung der Bekämpfungsmethoden ohne Schädigung der natürlichen Vegetation und des Wasserhaushaltes.

Nr. 81
Heim, Sepp

Orthopädie-technisches Versorgungszentrum und Ausbildungsstätte für Orthopädie-Techniker. Lomé/Togo

1979. 38 Seiten. Deutsch, Englisch, Französisch. ISBN 3-88085-071-2. DM 15,60.

Die deutsche Orthopädie-Technik nimmt in der Welt eine gewisse Leitposition ein. Die Broschüre vermittelt einen Einblick in ein orthopädie-technisches Versorgungszentrum und in eine Ausbildungsstätte, die in togoisch-deutscher Kooperation in Lomé errichtet wurden.

Nr. 82
Kopisch-Obuch, Friedrich-W., Ulsperger, Walter, Werner, Hans-J., Wieland, Thomas

Pesticide Residue Problems in the Third World, a Contribution of the GTZ Residue Laboratory in Darmstadt

1979. 64 pages. English. ISBN 3-88085-074-7. DM 14,50.
(See also No. 95).

Nr. 84
Frhr. v. Grotthuss, Otto (Redaktion)

Die Bundesrepublik Deutschland und die Forstwirtschaft der Dritten Welt

1980. 148 Seiten. Deutsch. ISBN 3-88085-099-2. DM 12,50.

Die Projektbeschreibungen dieser Publikation sind auf dem Stand vom 1.1.1980.

Nr. 86

Agricultural Engineering, Training and Advisory Centres as a Means of Promoting Agriculture in Developing Countries

1980. ISBN 3-88085-069-0. DM 16,50.

This book is the English translation of No. 75.

Nr. 90
Hayfa, Gisela und Korte, Rolf

Gesundheit, Bevölkerungsentwicklung, Ernährung

1979. 228 Seiten. Deutsch. ISBN 3-88085-084-4. DM 13,50.

Der Band informiert über 40 Projekte in 27 Entwicklungsländern im Bereich der Gesundheit, Bevölkerungsentwicklung und Ernährung. Als gemeinsamer Nenner der von der GTZ geförderten Maßnahmen kann das gesundheitspolitische Konzept gelten, das sein Schwergewicht auf die medizinische Versorgung möglichst breiter Bevölkerungsschichten, auf ländliche Gebiete, präventive Programme und auf eine Förderung der Eigeninitiative der Betroffenen legt.

Nr. 92
Rau, Norbert und Rau, Anke

Commercial Marine Fishes of the Central Philippines

1980. 608 pages. English. ISBN 3-88085-089-5. DM 38,-.

This work concentrates on the fish species landed in the Philippines and describes about 400 different species including some entirely new ones. Forty-two internationally recognized specialists for tropical fish species assisted in the species identification.

A short introduction, equally short instructions for use and a glossary are followed by a key to the fish families with brief illustrated descriptions. The main section contains the species descriptions classified according to families. The text of the descriptions of the individual species follows a uniform format; it is split up into feature-related paragraphs (body, head, colour, fins, scales, teeth), permitting rapid comparison between species. In addition a detailed drawing is given of almost every species. At the back are a bibliography and an index.

Nr. 94
Bandel, Klaus, and Solameh, Elias

Hydrochemical and Hydrobiological Research of the Pollution of the Waters of the Amman Zerka Area, Jordan

60 pages. 6 photographs and many ills. English. ISBN 3-88085-115-8. DM 17,95.

The publication is the result of extensive water research in the Amman-Zerka area, carried out in partnership between Jordan and Brunswick universities.

The Amman-Zerka area is semi-arid, making its very low rainfall very precious. In the train of increasing industrialization and the resultant strong growth in population the meagre water resources are over-burdened by pollution. The two author's study offers a precise analysis of the chemical and biological composition of the waters in this area, with special consideration of the Zerka River. Aim of the study is to provide information on the degree of pollution by identifying water quality so that counter measures can be taken in order to secure drinking water supplies long term.

Nr. 95

Problemas de Residuos de Pesticidas en el Tercer Mundo — la Contribución del Laboratorio de Análisis de Residuos de la GTZ en Darmstadt — Sus Actividades en el Extranjero (Pesticide Residue Problems in the Third World. A Contribution of the GTZ Residue Laboratory in Darmstadt)

1981. 67 pages. 35 ills. Spanish. ISBN 3-88085-057-7. DM 14,50.

Although the introduction of intensive cultivation methods in the developing countries has greatly increased area yields, the use of more plant protectants was indisponable. But there is sufficient control of the amounts of protectants used and other residues. The Federal Republic recognized the problem early and within its Technical Cooperation programme created a project dealing with the residue problems worlwide. This publication gives an insight of the work of the central GTZ laboratory in Darmstadt and many residue laboratories in Third World countries with which it collaborates closely.

Nr. 96

Problèmes de Résidus Résultants de la Protection des Végétaux dans le Tiers Monde. La Contribution du Laboratoire de Résidus de la GTZ à Darmstadt et ses Activités Extérieures

1982. 67 pages. 35 ills. French. ISBN 3-88085-127-1. DM 14,50.

Translation of No. 95.

Nr. 98
Heidemann, C. and Ries, H. O.

Regional and Urban Development. A Methodological Framework

1980. 48 pages. English. ISBN 3-88085-087-9. DM 11,25.

Proceeding from the experience gained with Technical Cooperation projects completed further to, the authors come to the conclusion that development promotion measures addressed only selectively or to a particular sector of a national economy can make only a small contribution to solving the social problems of Third World countries.

The concept presented by this publication aims to show the special characteristics of problem-related development projects, analyze the methodological concept appropriate to this project type, discuss the project approaches elaborated so far, note experiences derivable from practical implementation and translate them into action-related proposals for the preparation, management and evaluation of these projects.

Nr. 100
Hagenbrock, T. Pohl, H., Ries, H. O., Spanger, U., Springer, W.

Aufgaben und Chancen von Regionalentwicklungsprojekten in Entwicklungsländern Erfahrungen und Ergebnisse aus Südminas, einer Schlüsselregion für die Dezentralisierungspolitik Brasiliens

1979. 256 Seiten. Deutsch. ISBN 3-88085-092-5. DM 39,50.

Das Buch reflektiert den entwicklungspolitischen Stellenwert der Regionalplanung in Entwicklungsländern (Schwellenländer) unter Gesichtspunkten der technisch-wirtschaftlichen Zusammenarbeit und unter Gesichtspunkten eines räumlichen und implementationsbezogenen Planungsansatzes. Von besonderem methodischen Interesse ist, daß diesem Projekt ein problemund ressourcen-orientierter Ansatz zugrundegelegt wurde und auch während des Planungsablaufs beibehalten werden konnte. (Vgl. C. Heidemann - H. O. Ries Raumordnung, Regional- und Stadtentwicklung - Ein methodisches Konzept; GTZ Schriftenreihe Nr. 74, 1979.)

Damit wird auch ein Weg aufgezeigt, wie ergebnisorientierte Planungsverfahren dem Charakter von Projekten der Politikberatung weitgehend gerecht werden können und wie sie dazu beitragen, relativ schnell die auch politisch abgesicherte Formulierung von räumlich bezogenen Handlungsprogrammen zu finden.

Abschließend äußert sich die Autorengemeinschaft auch zu Fragen der Arbeitsorganisation, Arbeitsprogrammierung und zu Arbeitstechniken in Planungsprojekten.

Im Anhang findet sich schließlich eine Zusammenfassung des für die Region Südminas erarbeiteten Rahmenentwicklungsplanes und des in ihm eingegliederten, sektoral vertiefenden Industrieentwicklungsplanes.

Nr. 103

Centres de Formation et de Vulgarisation en Génie Agricole, un Moyen de Promotion de l'Agriculture dans les Pays en Voie de Développement (Agricultural Technology Training and Advisory Centres as a Means of Promoting Agriculture in Developing Countries)

1980. French. ISBN 3-88085-094-1. DM 24,50.

Because of widely varying cooperative, economic, ecological and social conditions there is as yet no generally valid development strategy for mechanizing agriculture in developing countries. This study looks into possibilities of introducing or improving mechanization in Thailand, Sri Lanka, Turkey and Morocco. The role of German training centres in the development of agriculture in these countries is also appraised. Then follow strategies for furthering agricultural mechanization and recommendations for projects in this field.

Nr. 105
Doppler, Werner

The Economics of Pasture Improvement and Beef Production in Semi-Humid West Africa

1980. 195 pages. English. ISBN 3-88085-101-8. DM 14,-.

Nr. 106

Studien, Gutachten, Berichte — Beiträge internationaler Zusammennarbeit im ländlichen Raum 1977-80

1981. 351 Seiten. Deutsch. ISBN 3-88085-102-6. DM 13,80.

Es handelt sich hierbei um den ersten Fortsetzungsband, der lückenlos an die gleichlautende Veröffentlichung von 1977 (GTZ-Schriftenreihe

Nr. 53) anschließt. Aufgeführt sind die im Auftrag der Hauptabteilung 1 - Landwirtschaft und ländliche Entwicklung - seit 1977 angefertigten Gutachten und Berichte.

Nr. 107
Kölbing, Alexander und Seifert, Kurt

Fisheries at Lake Assad, Syria — Pêche sur le lac Assad, Syrie — Fisherei am Assad-See, Syrien

1981. 32 Seiten. Deutsch, Englisch, Französisch. ISBN 3-88085-100-X. DM 16,25.

Am Beispiel eines neu entstandenen Stausees wird der Aufbau einer leistungsfähigen Fischerei unter besonderer Berücksichtigung lokaler Gegebenheiten dargestellt.

Nr. 108
Schmidt, Gerhard, Hesse, Friedrich-Wilhelm und Trost, Karl

Die Zuckerrohrkultur in Marokko

1981. 88 Seiten. 26 farbige Abbildungen, 13 Skizzen. Deutsch. DM 34,50.

Die marokkanische Regierung sieht vor, bis 1990 134.000 Hektar Zuckerrohr anzubauen. Dadurch könnte der hohe und weiter steigende Zuckerbedarf des Landes voll gedeckt werden. Während bis 1962 der gesamte in Marokko benötigte Zucker importiert werden mußte, konnte anschließend durch die Einführung der Zuckerrübe eine nicht unwesentliche einheimische Produktion eingeleitet werden. Obwohl die Rübenzuckererzeugung in den großen Bewässerungs- gebieten im Bewässerungs- und Regenfeldbau zu hohen Erträgen führte, reichten die Ernten zur Deckung des Gesamtbedarfes nicht aus. Versuchsergebnisse, die von der GTZ auf der Station Centrale de Plantes Sucrières am nationalen Agrarforschungsinstitut durchge- führt wurden, zeigten, daß auch Zuckerrohr insbesondere in den wenig frostgefährdeten Zonen des Landes für den Anbau geeignet ist. Inzwischen weitet sich der Zuckerrohranbau immer stärker aus. Eine weitere Ausdehnung hängt vom Ausbau der Bewässerungsanlagen und der Fabrikationskapazität ab. Derzeit wird in zwei marokkanischen Fabriken Zuckerrohr verarbeitet. Die Ausdehnung der Zuckerrohr- kultur hängt vom weiteren Ausbau der Bewässerungsanlagen und der Fabrikationskapazität ab. Diese Publikation faßt die Ergebnisse mit Zuckerrohr zusammen, die im Rahmen der landwirtschaftlichen Forschung in dem marokkanisch-deutschen TZ-Projekt zur Förde- rung der Zuckerrohrerzeugung erzielt wurden.

Nr. 111
Munzinger, Peter (Redakteur)

Handbuch der Zugtiernutzung in Afrika

1981. 496 Seiten. Über 100 Abbildungen. Deutsch. ISBN 3-88085-103-4. DM 60,-.

Initiator für die Erstellung dieses Handbuches war der Arbeitskreis "Internationale Agrartechnische Zusammenarbeit" der Max-Eyth-Gesellschaft für Agrartechnik e.V. Im Rahmen eines überregionalen Projektes der Technischen Zusammenarbeit wurde dann die Erarbeitung des Werkes ermöglicht.

Die Zugtierhaltung, lange Zeit als Relikt vergangener Epochen vernachlässigt, gewinnt seit wenigen Jahren langsam an Salonfähigkeit für entwicklungspolitische und fachliche Diskussionen. Die deutsche Technische Zusammenarbeit ist seit den frühen 60er Jahren sehr eng mit dem Bemühen verknüpft, nicht nur die Gespanntiernutzung zu fördern, sondern die Prüfung, Herstellung sowie den Vertrieb geeigneter Maschinen und Geräte zu unterstützen.

Wesentlichstes Ziel des aus drei Teilen bestehenden Buches ist die Vermittlung von Informationen zu Fragen der Zugtiernutzung unter afrikanischen Bedingungen. Es sollen so die Fachkräfte und Entwicklungsplaner nationaler und internationaler Organisationen der Entwicklungszusammenarbeit angesprochen werden, die sich in Afrika oder in anderen Teilen der Welt mit der Zugtiernutzung beschäftigen. Daneben richtet sich das Handbuch an Entscheidungsträger in nationalen Behörden.

Nr. 112
Schmidt, G., Hesse, F. W. and Trost, K.

La culture de la canne à sucre au Maroc

1981. 88 pages. French. ISBN 3-88085-105-0. DM 34,50.

This publication contains the experiences of a Moroccan-German project on sugar cane production.

See also No. 108.

Nr. 113

Problémes de post-Rècolte

1981. 292 pages. French. DM 15,40.

See No. 115.

Nr. 114
Drescher, Wilhelm and Crane, Eva

**Technical Cooperation Activities: Beekeeping.
A Directory and Guide**

1982. 166 pages. English. ISBN 3-88085-111-5. DM 39,50.

This is the first systematic presentation of information on beekeeping development programmes in tropical countries. It contains a large amount of material in a small space: information on 91 projects and 42 proposals or feasibility studies in altogether 85 countries.

Nr. 115

**Post Harvest Problems Documentation of an OAU/GTZ
Seminar**

1981. 292 pages. English. DM 14,50.

From March 9 to 22, 1980, the OAU/STRC Interafrican Phytosanitary Council (IAPSC) and GTZ held a joint International Seminar for training on control of post-harvest crop losse at Lomé, Togo. It was attended by scientists and development officers of 14 different African States. The papers given during the seminar are collated in this publication.

Nr. 116
Karbe, Eberhard and Komlanvi Freytas, Edem

Trypanotolerance. Recherche et Application. Research and Implementation

1982. 314 pages. French and English. ISBN 3-88085-158-1. DM 10,50.

Proceedings of an international workshop on trypanotolerance and animal production held from May 10 to 14, 1982, at Lomé and Aventonou, Togo.

Nr. 117
Jahn, Samia al Azharia

Traditional Water Purification in Tropical Developing Countries — Existing Methods and Potential Application

1982. 284 pages. 18 ills., 31 sketches. ISBN 3-88085-116-6. DM 35,-.

This publication covers aspects of water supply which — despite a host of efforts to find technical solutions ensuring a sufficient supply of good potable water — have frequently been overlooked. It refers to the traditional capability of man to manage this precious commodity under harsh natural conditions. The author has concentrated her efforts on finding ways of using water more efficiently with existing means without breaking with or destroying local traditions.

Nr. 118
Kopp, Erwin

Stickstoff-Problematik in Einem Semiariden Raum

1981. 276 Seiten. 25 farbige Abbildungen, 26 Grafiken, 38 Tabellen. DM 51,60.

Im Oberen Medjerdatal im Nordwesten Tunesiens wurden über einen Zeitraum von acht Jahren Stickstofof-Düngungsversuche auf unterschiedlichen Böden durchgeführt und die Abhängigkeit der Erträge von der N-Düngung ermittelt. Bei Weizen, Zuckerrüben, Futterrüben, Mais, Futterpflanzen u.a. wurde festgestellt, daß die N-Düngung hier kein ertragsbestimmender Faktor ist. Die Analysen berücksichtigen Bodenbeschaffenheit, Klima, Vegetationsperioden, Produktionstypen, Wasserqualität, Bewässerungstechniken etc.

Nr. 119
Jaritz, Günther

Amélioration des Herbages et Cultures Fourragères dans le Nord-Ouest de la Tunisie: Étude Particulière des Prairies de Trèfles Graminièes Avec Trifolium Subterraneum

1982. 340 pages. French. ISBN 3-88085-118-2. DM 52,50.

The author gives an account of research results and practical experience gained through eleven years of work on pasture improvement in north-west Tunisia with special reference to subterranean clover.

Nr. 120
Munzinger, Peter (editor)

Animal Traction in Africa

1982. 496 pages. English. ISBN 3-88085-133-6. DM 60,-.

See also Nos. 111 and 121.

The use of animal traction power — which is to a very large extent a renewable energy source — will in the future continue to be of enormous importance for many agricultural holdings in Africa and its significance will probably increase still further. This hypothesis is supported in particular by the permanent worldwide shortage of fossil energies, for the resultant increase in the cost of conventional types of energy is already having disastrous consequences for some African countries.

The major objective of this book is therefore to provide information on questions concerning the use of draught animals under African conditions. It is aimed at the specialists and development planners working for national and international development cooperation organizations and dealing with the use of draught animals in Africa or other parts of the world. The handbook is also intended for decision-makers within national authorities.

Nr. 121
Munzinger, Peter (editor)

La Traction Animale en Afrique

1982. 496 pages. French. ISBN 3-88085-148-4. DM 60,-.

See also Nos. 111 and 120.

Nr. 122
Philipp, Ottmar, Koch, Werner und Köser, Heinz

Utilization and Control of Water Hyacinth in Sudan

1983. 224 pages. English. ISBN 3-88085-184-0. DM 32,-.

This book compiles some of the main results of a Sudanese-German Project. It emphasizes techniques for converting this weed into animal feed, fertilizer and for energy production. Its aim is to contribute information to those working on aquatic weed management and those who have to make decisions in related politics.

Nr. 123

Fleisch aus Ferké. Ein Feedlot am tropischen Standort

1983. 112 Seiten. Deutsch. DM 14,60.

Die Elfenbeinküste ist seit veilen Jahren gezwungen, in zunehmendem Maße Fleisch aus Europa, Südamerika und dem südlichen Afrika zu importieren. Die hierzu erforderlichen Devisenaufwendungen belasten die Handels- und Zahlungsbilanz. Gleichzeitig strömt auf uralten Triebwegen Magervieh zum Schlachten aus Obervolta und Mali in den Norden des Landes. Neue Zuckerfabriken sind im letzten Jahrzehnt entstanden, die als Nebenprodukt Melasse — ein hochwertiges, nicht für die menschliche Ernährung geeignetes Futtermittel — herstellen. Es lag daher auf der Hand, dieses Magervieh vor dem Schlachten aufzumästen und damit die heimische Fleischerzeugung zu steigern. Volkswirtschaftlich bedeutet dies, Schaffung zusätzlicher Arbeitsplätze auf dem Lande, Erhöhung der Wertschöpfung und Reduzierung der Einfuhrabhängigkeit. Ein solches Vorhaben war technisch, zusammen mit dem zu errichtenden Schalachthof unter den vorhandenen traditionellen Strukturen, nur auf großbetrieblicher Grundlage zu realisieren. Die Broschüre zeichnet aber gleichzeitig auch auf, wie sich dieses Vorhaben einordnet in die Gesamtkonzeption des Landes zur Förderung der tierischen Erzeugung in einem traditionell ackerbaulich strukturierten Entwicklungsland, in welche auch der Kleinbetrieb einbezogen wird.

Nr. 124
Becker, E. W.

Mikroalgen. Ergebnisse aus drei Versuchsvorhaben

1982. 118 Seiten. DM 24,50.

Die Anfang der siebziger Jahre in den Ländern Thailand, Peru und Indien eingerichteten Projekte der Technischen Zusammenarbeit haben eindeutig gezeigt, daß die Massenproduktion von Mikroalgen in diesen Ländern möglich ist, die Qualität und der Nährwert dieser Algen sehr hoch sind und eine Verwendung als Nahrungsmittel durchaus möglich erscheint. In der vorliegenden Veröffentlichung sind die wesentlichen Projektergebnisse zusammengestellt.

Nr. 125
Gross, Rainer, Bunting, E. S.

Agricultural and Nutritional Aspects of Lupines

1982. 884 Seiten. Englisch. ISBN 3-88085-134-4. DM 39,-.

Proceedings of the first International Lupine Workshop, held in Cusco, Peru, April 1980.

Nr. 126

Smallholder Farming Systems with Yam in the Southern Guinea, Savannah, of Nigeria

1982. 340 Seiten. DM 36,-.

In this publication farm survey data are presented, calculated and evaluated. They enable the reader to get a comprehensive picture of yam cultivation both in economical and production technical matters.

Nr. 127

**Mas Granos Basicos para Honduras
(Mehr Getreide für Honduras)**

1982. 44 Seiten. 55 Farbabbildungen. Spanisch. DM 12,50.

Die Darstellung eines Selbsthilfeprogrammes bei FACACH, einer Spar - und Kreditgenossenschaft in dem mittelamerikanischen Staat Honduras.

Nr. 128
Kropp, E. (Redaktion)

**Ländliche Regionalentwicklung - ein
Orientierungsrahmen**

1983, 152 Seiten, Deutsch, Englisch, Französisch. ISBN 3-88085-221-9. DM 21,-.

Nr. 131
Thal, J., Taodi, M.

**Bedeutung, Biologie und Bekämpfung der Kartoffelmotte
in Marokko
Importance, Biology and Control of the Potato Moth in
Morocco
Importance et Biologie de la Tergue de la Pomme de
Terre, la Lutte Contre ce Ravageur au Maroc**

1982. 144 Seiten. Deutsch-Englisch, Französisch. 19 Abbildungen. ISBN 3-88085-139-5. DM 23,-.

In fast allen tropischen und substropischen Gebieten der Erde befällt der Großschädling Kartoffelmotte (Phythorimaea operculella Zeller) Solanaceen wie Kartoffeln, Tomaten oder Tabak. Seine genaue Heimat ist nicht bekannt. Mit Kartoffelimporten wurde dieser Schmetterling auch in den Mittelmeerländern verbreitet und konnte sich hier infolge milder Winter besonders gut vermehren. In Algerien entdeckte man diesen Schädling 1874 (nach Balachowsky 1966), in Zypern 1918 (Attia und Mattar 1939), in Ägypten 1939, in Marokko 1933, in Palästina 1943 (Mason), im Libanon und in Syrien 1961 (Thalhouk).

In Marokko befaßten sich seit dem Jahre 1974 Fachleute der Deutschen Gesellschaft für Technische Zusammenarbeit (GTZ) GmbH im Auftrag des Bundesministeriums für wirtschaftliche Zusammenarbeit (BMZ) gemeinsam mit den zuständigen marokkanischen Stellen damit, die Kartoffelmotte zu untersuchen und Möglichkeiten einer gezielten Bekämpfung zu erproben. Der wirtschaftliche Schaden, der in Marokko durch diesen Schädling ensteht, geht in die Millionen, da die befallenen Krollen in den Imporländern entweder zurückgewiesen oder begast werden.

Nr. 132
Weidelt, Hans Joachim und Banaag, Valeriano S.

Aspects of Management and Silviculture of Philippine Dipterocarp Forests

1982. 308 pages. English. ISBN 3-88085-157-3. DM18,-.

This report deals primarily with the Philippine dipterocarp forest which ranks among the most valuable rain forest formations of the world. It tackles, from the silviculturist's point of view, the various methods and approaches involved in harnessing the full potentials of dipterocarp forests. Thus, the report includes analysis of Philippine selective logging system, timber stand improvement and regeneration surveys among others. It is addressed to policy makers as well as to foresters in the field, and aims to provide some basic information necessary for the rational use of forest resources.

Nr. 134
Schlolaut, W. (Redaktion)

"Kompendium der Kaninchenproduktion unter Berücksichtigung der Verhältnisse in der Dritten Welt"

1984. 260 Seiten. Deutsch. ISBN 3-88085-215-4.

Dank seiner überragenden Flächenproduktivität und der nicht obligatorischen Nahrungskonkurrenz zum Menschen ist das Kaninchen eine Alternative zu den anderen Nutztierarten bei der Verbesserung der Proteinversorgung des Menschen und der Erzielung monetären Einkommens. In verstärktem Maße wurde die Kaninchenproduktion deshalb in laufende Projekte der Entwicklungshilfe einbezogen. Diese Publikation ist ein Versuch, das breite Spektrum der Produktionsintensität in möglichst kurzgefaßter und praxisorientierter Form zu behandeln. Alle Bereiche wie Fütterung, Haltung, Fortpflanzung und Krankheiten werden umfassend beschrieben.

Nr. 135
Rüdenauer, Michael

Production de viande be bovins trypanotolérants en savane guinée d'Afrique occidentale

1982. 344 Seiten. Französisch. ISBN 3-88085-144-1. DM 29,-.

Organisation, rentabilité et possibilités de développement compte tenu notamment des possibilités d'integration en petites exploitations, á l'exemple du Togo.

Systematische Erhebungen zum Weideertrag und zur Fleischerzeugung mit unterschiedlichen Rassengruppen (N'Dama, Kreuzungstiere) sowie Stichprobenerhebungen im dörflichen Milieu der Region um Avetonou/Togo bilden die Grundlage zu Überlegungen der Wirtschaftlichkeit der Fleischerzeugung über das Rind in unterschiedlich organisierten Betrieben.

Nr. 136
Merkle, Alfred

The Cost for Health of All. A Feasibility Study from Upper' Volta

1982. 101 pages. English. ISBN 3-88085-145-X. DM 20,-.

This case study from Upper Volta gives an estimate of the cost of a "Health for all"-programme. The estimated costs are based on practical experiences in the implementation of a rural basic health programme in the south of the country.

Nr. 137
Steiner, Kurt G.

Intercropping in Tropical Smallholder Agriculture with Special Reference to West Africa

1982. 312 Seiten. English. ISBN 3-88085-176-X. DM 41,50. French in print.

Intercropping as the major cultivation system is found throughout the humid and semiarid tropics. This publication basically evaluates the international literature hereto. Furthermore the author contributes substantially his own experience as well as the results of various farming system surveys.

Nr. 138
Hayfa, Gisela

Aufbau von Frauengruppen in Bangladesch. Familienplanung, Gesundheitserziehung, Einkommensverbesserung

1982. 92 Seiten. 12 Schwarzweiß-Abbildungen. 1 Standortkarte. ISBN 3-88085-155-7. DM 24,50.

Die Autorin schildert die Erfahrungen mit den sog. Mother Clubs-Programmen.

Nr. 139
Küper, W. (Editor)

Bildung und Wissenschaft in der Technischen Zusammenarbeit mit Entwicklungsländern

1982. 636 Seiten. Deutsch. ISBN-3-88085-182-4. DM 29,-.

Die Publikation gibt einen Überblick über die Maßnahmen der deutschen staatlichen Bildungs- und Wissenschaftsförderung in der GTZ-Abteilung 22 (Stand:31.12.1982).

Nr. 142
Frey, Franz, Yabar, Erik.

Enfermedádes y Plagas de Lupino en el Peru

1983. 80 pagos. Espagnol. ISBN 3-88085-188-3.

El trabajo aquí presentado es el fruto de la cooperación entre un investigador peruano y un alemán. Los resultados están bien ordenados, proporcionando una descripción de los sintomas y daños, de los agentes causales e indicando las posibilidades para evitar or controlar la enfermedad o plaga.

Nr. 143
Becker, E. W.

Microalgae. Findings of Three Experimentation Projects

1983. 94 pages. English. ISBN 3-88085-189-1. DM 24,50.

The projects, initiated in the early seventies in Thailand, India and Peru, have clearly demonstrated that in these countries mass production of microalgae is possible, the quality and nutritive value of the algae is high, and a utilization of algae as human food seems to be possible. This publication presents the major result of these projects.

Nr. 144

Technische Zusammenarbeit im ländlichen Raum — Was — Wo — Wie 1984

1984. 1048 Seiten. Deutsch. ISBN 3-88085-191-3. DM 32,50.

In der Publikation werden ca. 300 Projekte beschrieben, die von der GTZ-Hauptabteilung 1 Landwirtschaft, Gesundheit und ländliche Entwicklung betreut werden.

Nr. 145
Neumaier, Thomas (Redaktion); Wörner, Beate (Texte)

"Landtechnische Ausbildung in der Türkei"

1983. 72 Seiten. Deutsch. ISBN 3-88085-192-1. DM 21,-.

Die Förderung landtechnischer Ausbildung gehört seit über 20 Jahren zum Programm deutscher Zusammenarbeit mit den Ländern der Dritten Welt. Sie ist Bestandteil vieler Projekte der Deustchen Gesellschaft für Technische Zusammernabeit an den verschiedensten Standorten in Afrika, Lateinamerika und Europa. Zwei Projekte, die im Rahmen dieser Zusammenarbeit viele Jahre gefördert wurden, sind das deutsch-türkische Landmaschinenausbildungszentrum in Gökhöhük und das landtechnische Ausbildungszentrum Söke in der Türkei. Aus diesen beiden Projekten ist ein weiteres Vorhaben in Malatya hervorgegangen. Alle drei Projekte werden in dieser GTZ-Publikation vorgestellt.

Nr. 148
Heim Sepp, Kaphingst, Wieland

Tanzania Training Centre for Orthopaedic Technologists (TATCOT), Moshi/Tanzania.

1984. 36 pages. German. English. ISBN 3-88085-198.0. DM 27,-.

With the help of numerous photographs this publication documents the efforts made to train orthopaedic technologists. Being the first institute of this kind in English-speaking Africa TATCOT is of immeasurable value for the rehabilitation of the physically handicapped in a large number of countries.

Nr. 149
Weinberger, Peter, Park Geun-Je, Kwan, Du-Jung

Korean Woodlands (Im-Ya) as Resources for Grassland Development

1983. 228 pages. English. ISBN 3-88085-199-9. DM 29,50.

The authors report on an attempt in South Korea to establish a stock of dairy and beef cattle adequate enough to make self-sufficiency possible. The main problem was to provide adequate feed. In the existing agricultural structure in South Korea, which is based on rice farming, suitable niches and possibilities for dual usage had to be found for cattle farming. The mountainous wooded country is especially suited for these purposes and therefore it is on this that the attention of the book is focused. In order to avoid extensive ecological damage, the use of land for grazing should only be considered in the presence of certain vegetation indicators. Recommendations are also given for reforestation projects. These guidelines, which are based on 10 years of experience, can also be applied in other parts of East and Southeast Asia with a similar structure.

Nr. 150
Krause, R., Lorenz, F., Hoogmoed, W. B.

Soil Tillage in the Tropics and Subtropics

1984. 320 pages. Numerous black and white ills. English. ISBN 3-88085-200-6. DM 33,50.

This revised English-language version of the original German publication deals with all aspects of mechanical soil tillage in great detail and is an aid for the correct selection and use of agricultural machines and equipment in the tropics and subtropics.

Nr. 152
Domdey, Siegbert

Un feed-lot en milieu tropical (A Feedlot in a Tropical Environment)

114 pages. French. ISBN 3-88085-205-7. DM 27,-.

This publication points out the conditions under which large-scale animal husbandry projects take their place in the economy and society of a country, in this case the Ivory Coast. At the same time it describes how this project fits into the country's overall concept for promoting livestock production.

Nr. 154
Zech, Wolfgang

Etudes sur l'Ecologie des Ligneux d'Intérêt Forestier dans l'Afrique de l'Ouest Semi-Aride (Studies on the Ecology of Timber Species in Semi-Arid West Africa)

1984. 394 pages. 22 colour photographs, 2 b/w photographs, 31 ills., 52 tables incl. 7 foldouts. French. ISBN 3-88085-214-6. DM 30,50.

Nr. 161
Schmutterer, Heinz, Ascher, K. R. S.

Natural Pesticides from the Neem Tree (*Azadirachta indica* A. Juss) and Other Tropical Plants. Proceedings of the Second International Neem Conference, Rauischholzhausen, Federal Republic of Germany, 25-27, May 1983

1984. 604 Seiten, 7 Farbfotos. Zahlreiche sw-Fotos und Abbildungen. Englisch, ISBN 3-88085-156-5 DM.

Erscheint Januar 1985.

Nr. 166

Technical Cooperation in Rural Areas — Facts and Figures

1984. 460 Seiten. Englisch. ISBN 3-88085-229-4. DM 30,50.

In this publication some 300 projects are described which are implemented by the departement Agriculture, Health and Rural Development.

Special Issues

Afrass, Ahmed und Mosich, Lutz

Manuel Des Matériels De Traitement Des Cultures

1980. 70 Seiten. 21 × 14,8 cm.

Le manuel donne en premier lieu un sommaire général de tous les types de matériels usés en traitement ou Maroc dans la Section

Technique d'Application. Les particularités des pulverisateurs pou cultures basses, des pulvérisateurs et atomiseurs pour vignes sont expliqués en détail. En particulier on fait attention ou contrôle, à la vérification des matériels, ou calcul des volumes par hectare et aux concentrations et vitesses durant le traitement. Instructions de réparation, entretien et maintenance des máteriels aussi qu'une marche à suivre pratique pour le traitement sont données.

La manuel contient beaucoup d'illustrations.

Röttger, U. (editor)

Paddy Deterioration in the Humid Tropics

1982. 232 pages. 21 × 15 cm. ISBN 3-88085-152-2.

This book presents the results of the GASGA-Seminar held in the Philippines from October 11th to 18th 1981. An overview of different kinds of post-harvest loss in paddy is given (microorganisms, insects) followed by the presentation of a grading system. The construction of a warehouse is illustrated and warehouse management is explained. Also the different aspects of the impact of deterioration on milling of paddy is reported.

One chapter deals with training.

Weis, Norbert (editor)

Rodent Pests and Their Control

1981. 200 pages. 25,5 × 21 cm.

In the first part the guide contains descriptions of destructive rodent species, their biology and preservation instructions for collectors. Guidelines for the careful planning and evaluation of results of a campaign are followed by the presentation of various methods for controlling rodent pests in built-up areas and in open country. Advice on rodenticides and resistance to rodenticides is given. The correct application of proper bait and the construction of bait containers is illustrated. Also the use of proven appraisal methods in sugar cane, rice and maize is demonstrated. Included is advice on accident prevention and — for emergencies — first aid measures.

The rodent pest situation in various countries is reported.

Schäfer, Jürgen (editor)

Manual on Surveillance and Early-Warning Techniques for Plant-Pest and Diseases

1982. 228 pages. 22 × 19,5 cm.

This book presents standardized procedures to be used in an efficient plant protection service in the form guidelines for the implementation of a Surveillance and Early-Warning System. The coding system for reports is described, followed by detailed description of all the pests (insects, weeds, rodents, birds, nematodes) and diseases encountered. Advice for their control is given.

Schmutterer, H., Ascher, K. R. S., Remhold, H.

Natural Pesticides from the Neem Tree

1981. 297 pages. 21,5 × 15 cm.

In the proceedings of the First International Neem conference an overview is given of the research about Neem extracts used as natural compound in pest control.

The chemistry of the deterent is reported, followed by papers which deal with the physiology and behaviour of insects affected by the extract of the neem tree (*Azodirachta indica* A. Jass) and the Persian lilac (*Melia azedurach* L.). The authors present information about the effects on insects and mites in laboratory and field.

Becker, E. W. and Venkataraman, L. V.

Biotechnology and Exploitation of Algae — the Indian Approach

1982. 216 pages. 23,5 × 15,5 cm.

Das Buch gibt einen zusammenfassenden Überblick über die Ergebnisse eines durch das Bundesministerium für wirtschaftliche Zusammenarbeit geförderten Projektes zur Massenkultur von Mikroalgen in Indien.

Vergleichend mit anderen internationalen Unternehmungen gleicher Art werden die verschiedenen Produktionsverfahren beschrieben sowie chemische Zusammensetzung, ernährungsphysiologische Qualität und eventuelle toxikologische Risiken mehrerer Algen diskutiert, wobei auf die Vielzahl der Anwendungsmöglichkeiten der Algen unter Berücksichtigung der Produktionskosten eingegangen wird.

Midohoe, Kodjo und Hecht, H.
Hrsg. Institut für Landtechnik der J. L. Universität Gießen,
Fachbereich Agrartechnik

Arbeitszeitstudien und Mechanisierungsmodelle in kleinbäuerlichen Betrieben in Togo/Westafrika

1982. 214 Seiten. 21 × 15 cm. DM 15,-.

Die für Anbau, Ernte und Verarbeitung der wichtigsten Kulturen
benötigten Arbeitszeit (überwiegend Handarbeit) wurde in mehreren
charakteristischen kleinbäuerlichen Betrieben Zentraltogos ermittelt.
Ausgehend von der jeweils gegebenen Familienarbeitskraft wurden
Modelle für eine gezielte Mechanisierung zur Berechnung der limi-
tierenden Arbeitsspitzen entwickelt.

Thierolf, J. G.
en colaboración con Castillo H. y Konold, F.

Pastos Cultivados en la Sierra y Engorde Compensatorio de Ovinos

1981. 85 pages. 20,5 × 14,5 cm.

El libro es una guia para la instalación y el manejo de pastos cultivados
bajo riego en zonas alto andinas. Las instrucciones se basan en
técnicas desarrolladas en la Sierra Central del Perú desde 1970 con el
fin de activar reservas forrajeras en pisos ecológicos exclusivamente
ganaderas.

van Tienhoven, Icaza und Lagemann

Farming Systems in Jinotega Nicaragua

1982. 169 pages. 12 × 15 cm.

Description of agronomic practices, labor use, production-influencing
factors, production and productivity, income situation of a typical
smallholder area in Nicaragua. Recommendations of agricultural
research, extension and innovations.

von Platen, Rodriguez and Lagemann

Farming Systems in Acosta Puriscal Costa Rica

1982. 146 pages. 21 × 15 cm.

Description of agronomic practices, labor, production-influencing
factors, production and productivity, income situation of a typical
smallholder area of Costa Rica. Recommendations of agricultural
research, extension and innovations.

Rohrmoser, Klaus

Kompendium für Feldversuche

Eschborn 1984. 276 Seiten. 20,9 × 14,5 cm. ISBN 3-88085-108-5. DM 22,50.

Das Kompendium ist Nachschlagwerk und Anleitung für die Planung, Durchführung und Auswertung von Versuchen in der pflanzlichen Produktion. Das Buch ist speziell auf die Bedurfnisse und Probleme zugeschitten, die im Feldversuchswesen in Projekten der Technischen Zusammenarbeit auftreten. Die statistische Auswertung der gängigen Versuchsmethoden ist bewußt in leicht verständlicher Form dargestellt, um ein schnelles Einarbeiten in das Versuchswesen zu ermöglichen.

Die englische Ausgabe ist in Vorbereitung.

Keim, Ricardo, Korte, Rolf, Osinski, Petra (editors)

Family Planning Strategies in the 1980's

1983. 272 pages. English in print.

In this book papers are presented from the international conference: "Current Approaches to Population Problems" held from October 25-29 in Bonn-Niederbachem.

It gives a broad view of experience gained so far in the field of family planning. At the same time from this basis future development activities can be renewed and improved.

Al-Hubaishi, A., Müller-Hohenstein, K.

An Introduction to the Vegetation of Yemen

1984. 209 + 60 Seiten. 21 × 15 cm. 151 coloured photos. Bilingual english and arabic. ISBN 3-88085-248-0.

This publication depicts the plant world and its ecotopes for all types of landscape in the Yemen Arab Republic. The potential vegetation and the plant species currently to be found there are illustrated and classified scientifically as functions of the geological location, climate and the influence of man and his livestock.

Kotschi, J. und Adelhelm R.

Standortgerechte Landwirtschaft zur Entwicklung Kleinbaüerlicher Betriebe in den Tropen und Subtropen

1984. 108 + 107 Seiten. 21 × 15 cm. Deutsch. (Englische Übersetzung in Vorbereitung). ISBN 3-88085-264-2.

Mit dieser Studie legt die Hauptabteilung "Landwirtschaft, Gesundheit und ländliche Entwicklung" der GTZ einen Zwischenbericht zum Thema "Standortgerechte Landwirtschaft" vor. Er ist in zwei Teile gegliedert: Teil A ist zu verstehen als Beitrag zur Grundsatzdiskussion, Teil B gibt einen Überblick zu Aktivitäten Standortgerechter Landwirtschaft.

Im Anhang sind Literatur, Zeitschriften und Kontaktadressen zum Thema "Standortgerechte Landwirtschaft" aufgelistet.

Handbuchreihe Ländliche Entwicklung

Payr, G., Sülzer R., unter Mitarbeit von Albrecht H., Bergmann, H., Diederich, G., Hoffmann, V.

Landwirtschaftliche Beratung Band 1 — Grundlagen und Methoden

1981. 348 Seiten. 21 × 15 cm. ISBN 3-88085-114-X. Deutsch. DM 20,-.

Das Handbuch zur landwirtschaftlichen Beratung versucht, theoretische Grundlagen und wesentliche Zusammenhänge aus dem umfassenden Aufgabenfeld der Beratung zu verdeutlichen und daraus — wie auch aus der bisherigen praktischen Erfahrung — Vorschläge für die Gestaltung der Beratungsplanung und - durchführung zu machen. Auf keinen Fall möchte das Handbuch als "Rezeptbuch" verstanden werden, denn es kann keine Anweisungen für Einzelfälle geben, sondern lediglich Wege aufzeigen, wie man auf systematische Weise in spezifischen Situationen zu Problemlösungen gelangen kann.

Englisch-, Französisch- und Spanisch-Fassung in Vorbereitung.

Payr, G., Sülzer, R., unter Mitarbeit von Albrecht, H., Bergmann, H., Diederich, G., Hoffmann, V.

Landwirtschaftliche Beratung Band 2 —Arbeitsunterlagen

1981. 325 Seiten. 21 × 15 cm. ISBN 3-88085-121-2. Deutsch. DM 20,-.

Band 2 enthält eine Sammlung von Arbeitsunterlagen, die zur weiteren Verdeutlichung, Illustration und Ergänzung von Band 1 dienen sollen.

Viele lassen sich auch in Programmen zur Aus- und Fortbildung von Beratungskräften einsetzen. Sie sind nach Herkunft und Verwendungszweck geordnet.

Werner Warmbier, Detlev Böttcher unter Mitarbeit von Hans Gsänger, Winfried Muziol, Jochen Pfeiffer, Bernd Schubert, Karl-Ludwig Zils

Vermarktung Von Agrarprodukten Band 1: Grundlagen und Methoden

1984. 307 Seiten. 21 × 15 cm. Deutsch, ISBN 3-88085-224-3. DM 20,-.

Nach der Darstellung der Rolle der Agrarvermarktung in Entwicklungsländern werden Erfahrungen mit Agrarmarktförderungsmaßnahmen vorgestellt. Vorgehensweise und Instrumente der Förderung sowie das Management von Agrarmarktkomponenten sind Inhalt der nächsten Kapitel. Der Band schließt ab mit der Bewertung der Förderungsmaßnahmen.

Werner Warmbier, Detlev Böttcher unter Mitarbeit von Hans Gsänger, Winfried Muziol, Jochen Pfeiffer, Bernd Schubert, Karl-Ludwig Zils

Vermarktung Von Agrarprodukten
Band 2: Arbeitsunterlagen

1985. 338 Seiten. 21 × 15 cm. Deutsch. ISBN 3-88085-259-6. DM 20,-.

Band 2 enthält eine Sammlung von Arbeitsunterlagen, die zur weiteren Ergänzung, Verdeutlichung und Vertiefung von Band 1 dienen sollen. Einige von ihnen lassen sich zur Identifizierung von Problemstellen im Vermarktungsbereich verwenden und geben Hinweise zu deren Behebung. Sie sind jedoch nicht als fertige Rezepte zu verstehen, die in jeder Situation sicher wirken, sondern sollen die Entscheidung über "rotes" oder "grünes" Licht für bestimmte Projektaktivitäten unterstützen. Die meisten der Arbeitsunterlagen sind aus sich selbst heraus verständlich, ansonsten ergibt sich ihre Funktion aus dem Text von Band 1, von wo aus auf sie verwiesen wird.

Appropriate Technology

The following publications can be obtained from

GATE Deutsches Zentrum für Entwicklungstechnologien
In Deutsche Gesellschaft für Technische Zusammenarbeit
GTZ GmbH
Postfach 5180
D-6236 Eschborn/Taunus

or   Vieweg Publishing
P. O. Box 5829
D-6200 Wiesbaden

R. Kaupp

Gasification of Rice Hulls

1984. 322 pages. English. DM 64,-. Free for developing countries. ISBN
3-528-02002-4

The book analyses all existing problems in the field of rice hull
gasification. It is a comprehensive treatment of the physical and
chemical properties of rice hulls as well as general aspects of
gasification which do not only apply to rice hulls. The theoretical
aspects of gasification as well as the past and present praxis are
treated equally in this book.

Aprovecho Institute

Charcoal: Small Scale Production and Use

1984. 60 pages. DM 16,80. Free for developing countries. ISBN
3-528-02009-1.

This booklet presents an introduction into various aspects of small
scale charcoal production, such as economics of charcoal production
and use, charcoal making devices, kiln operation and improvement of
stove efficiency. It will help the extension worker to get involved in
charcoal production, but it does not pretend to give all information
available on charcoal.

Werner Roos/Ursula Rojczyk

Construction of Simple Kiln Systems

1984. Revised edition. 34 pages. DM 9,80. Free for developing countries. ISBN 3-523-0212-1.

There are various methods to produce charcoal: simple kiln and pile systems and a number of distillation plants in which by-products are recovered in addition to charcoal. Here only the simple systems shall be described. For every kind of kiln one process is presented in detail, for the other processes short construction and operating instructions are given.

Aprovecho Institute

Fuel-Saving Cookstoves

1984. 2nd edition. 168 pages. DM 29,80. Free for developing countries. ISBN 3-528-02007-5.

This book shows methods to develop appropriate cookstoves with local population. It provides basic information about the process of deforestation and related matters as well as hints for designing, selecting and testing stoves.

John Spiropoulos

Small Scale Production of Lime for Building

1985. 80 pages. DM 19,80. Free for developing countries. ISBN 3-528-02016-4.

Lime has been produced and used in many parts of the world for hundreds of years. It is a cementitious material which can be produced to an adequate quality by simple means and in sufficiently small quantities to suit the requirements and conditions of the rural areas of developing countries.

This handbook is a practical guide for the project planner and implementation officer covering the project investigation and implementation process, as well as technical and production aspects of low technology lime production on a small scale.

Ludwig Sasse

Biogas Plants

1984. 85 pages. DM 19,80. Free for developing countries. Available in English, Spanish and German language. ISBN 3-528-02004-0 (English). ISBN 3-528-02010-5 (Spanish). ISBN 3-528-02003-2 (German).

Simple biogas plants have been constructed in Third World Countries for about thirty years. Good and bad solutions are featured side by side without comment in articles and books. The same mistakes are repeated over and over again. This need not to be the case. The designer of a biogas plant must be able to distinguish between valid and invalid solutions. This book is intended to help him in this respect.

Helmut Muche/Harald Zimmermann

The Purification of Biogas

1985. 34 pages. DM 9,80. Free for developing countries. Also available in German and French. ISBN 3-528-02015-6.

This review attempts to set out the procedures for removing hydrogen sulphide.from biogas. Only an optimally applied purification agent can ensure a long life for the gas user, particularly engines, and avoid unnecessary repairs and maintenance on the plant equipment.

Herbert Bergmann

Primary School Agriculture
Vol. I: Pedagogy

1985. 144 pages. ISBN 3-528-02013-X. DM 24,50. Free for developing countries. 2nd revised edition.

Agriculture as a subject in Primary School is treated in a systematic way. A number of reasons for including Agriculture in the primary school curriculum are given, with a clear preference for a science-oriented approach. Curriculum design and the structure of a scheme of work are discussed and a number of very practical hints on school farm work are given. A number of lesson notes using various formats conclude the book.

The book is based on the author's practical experience in Africa.

Herbert Bergmann/Richard Butler

Primary School Agriculture
Vol. II: Background Information

1985. 190 pp. ISBN 3-528-02014-8. DM 29,80. Free for developing countries. 2nd revised edition.

This volume provides factual background for volume I. This is meant to serve as a reforce book for the teacher who wants to structure the teaching of agriculture according to the ideas expressed in volume I. Its emphasis is on African subsistence and tree crop farming in the humid tropics. It deals specifically with traditional farming and a number of variants of modern low-input-farming. The most important food crops and the tree crops cocoa and coffee are documented. Crop storage of grains and tubers is also discussed.

R. Kaupp, J. R. Goss

Small Scale Gas Producer-Engine Systems

1984. 284 pages. ISBN 3-528-02001-6. DM 56,-. Free for developing countries.

This book gives a comprehensive overview of all aspects of gasification of many biomass fuels. Special attention is given to the analysis of performance data for small scale gas producer-engine systems and their gas cleaning trains. It is intended as a guideline and information resource for planners and engineers working in this field.

Der Göpel — eine Alternative bei der Mechanisierung der Landwirtschaft

1983. 62 Seiten. 23,5 × 16,5 cm. kostenlos.

In dieser Broschüre sind Informationen zusammengetragen über die älteste "Renewable energy": die Tierkraft und ihre Nutzung im Göpel. Diese Veröffentlichung soll dazu beitragen, dem Göpel neue Einsatzmöglichkeiten in Ländern der Dritten Welt zu erschließen. Dies gilt sowohl für die Neueinführung dieser Technik, als auch für ihre Verbesserung und Modernisierung dort, wo sie bereits traditionell verbreitet ist.

Besug uns dusch GATE.

Hannah Schreckenbach, Jackson G. K. Abankwa

Construction Technology for a Tropical Developing Country

1983. 348 pages. ISBN 3-88085-186-7. DM 126,-.

246 illustrations, many detail drawings, maps, etc, Glossary, Timber Catalogue "Ghana Timber and Woody Plants," "Strength Properties of Superior and High Quality West African Timber Species," index; List of References to each Chapter.

Distribution by TZ-Verlagsgesellschaft mbH, Bruchwiesenweg 19, D-6101 Rossdorf 1, Federal Republic of Germany.

K. Vorhauer

Low Cost/Self Help Housing

1980, 152 pp., English. Free.

The techniques and methods presented are based on the great variety of traditional construction methods. The information given can be utilized not only in constructing new houses, but also in improving parts of simple houses already standing.

Obtainable only from GATE.

"gate" — questions, answers, information (Journal)

First edition 1982; quarterly, approx. 36 pages. 24 × 18 cm. ISSN 0723-2225. Free. Publisher: Deutsches Zentrum für Entwicklungstechnologien (GATE) in GTZ.

"gate" is a publication issued by GATE (German Appropriate Technology Exchange), a division of the German Agency for Technical Cooperation (GTZ).

"gate" hopes to be a forum for everybody involved and interested in appropriate technology; it aims to discuss problems and to suggest approaches to solutions.

INTSOY

International Soybean Program
Programme International pour le Soja
Programa Internacional de la Soya

University of Illinois at Urbana-Champaign
113 Mumford Hall
1301 West Gregory Drive
Urbana, Illinois 61801, USA
Telex: 206957
Cable: INTSOY
Telephone: (217) 333-6422

Introduction and Objectives

INTSOY seeks to improve human nutrition around the world through the use of soybeans, a legume rich in protein and calories. Cooperating with likeminded regional, national, and international organizations, we work toward this goal through research, cultivar testing and breeding programs, feasibility studies, publications, regional and international conferences, training courses, and study programs.

INTSOY provides timely information about cultivar adaptation, crop management and protection, seed storage, and the processing and use of soy food. We offer direct access to a worldwide collection of soybean germplasm, and conduct a breeding program under varied ecological conditions. Through training programs, conferences, and publications, we are extending practical and timely information to soybean workers.

The United States Agency for International Development (USAID) has provided the core funding since INTSOY's creation in April 1973. Special grants from FAO, UNDP, UNICEF, CARE, and IBPGR have supported a range of special projects.

INTSOY, el Programa Internacional de la Soya, está tratando de mejorar la nutrición humana en todo el mundo a través del uso de la soya, una léguminosa rica en proteína y calorías. Un programa de la Universidad de Illinois en Urbana-Champaign, INTSOY coopera con organizaciones nacionales, regionales e internacionales para la expansión del uso de la soya. Comuníquese con nosotros si desea mayor información sobre los programas de INTSOY.

INTSOY, un Programme International pour le Soja, cherche à améliorer l'alimentation humaine dans le monde par lusage du soja, un légume riche en protéines et en calories. INTSOY est un programme de l'Université d'Illinois en cooperátion avec des organisations nationales, régionales et internationales pour la généralisation de lutilisation du soja. N'hésitez pas à nous contacter pour tous renseignements complémentaires concernant les programmes d'INTSOY.

Monographs

S. W. Williams and K. L. Rathod

A Case Study of Expeller Production of Soybean Flour in India

1974. 12 pages. 21 × 27.5 cm. Paperback. English. (INTSOY Series No. 3)

This report is a case study of soybean processing operations of Agro Processors Pvt. Ltd., Nagpur, Maharashtra, India. This small, specialized soybean processing organization needed facilities to produce low-fat soybean flour of high quality with a limited capital investment. The facilities and procedures described illustrate how, at modest cost, improvements can be made on conventional techniques used in expeller processing.

Mattias von Oppen

Soybean Processing in India: A Location Study on an Industry to Come

1974. 55 pages. 21 × 27.5 cm. Paperback. English. (INTSOY Series No. 4)

The material presented in this study is based on the author's Ph.D. thesis entitled "Optimal Size and Location of Marketing Facilities for Soybean in India."

S.W. Williams, W.E. Hendrix, and M.K. von Oppen

Potential Production of Soybeans in North Central India

1974. 32 pages. 21 × 27.5 cm. Paperback. English. (INTSOY Series No. 5)

This study was undertaken to estimate potential production of soybeans and to delineate possible areas in Uttar Pradesh and Madhya Pradesh capable of producing enough soybeans for processing plants with daily capacity of 200 to 500 tons.

T. Hymowitz, S. G. Carmer, and C. A. Newell

Soybean Cultivars Released in the United States and Canada: Morphological Descriptions and Responses to Selected Foliar, Stem, and Root Diseases

1976. 31 pages. 21.5 × 28 cm. Paperback. English. (INTSOY Series No. 9)

Morphological descriptions and responses are given to selected foliar, stem, and root diseases of 331 soybean cultivars released in the U.S. and Canada.

T. Hymowitz, C. A. Newell, and S. G. Carmer

Pedigrees of Soybean Cultivars Released in the United States and Canada

1977. 23 pages. 21 × 27.5 cm. Paperback. English. (INTSOY Series No. 13)

This bulletin lists the pedigrees, maturity group, year introduced, and the year named or released for each cultivar.

A. I. Nelson, M. P. Steinberg, and L. S. Wei

Whole Soybean Foods for Home and Village Use

1978. 31 pages. 21 × 27 cm. Paperback. English. (INTSOY Series No. 14)

This booklet reports work by Professors Nelson, Steinberg, Wei, and their associates to develop concepts, methods, and processes for home use of the whole soybean directly for human food. The emphasis is on acceptability, quality of nutrition, and low cost of protein, calories, and food nutrients.

O. Tisselli, J. B. Sinclair, and T. Hymowitz

Sources of Resistance to Selected Fungal, Bacterial, Viral, and Nematode Diseases of Soybeans

1979. 134 pages. 21.5 × 28 cm. Paperback. English. (INTSOY Series No. 18)

This publication covers 19 diseases caused by fungi, bacteria, viruses, and nematodes. Information given for each disease includes: causal agent, importance, distribution, symptoms, resistance, photographs of advanced symptoms, and a bibliography.

R. S. Smith, W. H. Judy, and W. C. Stearn

International Inoculant Shipping Evaluation

1983. 25 pages. 21 × 27.5 cm. Paperback. English. (INTSOY Series No. 23)

Results presented here are from the International Inoculant Shipping Evaluation (IISE) trial, which evaluated the quality of granular soybean inoculant after it was exposed to shipping and storage conditions during international transport.

Conference, Workshop, and Symposia Proceedings

INTSOY

Proceedings of the Workshop on Soybeans for Tropical and Subtropical Conditions

1974. 21.5 × 28 cm. Paperback. English. (INTSOY Series No. 2)

Held February 4 to 6, 1974, at the University of Puerto Rico, Mayaguez Campus, this was the first INTSOY-sponsored regional conference on the production, protection, marketing, and use of soybeans.

D. K. Whigham (Editor)

Soybean Production, Protection, and Utilization. Proceedings of a Conference for Scientists of Africa, the Middle East, and South Asia

1975. 266 pages. 21.5 × 28 cm. Paperback. English. (INTSOY Series No. 6)

This conference was held October 14 to 18, 1974 in Addis Ababa, Ethiopia, sponsored by INTSOY and the Ethiopian Institute of Agricultural Research.

R. M. Goodman (Editor)

Expanding the Use of Soybeans. Proceedings of a Conference for Asia and Oceania

1976. 259 pages. 21.5 × 28 cm. Paperback. English. (INTSOY Series No. 10)

In February 1976, INTSOY, the Asian Vegetable Research and Development Center (AVRDC), and the Government of Thailand, Ministry of Agriculture and Cooperatives, sponsored this conference held in Chiang Mai, Thailand. Financial support was provided by the United States Agency for International Development.

R. E. Ford and J.B. Sinclair (Editors)

Rust of Soybean: Problems and Research Needs

1977. 110 pages. 21.5 × 28 cm. Paperback. English. (INTSOY Series No. 12)

This workshop, which was held in Manila, Philippines, from February 28 to March 4, 1977, was sponsored by the Government of the Philippines, INTSOY, the Asian Vegetable Research and Development Center (AVRDC), and the Philippine Council for Agriculture and Resources Research and Development (PCARRD).

W. H. Judy and J. A. Jackobs (Editors)

Irrigated Soybean Production in Arid and Semi-Arid Regions

1981. 193 pages. 21.5 × 28 cm. Paperback. English. (INTSOY Series No. 20)

This volume contains the proceedings of a conference held in Cairo, Egypt, August 31 to September 6, 1979. The Conference was sponsored by the Egyptian Ministry of Agriculture, Menoufeia University, and INTSOY, in collaboration with the Food and Agriculture Organization (FAO) of the United Nations and the U.S. Agency for International Development.

J. B. Sinclair and J. A. Jackobs (Editors)

Soybean Seed Quality and Stand Establishment. Proceedings of a Conference for Scientists of Asia

1982. 206 pages. 21.5 × 28 cm. Paperback. English. (INTSOY Series No. 22)

The Conference was held at the Agrarian Research and Training Institute, Colombo, Sri Lanka, January 25 to 31, 1981. Sponsors included the Sri Lanka Ministry of Agricultural Development and Research, the Seed Technology Laboratory at Mississippi State University, and INTSOY in collaboration with the United Nations Food and Agriculture Organization and the U.S. Agency for International Development.

B. J. Irwin, J. B. Sinclair, and Wang Jin-ling (Editors)

Soybean Research in China and the United States. Proceedings of the First China/USA Soybean Symposium and Working Group Meeting

1983. 194 pages. 21.5 × 28 cm. Paperback. English. (INTSOY Series No. 25)

Sponsored by the United States Department of Agriculture's Office of International Cooperation and Development, the Chinese Ministry of Agriculture, Animal Husbandry, and Fishery, and the University of Illinois at Urbana-Champaign, College of Agriculture, Office of International Agriculture in collaboration with INTSOY, this symposium was held at the University of Illinois at Urbana-Champaign from July 27 to 30, 1982. The proceedings provide the first comprehensive and formal presentation of the past and present research and development activities on soybeans in China, the soybean's center of origin. Also, the proceedings present a summary of the U.S. soybean research and development which led to the U.S. becoming the world's largest producer and consumer of soybeans.

Miscellaneous Reports and Brochures

Variety Trials

The International Soybean Variety Evaluation Experiment (ISVEX) is designed to: test adaptation of soybean cultivars under a wide range of environmental conditions; provide research workers with an opportunity to compare local and introduced cultivars; provide a source of new germplasm which the cooperator may use directly or incorporate into the national breeding program; identify areas of the world that have a potential for soybean production; and evaluate the response of soybeans to different environments.

D. K. Whigham

International Soybean Variety Experiment, First Report of Results, 1973

1975. 161 pages. 21 × 27.5 cm. Paperback. English. (INTSOY Series No. 8)

D. K. Whigham

International Soybean Variety Experiment, Second Report of Results, 1974

1976. 223 pages. 21.5 × 28 cm. Paperback. English. (INTSOY Series No. 11)

D. K. Whigham and W. H. Judy

International Soybean Variety Experiment, Third Report of Results, 1975

1978. 369 pages. 21.5 × 28 cm. Paperback. English. (INTSOY Series No. 15)

W. H. Judy and D. K. Whigham

International Soybean Variety Experiment, Fourth Report of Results, 1976

1978. 401 pages. 21.5 × 28 cm. Paperback. English. (INTSOY Series No. 16)

W. H. Judy and H. J. Hill

International Soybean Variety Experiment, Fifth Report of Results, 1977

1979. 285 pages. 21.5 × 28 cm. Paperback. English. (INTSOY Series No. 19)

W. H. Judy, J. A. Jackobs, and E. A. Engelbrecht-Wiggans

International Soybean Variety Experiment, Sixth Report of Results, 1978

1981. 305 pages. 21.5 × 28 cm. Paperback. English. (INTSOY Series No. 21)

J. A. Jackobs, M.D. Staggs, and D. R. Erickson

International Soybean Variety Experiment, Seventh Report of Results, 1979

1983. 211 pages. 21.5 × 29 cm. Paperback. English. (INTSOY Series No. 24)

J. A. Jackobs, C. A. Smyth, and D. R. Erickson

International Soybean Variety Experiment, Eighth Report of Results, 1980-1981

1984. 224 pages. 21.5 × 29 cm. Paperback. English. (INTSOY Series No. 26)

J. A. Jackobs, C. A. Smyth, and D. R. Erickson

International Soybean Variety Experiment, Ninth Report of Results, 1982

1984. 103 pages. 21.5 × 28 cm. Paperback. English. (INTSOY Series No. 27)

J. A. Jackobs, C. A. Smyth, and D. R. Erickson

International Soybean Variety Experiment, Tenth Report of Results, 1983

1985. English. (INTSOY Series No. 28). In preparation.

Bibliographies

J. B. Sinclair and O. D. Dhingra

An Annotated Bibliography of Soybean Diseases

1975. 280 pages. 21.5 × 28 cm. Paperback. English. (INTSOY Series No. 7)

This volume marks the first time the world literature on soybean diseases has been brought together and annotated in a single volume. There is a detailed index to the 2,250 citations dating from 1882 through 1974.

,I Kogan, D K Sell, R. E. Stinner, J. R. Bradloy, Jr., and M. Kogan

The Literature of Arthropods Associated with Soybean. V. A Bibliography of *Heliothis zea* (Boddie) and *H. virescens* (F.) (Lepidoptera:Noctuidae)

1978. 242 pages. 21.5 × 28 cm. Paperback. English. (INTSOY Series No. 17)

From the standpoint of potential crop loss, the genus *Heliothis Ochsenheimer* probably contains the single most important insect complex in the world. There is a detailed subject index to the 5,178 citations.

Periodicals

INTSOY Newsletter

Issued twice a year in English, Spanish, and French editions.

The Newsletter contains information on current INTSOY activities and research, and features news items of interest to soybean workers around the world. Names are added to the mailing list upon request.

ICIMOD

International Centre for Integrated Mountain Development

P.O. Box 3226, Kathmandu, Nepal
Cable: ICIMOD Kathmandu
Telex: NP 2245 SATA
Telephone: 522819, 521575

General Information

The primary objective of this newly established centre is to promote economically and environmentally sound development in the Hindu Khush-Himalayas. This region includes Afghanistan, Bangladesh, Bhutan, China, India, Nepal, and Pakistan.

The Centre will be a focal point for multidisciplinary documentation, training, and applied research, and will provide consultative services for mountain resource management and development.

The establishment of the Centre is based on an agreement between HMG Nepal and UNESCO. The Centre is presently sponsored by the government of Nepal, the Federal Republic of Germany, and Switzerland.

The Centre is located in Kathmandu with the status of an autonomous international organization under a board of governors from Nepal, UNESCO, the countries of the region, and the sponsors.

Proceedings

ICIMOD

Mountain Development 2000: Challenges and Opportunities (Proceedings of the First International Symposium and the Inauguration of ICIMOD, 15 December 1983)

1984. 123 pages. 21.5 × 28 cm. Paper cover, stitched. $10.00.

Proceedings of the First International Symposium and Inauguration of the International Centre for Integrated Mountain Development (ICIMOD), which was held in Kathmandu, Nepal in December 1983.

ICIMOD

1984. 12 pages. 15 × 23.5 cm. Stapled.

ICIMOD Brochure, describing composition, aims and objects, problems of development and ecology in the HinduKush-Himalaya, key factors of ecological degradations, obstacles to integrated development in the region, ICIMOD work programmes, and sponsorship and collaboration.

Periodicals

ICIMOD Newsletter No. 4

1985. 8 pages. 23 × 28.5 cm. Stapled.

Periodic approx. 4-monthly newsletter of ICIMOD programmes, conferences, workshops, organisation and staffing etc.

IIMI

International Irrigation Management Institute

Headquarters:
IIMI, Digana Village via Kandy, Sri Lanka
Telex: 22318 IIMIHQ CE
Telephone: (08) 74274

Liaison Office:
IIMI, P.O. Box 2075, 5A Schofield Place, Colombo 5, Sri Lanka
Telephone: (01) 589933

General Information

The International Irrigation Management Institute (IIMI) is an auto-nomous non-profit international organization chartered in Sri Lanka in 1984 to conduct research, provide opportunities for professional development, and communicate information about irrigation management in developing countries.

IIMI's research is conducted in two areas: systems operated by national governments and systems operated by farmers. Attention is given to problems with wide applicability. Researchers seek to understand the ways in which local conditions — physical and social — affect the performance of irrigation systems. Research is conducted on whole systems and representative portions of whole systems, and on problems which require both multidisciplinary and multi-institutional attention.

IIMI provides professional training through guided study tours, on-the-job research training, specialized workshops and conferences, and graduate student fellowships. The primary clients for training are irrigation system managers, and policy makers, educators, and researchers in irrigation management.

Information documentation and dissemination, and the development of communication networks among institutes and researchers involved in irrigation management are essential aspects of IIMI's program. IIMI invites anyone interested in improving irrigation management in developing countries to join its network. Please write for further information.

International Irrigation Management Institute

Informational Brochure

1984. 8 pages. 10.2 × 23 cm. Free. Available in English and French.

Includes a tear-off mailing list access card for those interested in being placed on IIMI's mailing list.

Conferences, Symposia, and Workshop Proceedings

International Irrigation Management Institute

Workshop on Research Priorities for Irrigation Management in Asia, January 6-11 1985

In press. Mimeos of two summary papers are available free on request. English.

Sponsored jointly by IIMI and Water Management Synthesis II Program of USAID, the workshop discussed critical issues in irrigation management with attention to the status of irrigation management research in nine Asian countries. The purpose was to identify priorities for future research.

Small, Leslie, E. 1975. "Research Priorities for Irrigation Management in Asia."

Barker, Randolph. 1975. "An Overview of Research in Irrigation Management in Asia."

International Irrigation Management Institute

Workshop on New Approaches and Methods in Irrigation Management, July 15-19 1985

In press. Mimeos of invited papers are available free on request. English.

This workshop met to assess the present state of knowledge about irrigation management relating to: rapid appraisal methodology, managing the rehabilitation process, managing main system water distribution, and improving management institutions and staff-farmer relations. Gaps in present knowledge were identified and research methods recommended.

Periodical

International Irrigation Management Institute

International Irrigation Management News (IIMN)

Number of pages varies. Quarterly. First edition in preparation/press. Free.

The International Irrigation Management News (IIMN) is a publication for IIMI's irrigation management network. Network members are invited to contribute concise summaries of research in progress, publication abstracts, and summary reports on conferences, workshops, and symposia relating to irrigation management issues. Contributions should be limited to less than two double-spaced typewritten pages which can include tables and figures. Edited copy will be returned to the author for approval before publication. English only please.

International Irrigation Management Institute

IIMI Newsletter

Number of pages varies. Quarterly. First edition in preparation/press. Free. English.

The IIMI Newsletter describes the Institute's research, professional training, and communication activities.

Addendum

International Rice Research Institute

The Rice Economy of Asia

1985. 324 pages. 21.59 × 27.94 cm. Paperback. ISBN 0-915707-15-2. HDC: send orders to Resources for the Future, 1616 P Street, N. W., Washington, D.C. 20036, USA; LDC US$7.75 plus airmail (US$20.00) or surface mail (US$2.25) postage.

The Rice Economy of Asia is about trends and changes in the Asian rice economy since World War II, and particularly since the introduction of new rice varieties and modern technology in the mid-1960s.

There is a vast amount of literature and statistical data on various aspects of the subject, but no single comprehensive treatment has previously has been prepared. *The Rice Economy of Asia* not only provides such a treatment but also presents a clear picture of some of the critical issues dealing with productivity and equity.

The volume should be interesting and useful to decision makers at national and international levels, to professionals, and to students of development. In addition to 18 chapters, it includes an extensive bibliography, 150 tables, and 50 charts.

The volume draws on the experience of the authors, two of whom served as IRRI agricultural economists.

KEYWORD INDEX

[Forests] Los limites socioecologicos del crecimiento agricola en ceja de selva **104**

Forests of South America/The (S) **473**

Fruit trees in tropical agroforestry systems **436**

Fruits/Directory of germplasm collections; 6: I tropical **145**

Fruits in the tropics and subtropics/Postharvest problems of vegetables and **391**

Fuel and fertilizer from organic wastes/Food **451**

Fuel for motor transport/Producer gas: another **453**

Fuel-saving cookstoves **505**

Fuels: options for developing countries/Alcohol **453**

Fuelwood in the lowland tropics/Potential contribution of Leucanea hedgerows intercropped with maize to the production of organic nitrogen and **433**

Fungal, bacterial, viral, and nematode diseases of soybeans/Sources of resistance to selected **512**

Fungal diseases of rice* **322**

Fungi and their control/Bean disease caused by* **49**

Gall midge and whorl maggot/Rice* **324**

Gall midge, *Orseolia oryzae,* in Asia/Reactions of differential varieties to the rice **309**

Gambia economy, February 1976/The Mac Carthy Islands irrigation project in the **384**

Gambian rice seed multiplication programme, February 1983/A review of the **379**

Gardening workshop-Thailand/Taiwan (April 21-27)/International **399**

Gardens: theoretical considerations on an old survival strategy/Household **102**

Gas: another fuel for motor transport/Producer **453**

Gas producer-engine systems/Small scale **507**

Gasification of rice hulls **504**

"Gate" — questions, answers, information (Journal) **508**

Gel electrophoresis/Detection of PSTV by* **91, 112**

Gene pools: spring X winter crosses in bread wheat/Probing the **123**

Gene rotation concept for rice blast control/Evolution of the **275**

Genebanks/A global network of (E, F, S) **143**

Genebanks/Institutes conserving crop germplasm: the IBPGR global network of **132**

Genebanks/Seed management techniques for **132**

Genetic analysis of traits related to grain characteristics and quality in two crosses of rice **308**

Genetic and sociologic aspects of rice breeding in India **307**

[Genetic breeding material] Multiplicacion acelerada de material genetico promisorio de yuca* **61**

Genetic conservation/Institutes working on tissue culture for **131**

Genetic conservation of rice **299**

Genetic conservation of rice germplasm for evaluation and utilization/Manual on **269**

Management variables in IRRIMOD/Sensitivity tests of the crop and **310**

Mangium and other fast-growing acacias for the humid tropics **452**

(*Manihot esculenta* Crantz)/Morphology of the cassava plant* (S, E) **57, 58**

Manpower and training plan for the agricultural research system in Kenya 1983-1987/A **348**

Manpower planning in a national agricultural research system **345**

Manual/Agroforestry species: a crop sheets **430**

Manual/Cowpea production training (E, F) **226**

Manual/Crop genetic resources field collection **141**

Manual for testing insecticides on rice **274**

Manual for use with programmable calculators/Fish population dynamics in tropical waters/A **413**

Manual/International bean yield nurseries: descriptive (S only) **16**

Manual/Livestock productivity and trypanotolerance, network training (E, F) **245**

Manual/Maize production (E, F) **227**

[Manual] Manuel des materiels de Traitement des cultures **497**

Manual/MULBUD user's **430**

Manual of common faba bean diseases in the Nile Valley/Field **151**

Manual of reforestation and erosion control for the Philippines **465**

Manual on surveillance and early-warning techniques for plant-pest and diseases **499**

Manual/The AVRDC vegetable preparation **398**

Manual/The tropicultor operators' (E, Telugu) **192**

Manual to major pests and diseases of wheat and barley/Field (A) **151**

Manual/Tuber and root crops production (E, F) **226**

Map of the Philippines/Agroclimatic* **304**

Mapping of tropical African rangelands/Evaluation and (E, F) **245**

Maps of South, Southeast, and East Asia/Agroclimatic and dry-season* **303**

Marine capture fisheries/Review of Indonesian **414**

Marine fishes of the central Philippines/Commercial **481**

Market: structure, conduct, and performance/The world rice **212**

Market: the case of wheat in India/Policy modeling of a dual grain **212**

Marketing and policy in Ghana, January 1981/Economic study of rice production (E, F) **385, 386**

Marketing and policy in Nigeria, July 1980/Rice production (E, F) **385, 386**

Marketing Bhutan's potatoes: present patterns and future prospects **104**

Marketing in Ghana/Economic study of rice production and (E, F) **367**

Marketing margins and pricing policies for wheat in developing countries/1985 world wheat facts and trends, report three: prices **122**

Marketing of milkfish in Taiwan, China: an economic analysis/Production and **421**